Insect Pheromones
in Plant Protection

Insect Pheromones in Plant Protection

Edited by

A. R. Jutsum
ICI Agrochemicals, Haslemere, UK

R. F. S. Gordon
ICI Agrochemicals, Bracknell, UK

A Wiley–Interscience Publication

JOHN WILEY & SONS

Chichester • New York • Brisbane • Toronto • Singapore

Library of Congress Cataloging-in-Publication Data:

Insect pheromones in plant protection.

"A Wiley–Interscience publication."
1. Semiochemicals.   2. Insect pests — Biological
control.   3.Insects — Behavior.   I. Jutsum, A. R.
II. Gordon, R. F. S.
SB933.5.C48 1989   632'.7   88-10726

ISBN 0 471 92019 3

British Library Cataloguing in Publication Data:

Insect pheromones in plant protection.
1. Insects. Behaviour. Chemical control
I. Jutsum, A. R.   II. Gordon, R. F. S.
595.705

ISBN 0 471 92019 3

Phototypeset by Dobbie Typesetting Limited, Plymouth, Devon
Printed in Great Britain by Biddles Ltd., Guildford, Surrey

Contents

Chapter 7.   Plastic laminate dispensers/
A. R. Quisumbing and A. F. Kydonieus

Chapter 8.   Hollow-fibre controlled-release systems/
D. W. Swenson and I. Weatherston

List of Contributors

DR A. BAKKE, *Norwegian Forest Research Institute, Box 61, 1432 As-NLH, Norway*

DR M. C. BIRCH, *Department of Zoology, University of Oxford, South Parks Road, Oxford OX1 3PS, UK*

DR D. G. CAMPION, *Overseas Development Natural Resources Institute, Central Avenue, Chatham Maritime, Chatham, Kent ME4 4TB, UK*

DR B. R. CRITCHLEY, *Overseas Development Natural Resources Institute, Central Avenue, Chatham Maritime, Chatham, Kent ME4 4TB, UK*

DR C. T. DAVID, *AFRC Insect Physiology Group, Department of Pure and Applied Biology, Imperial College of Science and Technology, London SW7 2AZ, UK*

DR R. F. S. GORDON, *ICI Agrochemicals, Jealott's Hill Research Station, Bracknell, Berks RG12 6EY, UK*

DR D. R. HALL, *Overseas Development Natural Resources Institute, Central Avenue, Chatham Maritime, Chatham, Kent ME4 4TB, UK*

DR M. HEBBLETHWAITE, *Overseas Development Natural Resources Institute, Central Avenue, Chatham Maritime, Chatham, Kent ME4 4TB, UK*

DR G. J. JACKSON, *ICI Agrochemicals, Fernhurst, Haslemere, Surrey GU27 3JE, UK*

DR A. R. JUTSUM, *ICI Agrochemicals, Fernhurst, Haslemere, Surrey GU27 3JE, UK*

PROFESSOR J. S. KENNEDY, *Department of Zoology, University of Oxford, South Parks Road, Oxford OX1 3PS, UK*

DR A. F. KYDONIEUS, *Hercon Laboratories Corporation, South Plainfield, New Jersey 07080, USA*

DR R. LIE, *Borregaard Industries Ltd, N-1701 Sarpsborg, Norway*

MR G. J. MARRS, *ICI Agrochemicals, Jealott's Hill Research Station, Bracknell, Berks RG12 6EY, UK*

MR L. MCVEIGH, *Overseas Development Natural Resources Institute, Central Avenue, Chatham Maritime, Chatham, Kent ME4 4TB, UK*

DR K. OGAWA, *Shin-Etsu Chemical Co. Ltd, Tokyo, Japan*

DR N. PUNJA, *ICI Agrochemicals Fernhurst, Haslemere, Surrey GU27 3JE, UK*

DR A. R. QUISUMBING, *Hercon Laboratories Corporation, South Plainfield, New Jersey 07080, USA*

DR C. J. SANDERS, *Great Lakes Forest Research Centre, Sault Ste Marie, Ontario, Canada*

DR D. W. SWENSON, *Morton Thiokol Inc, Alfa Products, Danvers, Massachusetts 01923, USA*

DR C. WALL, *Department of Animal Ecology, Ecology Building, University of Lund, S-223 62 Lund, Sweden*

DR I. WEATHERSTON, *Département de Biologie, Université Laval, Québec G1K 7P4, Canada*

DR A. YAMAMOTO, *Shin-Etsu Chemical Co. Ltd, Tokyo, Japan*

Foreword

This book is 'not another pheromone review', in the words of the editors. Rather, it chronicles—and in effect celebrates—a fact of historic importance: we have now passed the point where the practical, commercial feasibility of using pheromones to control insect pests could still be doubted. It has taken a long time, around 25 years, to achieve this two-pronged breakthrough in pest management technique and in environmental concern. Admittedly there are so far only a handful of cases that demonstrate the achievement, and it is possible to belittle their importance by pointing to the especially favourable circumstances in these cases. This sadly misses the point that deliberately choosing such circumstances to start with was correctly perceived as the only sensible strategy, when trying to develop a technique that was new and very tricky biologically, physicochemically, logistically, economically and socially. The many lessons learnt from these operations are the indispensable springboard for future moves into less favourable circumstances, in a stepwise progression that is acceptable commercially.

Of course, the pheromone breakthrough is a very modest one indeed by comparison with the great insecticidal breakthroughs since the Second World War. However successful the use of pheromones and other behaviour-modifying 'semiochemicals' may eventually become, nobody expects them to sweep the board as did the insecticides. At best they will become a valued addition to the pest manager's multiple tool kit. Nevertheless their development should prove significant in the long run because it is an object lesson. It shows how the development of new pest management techniques (including non-pheromonal techniques such as the use of insect growth regulators) is going to need a much larger *scientific content* than in the past.

Pest management using semiochemicals is a sophisticated procedure intellectually and practically. Its development demands solid research inputs from both pure and applied scientists of many kinds. It calls for deeper understanding of the insect 'enemy' itself, of the mode of action of the chemical weapon, of the agroecosystem and of the social context.

Underestimation of these research needs was undoubtedly the reason for the early setbacks and disappointments and the premature, faint-hearted prognosis that the future for pheromones lay in pest monitoring, not control. There has been one *non*-material, favourable circumstance without which the commercial success of pheromonal control could not have been achieved. It is the driving determination to achieve it which has united a small, scattered, multinational band of pure and applied scientists, academic, state-supported and industrial alike. This exceptional circumstance accounts for the remarkable perseverance of these people despite the setbacks and slow progress, not to mention the dire warnings that the main virtue of semiochemicals — specificity — must also severely restrict the market for any one of them. The perseverance paid off when it emerged that this economic objection need not be the Achilles heel of the whole enterprise which it once seemed: money *can* be made out of the proper application of these chemicals; moreover, their application has a telling promotional advantage in its defence of the biosphere. A study of the diverse motivations of the specialists contributing to the success of semiochemical pest control, and how they came to converge on it, would merit yet another book. Meanwhile the full gamut of their contributions is presented in this one.

JOHN KENNEDY
Oxford

Insect Pheromones in Plant Protection
Edited by A. R. Jutsum and R. F. S. Gordon
©1989 John Wiley & Sons Ltd

1

Introduction. Pheromones: Importance to Insects and Role in Pest Management

A. R. Jutsum

ICI Agrochemicals, Fernhurst, Haslemere, Surrey GU27 3JE, UK

R. F. S. Gordon

ICI Agrochemicals, Jealott's Hill Research Station, Bracknell, Berks RG12 6EY, UK

1.1 INTRODUCTION

Chemical communication is arguably the primary mode of information transfer in members of the class Insecta, and is well exemplified in the insect world. Chemicals—termed semiochemicals—are employed for both intraspecific and interspecific communication. Compounds which convey information between members of the same species are known as pheromones, and it is these chemical messengers and their use in plant protection which is the subject of this book.

Pheromones were originally defined over a quarter of a century ago by Karlson and Luscher (1959) as 'substances which are secreted to the outside by an individual and received by a second individual of the same species in which they release a specific reaction, for example, a definite behaviour or developmental process'. In the past two decades, the terms applied to different patterns of olfactory communication have proliferated. The term pheromone continues to describe intraspecific communication. An allelochemic is the analogous term for an interspecific effector, and it is subdivided into kairomones, which attract exploiters, and allomones, which are advantageous to the odour-releasing individuals.

1

1.2 SCOPE OF THE TREATISE

The subject of semiochemicals, and of pheromones in particular, has been reviewed before. A classic treatise was published over a decade ago by Shorey and McKelvey (1977). More recently Hummel and Miller (1984) have edited a valuable text on techniques employed in pheromone research and Prestwick and Blomquist (1987) surveyed pheromone biochemistry. The present book, however, is not another pheromone review. It is a text produced jointly by academia and industry which aims to combine scientific, practical and commercial approaches to exploiting pheromones in plant protection. The book is divided into four parts:

Part A provides an overview of the biological background to the subject, pulling together recent research findings in insect physiology and behaviour.

Part B deals with the most important existing and potential uses of pheromones in plant protection, and aims to be of value at the practical level to all those who use pheromones, whether in research or farming. It also provides an investigation of the strengths and limitations of these methods.

Part C gives a review of the synthetic chemistry of pheromones along with the controlled release formulations now employed, and their advantages and disadvantages.

Part D provides a unique commercial analysis of pheromones and a look into the future—as traditional insecticides will probably give way increasingly to alternative control methods, of which the use of pheromones may be one of the most important.

1.3 TYPES OF PHEROMONES

Pheromonal communication is ubiquitous in insects, pheromones being employed to mediate a wide variety of behaviour.

1.3.1 Sex pheromones

The most widespread and widely documented types of pheromone are those which are used to increase the probability of successful mating. Sex pheromones may be produced by females or by males, according to species. In some cases, both sexes contribute to the chemical communication involved in mating. In many cases, different pheromone components are responsible for the principal behavioural phases of mate location and courtship.

1.3.2 Aggregation pheromones

Aggregation pheromones have been reported for members of the Coleoptera, Dictyoptera, Hemiptera, Homoptera and Orthoptera, and function in many ways including mate selection, defence against predators, and overcoming host resistance by mass attack. They release behaviour leading to an increase in density of conspecifics in the vicinity of the pheromone source.

1.3.3 Alarm pheromones

Alarm pheromones which stimulate escape and other defensive behaviour have been observed in members of the Dictyoptera, Hemiptera and eusocial Hymenoptera. They are generally highly volatile, low molecular weight compounds that can be spread rapidly through a colony, act quickly and are short-lived.

1.3.4 Epideictic pheromones

Epideictic, dispersive or spacing pheromones elicit behaviour resulting in increased spacing between conspecifics and a reduction in intraspecific competition. Examples can be found in the Coleoptera, Diptera, Homoptera, Hymenoptera, Lepidoptera and Orthoptera and include oviposition deterrent pheromones.

1.3.5 Trail pheromones

Trail pheromones for recruitment or emigration are common in the Dictyoptera, Lepidoptera and eusocial Hymenoptera. They are used to recruit other insects in a colony to new food sources or are used to facilitate migration of a colony to a new site. In order to be effective, these pheromones are more stable and persistent than many other types of pheromone.

1.3.6 Other pheromones

There are many other pheromones which elicit additional effects. These include maturation pheromones, which influence physiological processes leading to development in the Coleoptera and Orthoptera, and pheromones affecting caste determination and ovarian inhibition in the Dictyoptera and eusocial Hymenoptera. Many social interactions such as trophillaxis, grooming and digging are also pheromonally mediated.

The chemical components of many insect pheromones have now been identified and are characterized by a considerable structural eclecticism in the different orders, as shown in Table 1.1. Pheromone chemistry ranges from the blends of aliphatic alcohols, aldehydes, esters and epoxides exploited by lepidopterous insects to alkenoic acids and aldehydes, branched alkanones, esters, monoterpene alcohols and aldehydes and a furanone employed by

TABLE 1.1 Examples of insect pheromones

Pheromone type	Compound	Order	Family	Comments
Sex pheromones	(Z)-11-Hexadecenal	Lepidoptera	Noctuidae	Produced by female
	(Z)-9-Tetradecenal	Lepidoptera	Noctuidae	Produced by female
	Methyl jasmonate[1]	Lepidoptera	Tortricidae	Produced by male
	Undecanal	Lepidoptera	Pyralidae	Produced by male
	(E,Z)-10,12-Hexadecadien-1-ol	Lepidoptera	Bombycidae	Produced by female
	(Z)-7,8-Epoxy-2-methylheptadecane[1]	Lepidoptera	Lymantriidae	Produced by female
	2-(Z-Isopropenyl-1-methylcyclobutyl)-ethanol	Coleoptera	Curculionidae	Produced by female
	10-Methyl-2-tridecanone	Coleoptera	Chrysomelidae	Produced by female
	(Z)-5-(1-Decenyl)dihydro-2(3H)-furanone[1]	Coleoptera	Scarabaeidae	Produced by female
	(Z)-9-Tricosene	Diptera	Muscidae	Produced by female
	Syringaldehyde	Hemiptera	Coreidae	Produced by male
	3-Methyl-6-isopropenyl-9-decen-1-ol acetate[1]	Homoptera	Diaspididae	Produced by female
Aggregation pheromones	Periplanone-B[1]	Dictyoptera	Blattidae	Produced by female
	3,7-Dimethylpentadecan-1-ol acetate[1]	Hymenoptera	Tenthredinidae	Produced by female
	2-Methyl-6-methylene-7-octen-4-ol	Coleoptera	Scolytidae	From gut-derived frass
	2,4-Dimethyl-5-ethyl-6,8-dioxabicyclo-(3,2,1)-octane	Coleoptera	Scolytidae	From gut-derived frass
Alarm pheromones	Germacrene-A[1]	Homoptera	Aphididae	From fat cells extend-ing into cornicles
	(E)-2-Hexanal	Hemiptera	Phrrhocoridae	From larval dorsal abdominal glands
Trail pheromones	4-Methyl-3-heptanone[1]	Hymenoptera	Formicidae	From mandibular glands
	Neocembrene-A[1]	Dictyoptera	Termitidae	From sternal gland
	Methyl-4-methylpyrrole-2-carboxylate	Hymenoptera	Formicidae	From poison gland of workers
	3-Ethyl-2,5-dimethyl pyrazine	Hymenoptera	Formicidae	From poison gland of workers
Ovarian inhibitor	9-oxo-(E)-2-decenoic acid	Hymenoptera	Aphididae	From mandibular glands
Phase transformer	2-Methoxy-5-ethylphenol	Orthoptera	Acrididae	From the crop

[1]Possession of a chiral centre(s) results in at least one enantiomer being present in the pheromone.

beetles. The chemistry and commercial production of pheromones is examined in detail in Chapter 6.

1.4 PHEROMONES AS TOOLS FOR PEST CONTROL

The global end-user insecticide market was estimated to be $6100 million in 1987 (Wood Mackenzie, 1988) and is dominated by conventional chemical control agents. The use of pheromonal-type compounds was examined in the 1940s in the USA when crude extracts of abdominal tips of female Gypsy moths were used in monitoring traps (Holbrook *et al.*, 1960). However, it was not until twenty years ago that the possibility of using pheromones in crop protection was accepted, and the first meaningful field trial on mating disruption of a lepidopterous pest was reported (Shorey *et al.*, 1967). Since then interest in pheromones has intensified, not only because of public pressure for more selective methods of pest control, but also because advances in the laboratory have enabled structural assignment to be made to many pheromones. This advance can be attributed to the exploitation of modern microtechniques such as chromatographic separation and nuclear magnetic resonance, infrared and mass spectrographic analysis. Although the pheromones of many major insect pest species have been identified and synthesized, and studies have been performed which have increased our understanding of their mode of action and biological significance to the species involved, progress towards widespread practical use of these chemicals has been slower than expected. This is mainly due to difficulties encountered in relation to understanding the biology of the insect and the use of pheromones in the field.

Since the first pheromone was identified in 1959 by Butenandt *et al.* (1959), pheromones have been identified for hundreds of species of insects. Many early studies concentrated on pheromones used by social insects, but subsequently attention has turned to non-social, economically important species, especially members of the Coleoptera and Lepidoptera. These include the boll weevil, *Anthonomus grandis* (Boheman); the cotton bollworm, *Heliothis zea* (Boddie); the tobacco budworm, *Heliothis virescens* (F.), the Old World bollworm, *Heliothis armigera* (Hübner), the Egyptian cotton leafworm, *Spodoptera littoralis* (Boisd.); the Eastern spruce budworm, *Choristoneura fumiferana* (Clem.) and the striped rice stemborer, *Chilo suppressalis* (Wlk.).

The value of pheromones in controlling these and other major pest insects has been examined during the last decade and successful products will no doubt emerge, but it is worth noting that research findings on commercial applications of pheromones in monitoring, mass trapping and disruption have been published for less than 5% of species for which pheromones have been identified. Nonetheless, it should be emphasized that all commercially important insect pests appear to use pheromonal communication in some way, often in more than one stage in the life cycle.

The objective of this book is to concentrate on pheromones which are, or are near to, being exploited commercially. By definition, this narrows the list of pheromones to those involved in sex and aggregation. For instance, sex pheromones and the behaviour associated with these semiochemicals are crucial in the life cycle, and are therefore excellent targets to interfere with to obtain pest control. However, field studies which have been performed with epideictic pheromones (e.g. Katsoyannos and Boller, 1976, 1980), with alarm pheromones (Dawson *et al.*, 1982) and with trail pheromones (e.g. Robinson *et al.*, 1982) have not been overlooked. Further reference to developments in such areas will be found in Chapter 14.

An outline history of pheromone development in the context of crop protection is shown in Table 1.2, along with further enabling actions awaiting

TABLE 1.2 History of pheromone development in crop protection

	Chemistry	Formulation	Biology/field studies
1960	First pheromone identified, *Bombyx mori*		
			First meaningful disruption field trial
1970		Hollow fibre dispenser conceived	First successful field trial with a microencapsulated formulation
1975		Plastic laminate flake dispenser conceived	First field trials with fibre and flake dispensers. First mating disruption product registered
1980		'Protected' microencapsulated product developed	First field trials with pheromone/insecticide mixtures
1990	Ensure 'commercially usable' pheromones are identified for all major pest species. Re-examine and confirm previous chemical identifications	Continue fundamental research on controlled release systems which are cheap, non-toxic and biodegradable, that can be applied using conventional application equipment	Undertake population studies/ identify economic thresholds. Examine behavioural responses of insects to pheromones in the field. Investigate relative importance of different communication systems in the life of the insect. Examine effects of meteorological and physicochemical factors on chemical behaviour. Perform field evaluations on a directed basis

completion under the headings of chemistry, formulation and biology/field studies of the pheromones thus discovered. In crop protection, to date, three main types of use can be recorded, as follows.

1.4.1 Monitoring and survey work

Sex pheromones employed as baits in traps can be used to capture insects of the opposite sex and thus provide the basis for monitoring a pest population. This technique can be used, for example, to detect early infestations of a pest insect, to monitor established populations and to assist in the timing of pesticide applications in relation to the build-up of populations to dangerous levels. The traps used are simple in construction and are easy to install and operate. Traps of many varieties have been devised for different pests, ranging from sticky boards and water traps to simple funnel traps. Even the pheromone-dispensing device, which is a key part of the trap, can be very simple, comprising impregnated papers, wicks or plastic vials.

1.4.2 Control by mass trapping

The rationale behind mass trapping is to concentrate pest insects into a restricted space where they can be killed easily and cheaply and with less environmental impact than with the use of conventional insecticides. It is especially applicable where the pest population is usually widely dispersed, where control by conventional pesticides is inapplicable, where resistance has developed to conventional insecticides and no other form of control is available or, most importantly, where it can be an economic form of pest control. Methods used to concentrate the insects are pheromone/attractant baits, e.g. sex attractant pheromones to trap lepidopteran adult males and aggregation pheromones to trap coleopteran adult males and females.

Pheromones for the most part confer species specificity on the control technique, a quality considered most desirable by some authorities. Once concentrated, there are numerous ways of killing the insects. Electric grids have been used, but usually a sticky surface or insecticide is used. In certain cases (e.g. coleopteran aggregation pheromones) where the insects are attracted to the general region of the trap as well as to the actual trap, spraying the area around the traps with conventional insecticide may be the most appropriate method of control.

The major costs of this technique are probably associated with placement of the traps in the field. The cost of the trap materials is probably minimal, except where traps need continual replacement of the sticky surface. As the cost-effectiveness of this technique is closely related to the trap density and deployment time, those 'crop' situations where protection is needed for more than one year, due to the continuity of the 'crop', are intrinsically better suited to this type of control. Orchards and forests are particular examples.

1.4.3 Disruption

Many species of insects use pheromonal communication to bring together the opposite sexes for mating. One sex, usually the female in the Lepidoptera, or either sex in the Coleoptera, will remain stationary and 'call', releasing the volatile pheromone. The other sex uses the resultant pheromone plume as a guide to finding a mate. Successful matings have been reduced by permeating the atmosphere around calling insects with a higher concentration of their own sex pheromone than they normally produce, or an anti-pheromone that prevents the perception of that pheromone. In Lepidoptera the high concentration of permeated chemical may prevent the male using the natural pheromone plume to locate the female. Alternatively, the high concentration of pheromone may lead to the habituation of the nervous system, preventing any response to the pheromone.

The disruption control strategy depends on reducing the number of successful matings and thereby the number of damaging larvae in the next generation and perhaps the number of adults in future generations. Knowledge of the time and place of mating is desirable, as gravid female immigration into the treatment area can negate any benefits gained through disruption of mating. The number of generations a year, their duration, synchrony and predictability determine the ease of timing and length of time the compounds must be permeated through the crop, and hence the overall cost.

A univoltine pest with a limited mating period is particularly susceptible to this type of control, as the population has less time to recover, and treatment is necessary only for a limited period during the season. A multivoltine, asynchronous pest would allow greater chances for population recovery, and would necessitate treatment over the whole season, which may be prohibitively expensive.

The longer the adult life, the greater the likelihood of chance encounters between males and females in pheromone or antipheromone permeated areas. This strategy is therefore less appropriate for controlling coleopteran pests such as boll weevil (*Anthonomus grandis*) or corn rootworm (*Diabrotica* spp.). As the likelihood of chance encounters leading to mating depends on the density of the pest and its mobility, the mating disruption technique would be best suited to low-density pest problems.

Thus, the greatest likelihood of success with mating disruptants will be in situations where there is only one important species, with a limited mating period, little immigration of females mated outside the treatment area, and where secondary pest problems arise through pesticide treatments directed at the key pest. In some instances better control may be achieved by adding small quantities of insecticide to the pheromone.

1.5 ADVANTAGES AND DISADVANTAGES OF PHEROMONES

Overall, the use of pheromones has a number of potential advantages. The compounds are naturally occurring, generally non-toxic and should not pollute the environment. Pheromones are insect-specific and their safety to beneficials makes them ideal components of integrated pest management systems. In addition, these various advantages of pheromones *may* obviate lengthy registration processes.

On the other hand, a number of problems are associated with the use of pheromones. The effectiveness of such compounds may be limited by the ratio of major and minor pheromone components used, by the timing of application and by insect population density. Many pheromones are also expensive to manufacture. Many are unstable, decomposing within minutes in the presence of light or oxygen. Some of these problems are now being surmounted. In particular, the use of controlled release formulations is highly applicable to unstable volatile compounds. Such formulations permit control of the duration of the chemical effect, deliver pheromone at target levels, maximize effectiveness with the minimum quantity of chemical and protect the active ingredient from light, oxidation and hydrolysis.

1.6 CONTROLLED-RELEASE PHEROMONE FORMULATIONS CURRENTLY AVAILABLE

The formulation requirements for pheromones are complex; the characteristics of a system required for use as a mating disruptant are shown in Table 1.3. At the present time, research and development effort is concentrated on four controlled release formulations: the hollow fibre, the twist-tie rope, the plastic laminate flake and microcapsules.

1.6.1 Hollow fibres

Controlled-release hollow-fibre formulations are produced by Scentry. The fibres comprise short lengths of thermoplastic tubing, sealed at one end and filled with liquid pheromone; the plastic is impermeable and non-reactive with the pheromone. Pheromone release from the fibres involves evaporation

TABLE 1.3 Formulation requirements for pheromones

1. Release a constant amount of pheromone per unit time, independent of temperature, humidity, light and crop
2. Have the ability to release different pheromones
3. Have the ability to provide different release rates
4. Protect the pheromone from degradation
5. Release the pheromone for a short or long time, depending upon the pest/crop situation
6. Release all the pheromone
7. Are relatively easy to apply, and suitable for aerial application
8. Are biodegradable, non-toxic and cheap

at the liquid–air interface, diffusion through the air column to the open end of the fibre and convection away from the end (Ashare *et al.*, 1975). In theory the emission rate is not constant but decreases steadily as the tube empties of pheromone. The biologically effective life of the formulation can be controlled by adjusting the length of the fibre filled. However, the formulation has to be dispensed by hand or as a highly viscous syrup comprising fibres and sticker using specialized equipment.

1.6.2 Twist-tie rope

A variety of plastic fibre, sealed at both ends and containing a hollow channel with pheromone and a wire 'spine', is produced by the Shin-Etsu Chemical Co. The rope is about 15 cm long and must be attached to the crop by hand. In comparison with the three other formulations, the rope places a very large initial reservoir of pheromone at one point, and the high dose per rope provides a relatively long persistence of release.

1.6.3 Laminate flakes

Three-layer plastic laminate flakes are produced by Hercon, the controlled release products business of Health-Chem Corporation. The flakes consist of two layers of vinyl sandwiching a central porous layer containing pheromone. The flakes are applied with sticker and thickening agent through specialized equipment, or by hand. The emission rate from the flakes is controlled by regulating the thickness of the layers, the edge–volume ratio, the concentration of chemical in the middle-layer reservoir, the stiffness of the plastic membranes and additives in the formulation (Kydonieus, 1977). The pheromone is emitted not only by diffusion through the membranes but also from the perimeter edge of the individual pieces (Bierl-Leonhardt *et al.*, 1979), yet emission under controlled conditions is relatively constant over long periods (Bierl *et al.*, 1976).

1.6.4 Microcapsules

Microencapsulated systems for pheromones are being developed and sold by ICI Agrochemicals. The microcapsules consist of small droplets (or particles) of pheromone protected by an outer shell of polymer. The system presently under development employs plastic polymer systems such as polyamide and polyurea prepared as capsules by interfacial polymerization. Emission of pheromone from these capsules is fast initially, as material which has migrated during storage is released. However, this is followed by a period of constant, sustained release which is diffusion controlled.

The ICI microcapsules are easily manufactured on a large scale using known technology (Nesbitt *et al.*, 1980) and are readily applied over large areas with conventional spray equipment. In addition, they possess numerous variables that can be manipulated to control the release characteristics; for example, wall thickness, wall composition, internal composition and capsule size.

These major formulations and other dispensers are discussed in Part C.

1.7 REGISTRATION, COMMERCIALIZATION AND ADOPTION OF PHEROMONES AS PEST CONTROL AGENTS

Technical efficacy is only one factor in the decision of an agrochemical company to develop a new product. Typically, in taking the decision to commit resources at the expense of other product possibilities, a company will undertake a thorough review of issues such as patent coverage, the expectation and cost of registration, the capital and output cost of manufacture, market size, the competitive environment, the likely build-up of sales and hence the return on investment. Chapter 11 describes the decision process in detail.

Beyond technical efficacy, which itself may be difficult to establish without the commitment of significant resources, potential pheromone products are seen in a generally favourable light from a registration viewpoint. But pheromones often appear in a poor light from other viewpoints, especially in comparison with conventional insecticide products. Table 1.4 summarizes the amenability of pheromone products for insect control in comparison with broad spectrum insecticides.

The repayment of registration costs is a major burden for all new agrochemical products. However, with pheromone products, it has been the case that such costs are relatively low, and registration timetables relatively short. These factors must be seen as essential if pheromone products are to be serious competitors for agrochemical company support.

TABLE 1.4 Relative amenability of pheromones to commercial development as mating disruptants compared with the broad spectrum insecticides

	Pheromones as mating disruptants	Broad-spectrum insecticides
Patentability	Patent protection difficult. Use, formulation and synthetic route are patentable	Patent protection afforded
Market size	Species-specific control agents have limited markets, but some may be large enough to recoup costs	Large market: costs are recouped and profit made
Manufacture	Scale of manufacture may be small	Generally, compounds are readily and economically manufactured in bulk
Packaging	Formulation technology is still advancing, but storage and packaging do not present major problems	Readily stored and packaged
Application	Microcapsules are applied through conventional equipment, but flakes and fibres require specialised application equipment	Reliably applied with standard application equipment
Treatment area	A large minimum treatment area may be necessary to obtain acceptable control	Protection can be provided to small areas surrounded by large reservoirs of pests
Biological effect	Essentially prophylactic, acting on adults. Therefore, need to act over at least one generation	Active primarily on damaging larval stages. Can dramatically suppress pest populations

Chapter 12 describes the background to this distinction between conventional agrochemicals and pheromones, taking the USA Environmental Protection Agency regulations as a model.

A significant factor in the success of a pheromone product, and hence in the choice of a target for pheromone research, is the extent to which the farmer or government pest control agency will adopt pheromone technology, with its attendant re-education needs and inevitable initial risks. Chapter 13 analyses the issue of adoption in terms of farm type, pest type and the nature of pest-control decision-making found in different countries.

Thus, Section D takes a close look at the commercial aspects of pheromone use, focusing on pheromones for pest control and using as models the major examples of successful pheromone implementation. The section concludes with a view of future pheromone research needs, aimed at technical efficacy in particular, and considers the pheromones other than sex and aggregation pheromones which may offer alternative pest control products.

1.8 REFERENCES

Ashare, E., Brooks, T. W. and Swenson, D. W. (1975) Controlled release from hollow fibres. In *Proc. 1975 Controlled Release Pesticide Symp.*, F. W. Harris (Ed), 42–49.

Bierl, B. A., Devilbiss, E. D. and Plimmer, J. R. (1976). Use of pheromones in insect control programs: slow release formulations. In *Controlled Release Polymeric Formulations*, D. R. Paul and F. W. Harris (Eds), American Chemical Society, Washington, 265–272.

Bierl-Leonhardt, B. A., Devilbiss, E. D. and Plimmer, J. R. (1979). Rate of release of disparlure from laminated plastic dispensers. *J. Econ. Entomol.*, **72**, 319–321.

Butenandt, A., Beckmann, R., Stamm, D. and Hecker, E. (1959). Uber den Sexual-Lockstaff des Seidenspinners *Bombyx mori*. Reidanstellung und Konstitution. Z. Naturforsch. B, **14**, 283–284.

Dawson, G. W., Gibson, R. W., Griffiths, D. C., Pickett, J. A., Rice, A. D. and Woodcock, C. M. (1982). Aphid—*Myzus persicae*—alarm pheromone derivatives affecting settling and transmission of plant viruses. *J. Chem. Ecol.* **8**, 1377–1388.

Holbrook, R. F., Beroza, M. and Burgess, E. D. (1960). Gypsy moth (*Porthetria dispar* (L)) detection with the natural female sex lure. *J. Econ. Entomol.*, **53**, 751.

Hummel, H. E. and Miller, T. A. (1984). *Techniques in Pheromone Research*, Springer-Verlag, New York, 464 pp.

Karlson, P. and Luscher, M. (1959). 'Pheromones' A new term for a class of biologically active substances. *Nature*, **183**, 55–56.

Katsoyannos, B. I. and Boller, E. F. (1976). First field application of oviposition-deterring marking pheromone of European cherry fruit fly. *Environ. Entomol.*, **5**, 151–152.

Katsoyannos, B. I. and Boller, E. F. (1980). Second field application of oviposition-deterring pheromone of the European cherry fruit fly, *Rhagoletis cerasi* L. (Diptera: Tephritidae) Z. *Ang-Ent.*, **89**, 278–281.

Kydonieus, A. F. (1977). The effect on some variables on the controlled release of chemicals from polymeric membranes. In *Controlled Release Pesticides*, H. B. Scher (Ed), American Chemical Society, Washington DC, 152.

Nesbitt, B. F., Hall, D. R., Lester, R. and Marrs, G. J. (1980). UK Patent 8 007 581.

Prestwick, G. D. and Blomquist, G. J. (1987). *Pheromone Biochemistry*, Academic Press Inc., London.

Robinson, S. W., Jutsum, A. R., Cherrett, J. M. and Quinlan, R. J. (1982). Field evaluation of methyl 4-methylpyrrole-2-carboxylate, an ant trail pheromone, as a component of baits for leaf-cutting ant (Hymenoptera: Formicidae) control. *Bull. Ent. Res.*, **72**, 345–356.

Shorey, H. H. and McKelvey, J. J. (1977). *Chemical Control of Insect Behaviour*, Wiley–Interscience, Chichester.

Shorey, H. H., Gaston, L. K. and Saario, C. A. (1967). Sex pheromones of Noctuid moths. XIV. Feasibility of behavioural control by disrupting pheromone communication in cabbage loopers. *J. Econ. Entomol.*, **60**, 1541–1545.

Wood Mackenzie & Co. Ltd (1988). Agrochemical Service, March 1988.

PART A

Background

Insect Pheromones in Plant Protection
Edited by A. R. Jutsum and R. F. S. Gordon
©1989 John Wiley & Sons Ltd

2

Pheromones and Insect Behaviour

C. T. David

*AFRC Insect Physiology Group, Department of Pure and Applied Biology,
Imperial College of Science and Technology, London SW7 2AZ, UK*

M. C. Birch

*Department of Zoology, University of Oxford, South Parks Road,
Oxford OX1 3PS, UK*

2.1 INTRODUCTION

Almost any behavioural modification induced by a chemical compound can be used to manipulate some aspect of an insect's relationship with plants. For instance, populations of a pest insect can be reduced by mass-trapping using the insect's own sex pheromone as a lure, and as a result less of a particular crop is damaged. The same result could be achieved by spraying the crop with a feeding deterrent, making those plants inedible, or with an oviposition deterrent, stopping the pest from laying eggs on it.

Consideration of the action of a pheromone on a target insect's behaviour will lead to insights into how to use such chemicals in plant protection. Immediately, however, it will more often highlight areas where more research is needed before chemicals can be applied intelligently. For instance, much effort has been put into trying to disrupt the mating of pest insects by applying behaviourally active chemicals in the field, as with the 'confusion' technique (see Chapter 5), but very few of the attempts have been successful (Wall, 1985). At least five possible mechanisms by which confusion might work have been proposed (Bartell, 1982). Knowledge of the actual mechanism or mechanisms would surely suggest more effective ways of applying the chemical. Yet few studies have attempted to discover that mechanism, for example by establishing common factors in those cases where confusion does work. Is receptor adaptation the same in all cases, and is there any remaining response of male insects to female pheromone plumes?

17

There are still large gaps in our understanding of the behaviour we are trying to disrupt. We do not even know completely how flying insects orient to the sources of attractant odour in the laboratory, let alone in nature, although this problem has been studied since the days of Fabre.

In this chapter we have chosen to concentrate on how flying insects orient to a source of odour, since many plant protection applications of pheromones are believed to work through manipulating the orientation of flying insects. The chapter is divided into three sections. The first covers pheromone components. The second deals with how insects reach a source of attractant odour, what mechanisms they use and how these are different between species and in different environmental conditions. Lastly, ways will be considered in which pheromones can be used to monitor insect numbers, or to control them by trapping and confusion. In each case we will outline what is known and point out where more research needs to be done.

2.2 PHEROMONE COMPONENTS

An important area of research is the study of how the precise composition of the pheromone stimulus affects orientation behaviour, and also the other types of behaviour that lead to mating or aggregation. Insects not only synthesize pheromonal components to a high degree of purity by specific biosynthetic pathways, but also precisely control the geometrical and optical isomerism of the molecules, and the ratios in which they are produced (Löfstedt and Odham, 1984; Löfstedt et al., 1982, 1985a,b).

Different isomers can evoke very different behaviour patterns in a responsive insect, so discrimination between these molecules must occur in the antennae and in the central nervous system. In insects, olfaction is the dominant sensory system for pheromone detection, and insect pheromones are much simpler than mammalian ones. Typically an insect pheromone comprises a few compounds whose composition is well defined and generally very consistent from individual to individual. In this section we will give a simple overview of pheromone components and their functions.

The simplest way for two lepidopteran species to achieve reproductive isolation through chemical differences in their pheromones would be for each to employ a different compound. The sex pheromones of most Lepidoptera are a mixture of several components, not all of which may be necessary to stimulate upwind flight in the male. Some may stimulate close-range orientation or courtship behaviour. Two-component pheromones illustrate some of the principles of multicomponent pheromones in reproductive isolation (Tanaki, 1985).

Two tortricid moths which co-exist in nature, with apparently identical flight activity and response periods and with the same two component sex pheromone, are the leaf-roller *Clepsis spectrana* (Treitschke) and the summer fruit tortrix *Adoxophyes orana* (Fischer von Roslerstamm). Both females produce two components, (Z)-9- and (Z)-11-tetradecenyl acetates, but the males respond to different critical ratios (Minks *et al.*, 1973). Males of *A. orana* are most attracted to (Z)-9- and males of *C. spectrana* will go to (Z)-11-, and the optimal blend for the latter species is a 1 : 3 mixture of (Z)-9- : (Z)-11-. This is close to the ratio produced by the female.

In a blend of four components, males are attracted to more complex ratios: the blend ratio in the tortricid fruit-tree leaf-roller moth *Archips argyrospilus* (Walker) is 60 : 40 : 4 : 200. In the sibling species *A. mortuanus* Kearfoot, the blend ratio is 90 : 10 : 1 : 200. The components are (Z)- and (E)-11-tetradecenyl acetates, (Z)-9-tetradecenyl acetate and dodecanyl acetate. Shifts in these optimal ratios cause a drastic reduction in trap catches (Cardé *et al.*, 1977).

In another type of system, the redbanded leaf-roller, *Argyrotaenia velutinana* (Walker), the female produces a pheromone of three components (Roelofs *et al.*, 1975). The (Z) : (E) ratio of -11-tridecenyl acetate is a crucial factor in eliciting the response of males moving upwind towards the female (Cardé *et al.*, 1975; Baker *et al.*, 1976). The third component, dodecanyl acetate, seemed to lead to wing fanning by the male and may reflect an effect on the male's close-range behaviour towards the female.

The three major components of the sex pheromone of the pine beauty moth, *Panolis flammea* (Den. & Schiff), are (Z)-11-tetradecenyl acetate, (Z)-9-tetradecenyl acetate and (Z)-11-hexadecenyl acetate, which are present in the ratio 100 : 5 : 1 (Baker *et al.*, 1982). When all three compounds were released in a plume in a wind tunnel, a high percentage of males flew upwind, landed and attempted copulation (Bradshaw *et al.*, 1983). In the absence of (Z)-11-tetradecenyl acetate, the percentage of landings decreased markedly. A high percentage of landings occurred in the absence of (Z)-11-hexadecenyl acetate but few copulations were attempted. The (Z)-9-tetradecenyl acetate appears to activate the males. Thus, when the sex pheromone is produced, male moths are stimulated to fly upwind. The higher concentration of (Z)-11-tetradecenyl acetate occurring near the female, where turbulence has had less time to dilute it, increases the landing rate. Then (Z)-11-hexadecenyl acetate releases attempts at copulation. Thus the pheromone components in this case are critical factors in the control of a behavioural sequence.

Priesner and Witzgall (1984) have studied the modification of the response of the larch casebearer moth *Coleophora laricella* (Hubner) to its single-component pheromone by a compound produced by the females after mating, which stops the males from being attracted. This compound modifies the counterturns of a responding male, making the zigzags sharper. In higher doses it acts as a repellent. Interestingly, the responses of males are quite different if the pheromone and the 'inhibitor' are separated in space. When the two sources are across wind from each other the inhibitor has little effect if it is more than 20 cm from the attractant. If the inhibitor is downwind from the attractant it blocks the approach of moths to the attractant, but if it is upwind males can still reach the attractant if the inhibitor is more than 0.5 m from it. However, their flight tracks are characteristically altered: the zigzagging amplitude is reduced and the tracks appear canalized. Priesner and Witzgall consider that their results can best be explained on the assumption that male moths avoid concentrations of the 'inhibitor' that rise faster than concentrations of the attractant pheromone.

In other insect orders, pheromone components can act differently through time and space, as in the weaver ants (*Oecophylla* spp.). Their alarm pheromone is produced by paired glands at the base of the mandibles. There are thirty compounds in this secretion, and the behavioural effects of four of them have been identified (Bradshaw *et al.*, 1979). Hexanal alerts the major workers, which then rush around with their jaws open. 1-Hexanol attracts the ants, which may then circle the source, and the less volatile 3-undecanone and 2-butyl-2-octenal attract at short range and elicit prolonged biting responses. Thus the four components release a sequence of behaviour, proceeding from alerting to orientation and biting (Howse *et al.*, 1986).

In beetles, the role of aggregation pheromones in reproductive isolation is seen most clearly in the *Ips* species of bark beetle. In this group the pheromones comprise a few compounds manufactured by the insects plus plant compounds sequestered by them. The aggregation pheromone increases beetle density in the vicinity of the pheromone source on the host tree. Such aggregation increases the probability of successful copulation, so it is clear that there is no good distinction between aggregation and sex pheromones.

A blend of ipsenol, ipsdienol and *cis*-verbenone was the first pheromone isolated from a beetle, the California five-spined engraver beetle, *Ips paraconfusus* Lanier (formerly *I. confusus* Le Conte) (Wood *et al.*, 1967). Ipsenol and ipsdienol contain asymmetric carbon atoms in their molecules, each exists as two enantiomers. Thus these compounds are more accurately (*R*)-(−)-ipsenol, (*S*)-(+)-ipsdienol and (*R*)-(−)-ipsdienol. All *Ips* species whose pheromones have been identified use combinations of these pheromone components, and in only one species are additional compounds known (Birch, 1984). In northern California, *I. paraconfusus* and *I. pini* (Say) are sympatric; both attack and infest fallen ponderosa pine. Although both share the same habitat and are active at the same time of year they never breed in the same piece of host material. Only *I. paraconfusus* is attracted to logs containing *I. paraconfusus*, only *I. pini* is attracted to logs containing *I. pini*. The pheromone of *I. pini* in California is the single component (*R*)-(−)-ipsdienol (Birch *et al.*, 1980). This attracts *I. pini* but prevents the response of *I. paraconfusus* to its pheromone. Similarly the *I. paraconfusus* pheromone of (*R*)-(−)-ipsenol and (*S*)-(+)-ipsdienol prevents the response of *I. pini*. More than 2–3% of (+)-ipsdienol will completely shut off the response of *I. pini* to its pheromone (−)-ipsdienol. Pheromonal specificity is a principal isolating mechanism among sympatric species of *Ips* (Birch, 1984). The first species to colonize a new host effectively reserves that host for conspecifics. Interruption also prevents the reduction in brood that occurs when both species colonize a host together: this brood reduction is presumably a powerful selection pressure which leads the beetles to avoid colonizing already colonized material.

This section has shown how pheromone components and behaviour are closely intertwined. In the early days of pheromone studies it was thought that pheromones were single-compound substances in most species. Now it is known that the majority of species use blends of compounds and that the behaviour they induce is more complicated than simply flight or movement towards a source. In fact we now know that quite complex responses may be elicited and that these are a function of the context of pheromone emission.

2.3 IN-FLIGHT ORIENTATION TO ODOUR SOURCES

What do flying insects do when they orient to distant sources of chemical? Most work on answering this question has been done in wind tunnels in the laboratory, although some methods have been devised for studying larger species under more natural conditions in the field. The basic strategy which leads insects to arrive at the source of an attractant chemical is to fly upwind (as long as there is any wind) when stimulated by the chemical, and to land or to fly cross-wind when they lose the chemical. There appear to be two different ways of achieving this orientation to wind.

Most insects can orient in relation to the wind whilst airborne. This means that they can determine the particular values of airspeed and heading which will overcome the speed and direction of the wind. An airborne insect cannot determine these values through mechanoreceptive reactions to the direction of the air flow past itself because this flow depends on its own flight direction and airspeed, and not on the direction or speed of movement of the air over the ground. Instead, as first shown by Kennedy (1940), insects rely on visual reactions to the relative movement of fixed features on the ground to orient in relation to the wind. Solving the problem of flying upwind is relatively simple. If an insect is flying straight upwind, fixed objects on the ground underneath it will appear to pass along the body axis from front to back. If the insect is not flying fast enough to overcome the wind the fixed objects will appear to pass from back to front. Consequently, in order to fly upwind the insect should accelerate if it sees visual flow in this direction. This is precisely what insects do in a wind tunnel when the floor is moved forwards underneath them, simulating a wind that is blowing the animal backwards (Kellog *et al.*, 1962; Kennedy and Thomas, 1974; David, 1978, 1982b, 1985, 1986). So visual reactions to the ground do control the airspeed of the animal.

Obviously the insect must also steer in the right direction so that the visual flow is from front to back as well as control its airspeed. This is also accomplished visually. As long as the insect is flying upwind the objects on the ground will pass from front to back beneath it. As soon as it deviates from the upwind direction the insect will be drifted sideways by the wind (Fig. 2.1) and the objects on the ground will no longer pass from front to back but obliquely from the side. Kennedy (1940) showed that in response to such a visual flow mosquitoes turned in the upwind direction, towards the direction of the flow, and also demonstrated that such a reaction would always result in upwind flight. Similar reactions have been found in other flying insects (David, 1982a, 1986).

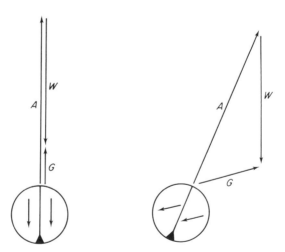

Fig. 2.1 Visual stimulus for steering upwind. The figure shows the ground track direction and groundspeed (G) produced by an insect flying with airspeed and heading (A) in a wind with speed and direction (W). If the insect is steering straight upwind (left) its ground track direction is aligned with its body axis, and consequently the visual flow past its eyes (circle at bottom) is also aligned with its body axis. If it deviates from the upwind direction (right), its ground track will no longer be aligned with its body axis and the visual flow past its eyes will deviate from its body axis. To return to upwind flight the insect should turn to the left when it sees visual flow towards its left, and vice versa (Kennedy, 1940)

2.3.1 Counterturning

Some insects, such as *Drosophila*, fly straight upwind in attractant chemical plumes as if the above two reactions to visual flow explained all of their orientation. Most insects, however, zigzag when flying up such a plume (Fig. 2.2). This means that most of the time they are flying across the wind and not straight upwind. As in the case of upwind flight, these cross-wind zigzag legs are, at least partially, visually steered and an insect's track can be controlled in a wind tunnel by moving the floor (Kennedy and Marsh, 1974). However, precisely how they are steered remains a matter for speculation and experiment (David, 1986). It is helpful to look at the evidence on what makes insects zigzag before considering its steering.

The first theory regarding the causes of zigzagging was that the loss of odour as the insect reached the edge of the plume either produced each turn or led to a sequence of turns. The insect supposedly flew obliquely upwind as long as the odour stimulus from the plume was increasing and switched to the other tack, flying obliquely the other way across the plume, as soon as the odour stimulus decreased in strength. This story is no longer generally believed. Kennedy *et al.* (1980, 1981) showed that zigzagging continued in *A. orana* even when there were no falls in the odour concentration as might occur at the boundaries of a pheromone plume. They uniformly permeated an area of the wind in a wind tunnel with pheromone and moths entering this area, whilst casting cross-wind after losing a normal odour plume, carried on zigzagging, but with decreased intervals between turns; however, they made no upwind progress after a brief surge as they entered the pheromone. Thus they behaved rather as if they were still casting cross-wind after the loss of pheromone odour: they did not advance upwind, but more as if they were in a plume in the amplitude of their zigzagging. So, far from a fall in pheromone concentration being necessary to produce a sequence of turns as had been thought, turns were induced by the presence of pheromone. A fall in pheromone concentration in fact led to longer intervals between turns. This is clearly shown in the typical widening of the zigzag track of a moth which has been in a plume into cross-wind casting when it loses it. These results have since been partially confirmed in the oriental fruit moth *Grapholitha molesta* (Busck) (Willis and Baker, 1984). The conclusions drawn by Kennedy *et al.* (1980, 1981) were that the zigzagging is produced by an internal counterturning mechanism which manifests itself both in the zigzag track of moths flying up an odour plume and in the turns at the ends of the cross-wind casts made by moths that have lost the plume. The frequency of the zigzag turns is modulated by the odour stimulus, increased by increases in odour concentration and decreased by decreases. If the odour stimulus is unchanging, either with no odour present or with

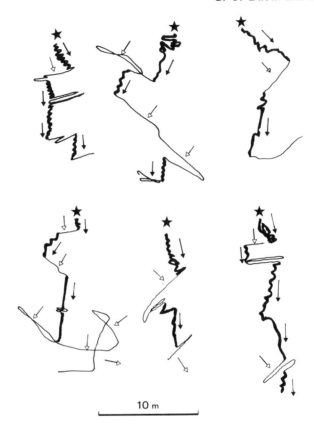

10 m

Fig. 2.2 Tracks of male gypsy moths approaching a source of 100 ng of artificial female pheromone ((+)-Disparlure) in the field. Soap bubbles were released at the pheromone source (star) and marked the position of the odour plume (David *et al.*, 1983). Moths were video-recorded from above and to the side as they flew through an open experimental area marked with numbered wind vanes and streamers, and the tracks subsequently plotted on an orthogonal grid removing the effects of perspective. Thick lines show the tracks of the male moths while they were surrounded by bubbles and dark arrows show the corresponding wind directions. The moths flew upwind on a more or less zigzag path when stimulated by the odour. Thin lines show the tracks of the moths when they were away from the bubbles and the open arrows show the corresponding wind directions. The moths flew across the wind when they were away from the odour plume. The moths frequently lost contact with the plume of bubbles, most often when the wind changed direction and blew the plume away from the moth

a uniform fog of odour, the moth makes no resultant upwind progress through its counterturning. Upwind progress is only made when the odour stimulus is constantly changing. This result was also confirmed by Willis and Baker (1984). Moths will orient in narrow zigzags up an odour plume superimposed on the uniform background fog of odour (Kennedy *et al.*, 1980,

1981; Willis and Baker, 1984). This observation has important implications for the mechanism of communication disruption, as will be seen later.

Baker and Kuenen (1982) showed that zigzagging continued in *G. molesta* even when the wind stopped: for many seconds the overall track of the moth was indistinguishable from that in a wind, even to the extent that it switched to cross-tunnel casting, across the old wind direction, if it lost the pheromone. To perform these feats, the moth must almost instantly have adjusted its airspeed, its cross-wind headings and its extent of turning at the end of each zigzag leg. This important result sheds some light on the mechanism of zigzagging. The zigzagging continues when there is no wind, and therefore no wind-induced visual drift. The moth is neither simply flying at an angle to the wind because it maintains the visual drift at a fixed value nor zigzagging by turning its other side to the wind at the same angle, and with the same visual drift. Similarly the zigzagging is unlikely to be noise in the control loop of the steering response to visual drift, as has been suggested by Preiss and Kramer (1986).

The amazing regularities in the zigzagging of moths at different wind speeds and in different pheromone concentrations also refute this hypothesis. Marsh *et al.* (1978) showed that *Plodia interpunctella* (Hubner) follow remarkably similar zigzag tracks in three different wind speeds. Just as in *G. molesta* (Baker and Kuenen, 1982) this must entail precise adjustments both in the angle through which they turn at each zigzag and in their airspeeds. In *G. molesta*, Kuenen and Baker (1982) showed that at different pheromone concentrations the zigzag tracks maintain a constant angle to the wind, but that groundspeed varies. Zigzag turns and airspeed are therefore under precise control. Zigzagging also occurs when moths that have never experienced wind are stimulated by pheromone, although in this case the zigzags are not oriented relative to the wind tunnel. As soon as wind occurs, the zigzags become oriented relative to the tunnel (Baker *et al.*, 1984). This is presumably a reaction to the induced visual drift, since similar steering can be evoked by moving the floor (David and Kennedy, unpublished). Some individuals are able to fly along a plume into which they had flown in still air without ever having been in wind (Baker *et al.*, 1984), but these may just have been flying in that direction by chance, rather than chemotactically steering their zigzags by the plume (but see Section 2.3.3).

In summary, this is how it is believed that a moth steers in an odour plume. Pheromone odour switches on an internal programme which produces flight on a zigzag track. The track is oriented relative to the wind through visual reactions to the image flow produced by the wind drift. The structure and concentration of the pheromone stimulus determine the frequency of the zigzagging and also the angle of the legs to the wind. At higher, changing concentrations the angle is steeper and the legs are short, so that the moth moves upwind along the plume. At very low concentrations, or if the stimulus is unchanging, it casts at right angles to the wind direction without advancing.

Several problems still remain. For instance, how do insects keep in the centre of a plume? As explained above, the theory that they turn to re-enter the plume when they experience a decrease in pheromone concentration is no longer valid. On the contrary a moth turns more frequently when stimulated by the odour, thus flying into the wind so long as it is in the plume. But this behaviour alone would not ensure that a moth found the centre of a wide plume, as they do in wind tunnels (although it is not known that they do so in the field). Some form of chemotaxis must be involved. More information comes from observations on the behaviour of moths in natural odour plumes. As Wright (1958) observed, plumes are structured and contain clean air inside them. This description has been quantified by Murlis and Jones (1981). If moths turned on losing the odour they would turn as frequently in the middle of a plume as they would at its edges. Odour plumes in the field rarely stay in one place for any length of time because the wind direction continually changes (David et al., 1982). Consequently insects flying in an odour plume at some distance from the source frequently lose it (David et al., 1983) (Fig. 2.2). Near the source, however, gypsy moths are amazingly good at following a plume, appearing to move sideways with the plume as the wind swings. This could be explained by the moths casting as they lose the odour, with each cast much longer than the one before (David, unpublished). Cardé and Charlton (1985) have mimicked this situation in a wind tunnel by moving the source of odour sideways. They believe that the simple modulation of an internal turning system in response to odour concentration, as proposed by Kennedy et al. (1980), is insufficient to explain the moths' behaviour and that some form of cross-wind chemotaxis is involved, although they have not clearly ruled out the casting explanation given above. The conclusion is that we still do not completely understand how insects find the source of an attractant odour in the field.

2.3.2 Non-visual orientation to wind

We mentioned that there are two ways of orientating upwind in response to odour. One, described above, is through in-flight reactions to visual flow. The other is to detect the wind direction with mechanoreceptors while sitting in some fixed position, then to orientate in that direction in flight and to fly forwards until the odour is lost. This behaviour has been reported from cabbage root flies (Hawkes and Coaker, 1979), onion flies (Dindonis and Miller, 1981) and tsetse flies (Bursell, 1984), all responding to host odours and not to pheromones, but this is probably a coincidence. Other insects, such as *Drosophila* responding to banana odour, respond to host odours with in-flight course corrections induced by visual drift (Kellog *et al.*, 1962; David, 1982a,b). Tsetse flies respond to the loss of the odour by turning, either through 180° (Bursell, 1984), or about 140° (Gibson and Brady, 1985). It is possible that non-visual anemotactic behaviour occurs when the flight speed is much faster than the wind speed, so that visual drift cues for anemotaxis are weak (although this is not necessarily so if the insects resolve their visual flow into components) (Marsh *et al.*, 1978; David, 1985, 1986). The insect would be keeping in the plume by a form of longitudinal klinotaxis (Kennedy, 1983; Bursell, 1984), turning in relation to its previous path on the loss of odour stimulation, rather than reorientating in relation to the wind. In slow winds this form of orientation may be as reliable and quick a way of reaching the source as the better-studied anemotactic mechanism.

2.3.3 Sensory physiology and counterturning

A great deal of work has been done on the electrophysiology of the sense organs that detect the chemical stimulus. Good recent reviews can be found in Bell and Cardé (1984) and in Payne *et al.* (1986). An approach has also been made to the understanding of the neural basis of zigzagging. As pointed out above this is believed to be an internally generated behaviour modified by pheromone input. Such modifications might be carried out by flip-flopping interneurons, such as those discovered by Olberg (1983), which produce course changes in silkworm moths in response to changes in pheromone concentration. It is clear that these cannot be the whole control system, as Kramer (1975) showed that silkmoths walking in uniform pheromone odour turned spontaneously, while the flip-flopping interneurons apparently only change state in response to a change in the odour stimulus.

There is much speculation on the evolution of pheromonal systems. It is hard to explain why the quantities of pheromone emitted by many female moths are so minute, and the males so exquisitely sensitive. These explanations often rely on sexual selection. It is interesting that Cardé and Hagaman (1984) have shown that the female gypsy moth is amazingly unchoosy in taking a mate once he finds her; any female choice is thus for the searching ability of the males and not for any other character. Another evolutionary problem is why so many insects reach odour sources by a combination of upwind flight and zigzagging. The answer can perhaps be derived from field observations of their behaviour, but these can be very difficult because of the small size and rapid movement of the insects and of the difficulty of knowing when they are detecting the invisible odour and when they are not. David *et al.* (1983) devised a means of making the gross distribution of odour visible by releasing soap bubbles at the same point as the odour. The soap bubbles are carried downwind with the odour and are easily visible at a distance from the source. Individuals of the large, day-flying gypsy moth *Lymantria dispar* (L.) are released in the plume of odour and bubbles and followed as they fly towards the source. Happily the behaviour is as expected from wind tunnel observations, although on a much larger scale (Fig. 2.2). The moths fly upwind in zigzags when they are among the bubbles and presumably in the odour. The zigzags are sometimes a metre across, whereas in a wind tunnel they are usually only centimetres. As in the wind tunnel the moths fly in wide casts across the wind when they lose the plume of bubbles, but some of these cross-wind casts are as long as 20 m. Presumably these types of behaviour are on a reduced scale in the wind tunnel because the moths have to turn to avoid hitting the walls.

One significant discovery was that the moths lose the plume of bubbles most often when the wind changes direction and blows the plume away from them, and seldom through bad steering. Their subsequent casting flight usually leads them to rediscover the plume closer to the source. This is because of the way the wind behaves in open areas: individual packets of air move in straight lines for long distances (David *et al.*, 1982) (Fig. 2.3). A consequence of this is that to and fro cross-wind flight on loss of the pheromone leads the moth alternately nearer to and further from the source, but they only refind the odour plume when they are nearer to it. In nature gypsy moths live in woods and the wind there behaves in a different way than in open fields (Cardé, 1986). However, the cross-wind casting on losing the odour would still serve to refind the plume, and even in those conditions could be better than flying in any other direction. Sabelis and Schippers (1984) have produced theoretical predictions of the direction insects should fly to find new odour plumes in winds that differ in reliability of direction. If the wind direction is very unstable, downwind flight is best. The same could apply to the refinding of a plume an insect has just lost. From time to time gypsy moths make downwind 'casts' in the field (David, unpublished), and Traynier (1968) has reported a sequence of cross-wind, then downwind, then cross-wind casts in moths that had lost a plume in a wind tunnel. *Drosophila* do the same (David, unpublished).

2.4 BEHAVIOUR AND TRAP DESIGN

There are two main ways in which pheromones are now being introduced into agricultural and forest pest management. One is the confusion technique, which is discussed in the next section, the other is in the use of traps. Traps are used in two ways: first, to detect and monitor pest populations — this allows other methods of regulating pest populations to be used more effectively; second, large numbers of pheromone traps can be used to attract and remove as many of the reproductively active pest population as possible.

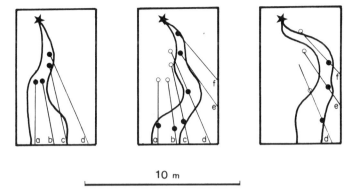

10 m

Fig. 2.3 Plans of an odour plume at 2 s intervals in the same area as Fig. 2.2. Black spots mark the positions of individual bubbles, thin lines show their trajectories, open circles their positions in the preceding plan and thick black lines the boundaries of the area enclosing all the bubbles. The plume of bubbles meanders over the ground even though the track of each individual bubble is straight (see David *et al.*, 1982). A moth flying upwind inside the plume of bubbles will frequently lose the plume since the wind direction is not always aligned with the long axis of the plume. In a wood, the movement of air carrying an odour plume is different, but the wind is still seldom aligned with the plume's long axis (Cardé, 1986)

Most traps have been used with moths and beetles, and a few with flies. The development of a trap suitable for the particular insect is vital. The behaviour of insects as they approach, enter and try to escape from traps can be very different and therefore affect the design. There is no way to assess the effects of a particular trap design on the behaviour of a given insect without seeing it in action. This will receive particular attention in Chapters 3 and 4. However, certain conclusions can be drawn: (1) with all pheromone trap studies it is essential to have as full an understanding as possible of the appropriate behavioural components of the target species; (2) when the behaviour is understood the appropriate trap can be designed; (3) traps can then be modified either for the detection or monitoring of the pest or for mass trapping in control programmes. Clearly trap design has to be tailored to the precise function required.

2.5 COMMUNICATION DISRUPTION

Bartell (1982) has recently reviewed the mechanisms by which communication disruption might work, and this is the subject of Chapter 5. In this section we will not review these again, rather we will report on two pieces of work in which we have been involved. One is an example of the experimental use of disruption in the field, the other sheds light on a possible mechanism of disruption and also on the natural behaviour of male moths in a pheromone plume. One way to study communication disruption in the field is to establish small-scale plots each with a trap baited with virgin female moths or with synthetic pheromone surrounded by baits emitting sex pheromone (Roelofs, 1979). The catch of male moths in the central trap can be compared with different baits, and with controls with no surrounding baits.

In the artichoke plume moth, *Platyptilia carduidactyla* (Riley), there is only one chemical component in the female sex pheromone, (Z)-11-hexadecanal (Klun *et al.*, 1981; Haynes and Birch, 1984). Mass trapping of this species appears to perform well at high population densities, although these traps are not necessarily the most practical integration into the already complex agricultural practices used in the management of artichoke fields. However, small-scale experiments indicate that the pheromone has the potential to disrupt mating in artichoke plume moths. In Castroville, California, small grids of pheromone lures (10 Hercon laminate flakes; $22 \mu g \, d^{-1}$) were set up with a trap in the centre of each grid. Each grid consisted of 4×4 lures with the lures spaced 3 m apart; control grids were the same but lacked lures. The central trap consisted of three, day-old, virgin females, or a lure, or three virgin females on a mating station consisting of a pan containing the females, each with one wing clipped (Haynes, 1982). The results of the experiments demonstrated a 97% reduction of males trapped compared with controls when lures were used in the central traps, and a 94% reduction with virgin females in the traps. Mating was completely disrupted on the mating stations. This may be the ideal crop–insect system to test the feasibility of disrupting mating behaviour, since there is little more than 4000 ha of artichokes to protect.

One of the possible mechanisms of disruption is that the pheromone artificially placed in the field produces plumes of pheromone. Male moths encountering these may respond to them as if they were plumes coming from an attractive female by flying upwind in them, thus reducing the mating rate. This waste of time would be increased if there were some way to habituate or adapt the male to the odour while it is flying in it. Van der Kraan (unpublished) suggested that this was what might occur if the pheromone plume from the artificial source could be made significantly stronger than that from the real females. In many cases, if a plume is stronger than that produced by a female, male moths will only fly in it for a short time before flying away. There is an upper as well as a lower limit to the strength of a pheromone plume that will induce males to fly upwind. However, it was not known what caused the moths to stop following such a strong plume, nor whether the moths were sensorally adapted or centrally habituated by the high pheromone concentrations they had just been experiencing so that they would not follow a weaker plume from a female.

It was decided to investigate whether such a phenomenon would occur by experiments using gypsy moths both in the field and in a wind tunnel (David, in preparation). In the wind tunnel male gypsy moths were flown in plumes of pheromone of similar strengths as would be found at distances of 1–20 m downwind from a calling female. They were then flown in a plume that was 10–1000 times stronger than the one they were previously following for lengths of time varying from 10 s to 1 min. At the end of this time the stronger plume was removed and they were once more exposed to the weaker plume they had been following before. This drop in pheromone concentration caused the moths to cast across the wind tunnel, passing repeatedly through the weaker plume. After a length of time that depended directly on the strength of the stronger plume and the length of time they were exposed to it, the moths 'locked on' again to the weaker plume and commenced to fly upwind. Thus it was quite clear that the moths were adapted or habituated by the stronger plume so that they no longer responded to the weaker one. This lack of response lasted about 1 min when the weaker plume was from a source containing 1 ng disparlure, and the stronger one had 1000 ng disparlure to which the moths were exposed for 1 min.

In the field moths were presented with two plumes of different strengths 5 m apart cross-wind. When the plumes differed by a factor of 20 in strength the large majority of the moths arrived at the stronger source, regardless of which plume they had been released in 20 m downwind. This was because as the wind swung they frequently lost the plume they were following. They then cast across the wind and were likely to meet the stronger plume. Once they had done so they followed it. However, if they met the weaker plume they were less likely to lock on to it even if it was the plume they had just lost, and they never locked on to it if they had just lost the stronger plume. The majority of male moths also arrived at an artificial pheromone source rather than at a calling female moth when the plume from the artificial source was only five times as strong as the one from the female.

These results show how moths manage to find one female when there are many calling females close by. One would have imagined that as the wind swings frequently, and a male is constantly losing the plume that it is following, it would then lock on to the first plume it discovers. This would be most likely to be a plume from a female further upwind than the female whose plume he has just lost, since a plume that has come from further will be broader and more likely to be discovered by the male. In consequence, the male would progress upwind through a dense population of female moths but seldom stopping at any of them. In fact this will not happen, since a plume that has come from further upwind will be weaker as well as broader (Murlis and Jones, 1981). A male moth will not lock on to a weaker plume for a time after he has lost a stronger one. He will keep on casting and not be 'seduced' by the weaker plume until he has 'searched' thoroughly for the stronger plume. And he will search for longer periods as the concentration difference between the two plumes increases. For confusing the male moth, therefore, a good strategy would be to place the pheromone in dispensers, each producing a plume as strong as is possible. These should cause the males to ignore the females' plumes unless they happen to come very close to her. This will only happen by chance, more often in high than low female densities, and will be less efficient than the natural way of reaching females by following their pheromone plumes. Despite the above evidence favouring competitive point-sources of pheromone, in fact it seems quite possible to perfect confusion systems using microencapsulated pheromone. It may be that fogs are not produced by such a system because of air turbulence mixing in clean air from above and around the treated area. However, it is obvious that this is an important area where much more research is needed.

Acknowledgements

We are grateful to Drs M. L. Birch, C. Rechten and J. E. Moorhouse for their helpful comments on the manuscript.

2.6 REFERENCES

Baker, R., Bradshaw, J. W. S. and Speed, W. (1982). Methoxymercuration–demercuration and mass spectrometry in the identification of the sex pheromones of *Panolis flammea*, the pine beauty moth. *Experientia*, **38**, 233–234.

Baker, T. C. and Kuenen, L. P. S. (1982). Pheromone source location by flying male moths: a supplementary non-anemotactic mechanism. *Science*, **216**, 424–427.

Baker, T. C., Cardé, R. T. and Roelofs, W. L. (1976). Behavioural responses of male *Agyrotaenia velutinana* (Lepidoptera: Tortricidae) to components of its sex pheromone. *J. Chem. Ecol.*, **2**, 333–352.

Baker, T. C., Willis, M. A. and Phelan, P. L. (1984). Optomotor anemotaxis polarizes self-steered zigzagging in flying moths. *Physiol. Ent.*, **9**, 365–376.

Bartell, R. J. (1982). Mechanisms of communication disruption by pheromone in the control of Lepidoptera: a review. *Physiol. Ent.*, **7**, 353–364.

Bell W. J. and Cardé, R. T. (Eds) (1984). *Chemical Ecology of Insects*, Chapman & Hall, London.

Birch, M. C. (1984). Aggregation in bark beetles. In *Chemical Ecology of Insects*, W. J. Bell and R. T. Carde (Eds), Chapman & Hall, London, pp. 331–354.

Birch, M. C., Light, D. M., Browne, L. E., Silverstein, R. M., Bergot, B. J., Ohloff, G., West, J. R. and Young, J. C. (1980). Pheromonal attraction and allomonal interruption of *Ips pini* in California by the two enantiomers of ipsdienol. *J. Chem. Ecol.*, **6**, 703–717.

Bradshaw, J. W. S., Baker, R. and Howse, P. E. (1979). Multicomponent alarm pheromones in the mandibular glands of the African weaver ant, *Oecophylla longinoda*. *Physiol. Ent.*, **4**, 15–25.

Bradshaw, J. W. S., Baker, R. and Lisk, J. G. (1983). Separate orientation and releaser components in a sex pheromone. *Nature*, **304**, 265–267.

Bursell, E. (1984). Observations on the orientations of tsetse flies (*Glossina pallidipes*) to wind-borne odour. *Physiol. Ent.*, **9**, 133–137.

Cardé, R. T. (1986). In *Mechanisms in Insect Olfaction*, T. C. Payne, M. C. Birch and C. E. J. Kennedy (Eds), Oxford University Press, Oxford.

Cardé, R. T. and Charlton, R. E. (1985). Olfactory sexual communication in Lepidoptera: Strategy, sensitivity and selectivity. In *Insect Communication*, T. Lewis (Ed.), Academic Press, London, pp. 241–265.

Cardé, R. T. and Hagaman, T. E. (1984). Mate location strategies of gypsy moths in dense populations. *J. Chem. Ecol.*, **10**, 25–31.

Cardé, R. T., Baker, T. C. and Roelofs, W. L. (1975). Behavioural role of individual components of a multicomponent attractant system in the Oriental fruit moth. *Nature*, **253**, 348–349.

Cardé, R. T., Cardé, A. M., Hill, A. S. and Roelofs, W. L. (1977). Sex attractant specificity as a reproductive isolating mechanism among the sibling species *Archips argyrospilum* and *mortuanus* and other sympatric tortricine moths. *J. Chem. Ecol.*, **3**, 71–84.

David, C. T. (1978). The relationship between body angle and flight speed in free-flying *Drosophila*. *Physiol. Ent.*, **3**, 191–195.

David, C. T. (1982a). Competition between fixed and moving stripes in the control of orientation by flying *Drosophila*. *Physiol. Ent.*, **7**, 151–156.

David, C. T. (1982b). Compensation for height by *Drosophila* in a new 'barber's pole' wind tunnel. *J. Comp. Physiol.*, **147**, 485–493.

David, C. T. (1985). Visual control of the partition of flight force between lift and thrust in free-flying *Drosophila*. *Nature*, **313**, 48–50.

David, C. T. (1986). Mechanisms of directional flight in wind. In *Mechanisms in Insect Olfaction*, T. L. Payne, M. C. Birch and C. E. J. Kennedy (Eds), Oxford University Press, Oxford.

David, C. T., Kennedy, J. S., Ludlow, A. R., Perry, J. N. and Wall, C. (1982). A re-appraisal of insect flight towards a distant point source of wind-borne odor. *J. Chem. Ecol.*, **8**, 1207–1215.

David, C. T., Kennedy, J. S. and Ludlow, A. R. (1983). Finding of a sex pheromone source by gypsy moths released in the field. *Nature*, **303**, 804–806.

Dindonis, L. L. and Miller, J. R. (1981). Host-finding behaviour of onion flies. *Hylema antiqua*. *Environ. Ent.*, **9**, 769–772.

Gibson, G. and Brady, J. N. (1985). 'Anemotactic' flight paths of tsetse flies in relation to host odour: a preliminary video study in nature of the response to loss of odour. *Physiol. Ent.*, **10**, 395–406.

Hawkes, C. and Coaker, T. H. (1979). Factors affecting the behavioural responses of the adult cabbage root fly, *Delia brassicae*, to host plant odour. *Ent. Exp. Appl.*, **25**, 45–58.

Haynes, K. F. (1982). The role of chemical communication in the mating behavior of the artichoke plume moth, *Platyptilia carduidactyla*. Doctoral Dissertation, University of California, Davis.

Haynes, K. F. and Birch, M. C. (1984). Mate location and courtship behaviors of the artichoke plume moth, *Platyptilia carduidactyla* (Lepidoptera: Pterophoridae). *Environ. Ent.*, **13**, 399–408.

Howse, P. E., Lisk, J. C. and Bradshaw, J. W. S. (1986). The role of pheromones in the control of behavioural sequences in insects. In *Mechanisms in Insect Olfaction*, T. L. Payne, M. C. Birch and C. E. J. Kennedy (Eds), Oxford University Press, Oxford.

Kellog, F. E., Frizel, D. E. and Wright, R. H. (1962). The olfactory guidance of flying insects IV. *Drosophila. Can. Ent.*, **94**, 884–888.

Kennedy, J. S. (1940). The visual responses of flying mosquitoes. *Proc. Zool. Soc. Lond. A*, **109**, 221–242.

Kennedy, J. S. (1983). Zigzagging and casting as a programmed response to wind-borne pheromones: a review. *Physiol. Ent.*, **8**, 109–120.

Kennedy, J. S. and Marsh, D. (1974). Pheromone regulated anemotaxis in flying moths. *Science*, **184**, 999–1001.

Kennedy, J. S. and Thomas, A. A. G. (1974). Behaviour of some low-flying aphids in wind. *Ann. Appl. Biol.*, **76**, 143–159.

Kennedy, J. S., Ludlow, A. R. and Sanders, C. J. (1980). Guidance system used in moth sex attraction. *Nature*, **288**, 475–477.

Kennedy, J. S., Ludlow, A. R. and Sanders, C. J. (1981). Guidance of flying moths by wind-borne sex pheromone. *Physiol. Ent.*, **6**, 395–412.

Klun, J. A., Haynes, K. F., Bierl-Leonhardt, B. A., Birch, M. C. and Plimmer, J. R. (1981). Sex pheromone of female artichoke plume moth, *Platyptilia carduidactyla. Env. Ent.*, **10**, 763–765.

Kramer, E. (1975). Orientation of the male silkmoth to the sex attractant Bombykol. *Olfaction and Taste*, **5**, 329–335.

Kuenen, L. P. S. and Baker, T. C. (1982). The effects of pheromone concentration on the flight behaviour of the oriental fruit moth, *Grapholitha molesta. Physiol. Ent.*, **7**, 423–434.

Löfstedt, C. and Odham, G. (1984). Analysis of moth pheromone acetates by selected ion monitoring using electron impact and chemical ionization mass spectrometry. *Biomed. Mass Spectrom.*, **11**, 106–113.

Löfstedt, C., Van Der Pers, J. N. C., Lofqvist, J., Lanne, B. S., Appelgren, M., Bergstrom, G. and Thelin, B. (1982). Sex pheromone components of the turnip moth, *Agrotis segetum*: chemical identification, electrophysiological evaluation and behavioural activity. *J. Chem. Ecol.*, **10**, 1305–1321.

Löfstedt, C., Lanne, B. S., Lofqvist, J., Appelgren, M. and Bergstrom, G. (1985a). Individual variation in the pheromone of the turnip moth *Agrotis segetum. J. Chem. Ecol.*, **11**, 1181–1196.

Löfstedt, C., Linn, C. E. Jr and Lofqvist, J. (1985b). Behavioural response of turnip moth males *Agrotis segetum* to sex pheromone in a flight tunnel and in the field. *J. Chem. Ecol.*, **11**, 1209–1222.

Marsh, D., Kennedy, J. S. and Ludlow, A. R. (1978). An analysis of anemotactic zigzagging flight in male moths stimulated by pheromone. *Physiol. Ent.*, **3**, 221–240.

Minks, A. K., Roeloffs, W. L., Ritter, F. J. and Persoons, C. J. (1973). Reproductive isolation of two tortricid moth species by different ratios of a two-component sex attractant. *Science*, **180**, 1073–1074.

Murlis, J. and Jones, C. D. (1981). Fine-scale structure of odour plumes in relation to insect orientation to distant pheromone and other attractant sources. *Physiol. Ent.*, **6**, 71–86.

Olberg, R. M. (1983). Pheromone-triggered flip-flopping interneurons in the ventral nerve cord of the silkworm moth, *Bombyx mori. J. Comp. Physiol.*, **152**, 297–307.

Payne, T., Birch, M. C. and Kennedy, C. E. J. (Eds) (1986). *Mechanisms in Insect Olfaction*, Oxford University Press, Oxford.

Preiss, R. and Kramer, E. (1986). In *Mechanisms in Insect Olfaction*, T. C. Payne, M. C. Birch and C. E. J. Kennedy (Eds), Oxford University Press, Oxford.

Priesner, E. and Witzgall, P. (1984). Modification of pheromonal behaviour in wild *Coleophora laricella* male moths by (*Z*)-5-decenyl acetate, an attraction-inhibitor. *Z. Ang. Ent.*, **98**.

Roelofs, W. L. (Ed.) (1979). *Establishing Efficacy of Sex Attractants and Disruptants for Insect Control*, Entomological Society of America.

Roelofs, W. L., Hill, A. and Cardé, R. T. (1975). Sex pheromone of the redbanded leaf-roller, *Argyrotaenia velutinana* (Lepidoptera: Tortricidae). *J. Chem. Ecol.*, **1**, 83–89.

Sabelis, M. W. and Schippers, P. (1984). Variable wind directions and anemotactic strategies of searching for an odour plume. *Oecologia*, **13**, 225–228.

Tanaki, Y. (1985). Sex pheromones. In *Comprehensive Insect Physiology, Biochemistry and Pharmacology* (Vol. 9), G. A. Kerkut and L. I. Gilbert (Eds), Pergamon Press, Oxford.

Traynier, R. M. M. (1968). Sex attraction in the Mediterranean flour moth, *Anagaster kuhniella*: location of the female by the male. *Can. Ent.*, **100**, 5–10.

Wall, C. (1985). The exploitation of insect communication by man. In *Insect Communication*, T. Lewis (Ed.), Academic Press, London, pp. 379–400.

Willis, M. A. and Baker, T. C. (1984). Effects of intermittent and continuous pheromone stimulation on the flight behaviour of the Oriental fruit moth, *Grapholitha molesta. Physiol. Ent.*, **9**, 341–358.

Wood, D. L., Stark, R. W., Silverstein, R. M. and Rodin, J. O. (1967). Unique synergistic effects produced by the principal sex attractant compound of *Ips confusus* (Le Conte) (Coleoptera; Scolytidae). *Nature*, **215**, 206.

Wright, R. H. (1958). The olfactory guidance of flying insects. *Can. Ent.*, **90**, 81–89.

PART B

Evaluation and Use of Behaviour-modifying Chemicals

Insect Pheromones in Plant Protection
Edited by A. R. Jutsum and R. F. S. Gordon
©1989 John Wiley & Sons Ltd

3

Monitoring and Spray Timing

C. Wall
*Department of Animal Ecology, Ecology Building,
University of Lund, S-223 62 Lund, Sweden*

3.1 THE PURPOSE OF MONITORING

At first glance the objectives of monitoring pest insects seem straightforward: to determine if and when a pest is present in order to decide the need for, and timing of, control measures such as conventional insecticides. However, there are many ways in which information from a monitoring scheme or method can be used and it is therefore important that the objectives of any one scheme should be clearly and precisely stated so that the details of the method can be designed to achieve those objectives. To this end and in the context of monitoring traps containing behaviour-modifying chemicals (BMCs) it is possible to classify the information extracted from the trap catches and the use to which it is put (Table 3.1). This classification is not intended to be comprehensive, but it indicates the complexity of the approach and a framework on which to discuss it.

TABLE 3.1 Main uses of pheromone-monitoring traps

Information from trap catch	Application
Detection	Early warning
	Survey
	Quarantine
'Threshold'	Timing of treatments
	Timing of other sampling methods
	Risk assessment
Density estimation	Population trends
	Dispersion
	Risk assessment
	Effects of control measures

3.2 BEHAVIOUR-MODIFYING
CHEMICALS (BMCs) USED IN MONITORING TRAPS

The most commonly used BMCs are insect pheromones, particularly sex pheromones. To date the most widely investigated group of insects is the Lepidoptera, and in this case a synthetic version of the olfactory female sex pheromone (or a chemical analogue of it) is used to attract and catch conspecific males. This has the major disadvantage that the trap catches of males must usually be interpreted in terms of the behaviour of the females, thus adding to the complexity of that interpretation. In those insects in which olfactory aggregation pheromones play an important role, traps containing synthetic pheromone can be used to catch both sexes (Hardee *et al.*, 1969; Levinson and Levinson, 1973; Blight *et al.*, 1984), thus providing a much better opportunity for successful monitoring. So far, such aggregation pheromones have been found mainly in the Coleoptera (bark beetles and some weevils). BMCs not of insect origin can be valuable in monitoring traps. For many years various Diptera have been trapped by combinations of food and general attractants (Chambers, 1977); in addition, host plant volatiles can enhance captures in sex-attractant traps as with *Sitona lineatus* L. (Blight *et al.*, 1984), as can food attractants (Wanabe *et al.*, 1982).

3.3 APPLICATIONS OF PHEROMONE TRAPS

3.3.1 Detection

3.3.1.1 Early warning of pest incidence

Pests which occur sporadically during a season or from year to year can be monitored qualitatively to determine the start of each flight period with great effect. When this was done in Cyprus for the Egyptian cotton leafworm, *Spodoptera littoralis* (Boisd.), it was also possible to show that infestations arose from localized populations within the island rather than from immigrants (Campion *et al.*, 1977).

3.3.1.2 Survey to define distribution and areas of infestation

This application is particularly relevant to forest pests. Enormous areas of continuous forest may have to be checked for the presence of a pest. Grids of pheromone traps help to define those areas in which the pest is present, and may even indicate local increases in density against an otherwise low-level background. This has been the approach adopted for the spruce budworm, *Choristoneura fumiferana* (Clem.) (Sanders, 1978, 1984), the gypsy moth, *Lymantria dispar* (L.), (Schwalbe, 1981), and also *Lymantria monacha* L.

and *Panolis flammea* (D. & S.) (Bogenschutz, 1982). Pheromone trap grids are used in food stores to identify locations with highest catches; these can then be used either as future monitoring sites (thus reducing the number of traps required) or as foci for control measures (Cogan and Hartley, 1984). Pheromone detection trapping has greatly extended the known ranges of two lepidopterous forest pests in the western United States (Livingston and Daterman, 1977; Sartwell *et al.*, 1980). A novel use of detection traps was employed by Arciero (1979) to determine whether beetles of the smaller European bark beetle, *Scolytus multistriatus* (Marsham), carrying spores of the Dutch elm disease fungus, *Ceratocystis ulmi* (Buisman), occurred outside known infection sites.

3.3.1.3 Quarantine inspection

Perhaps the most useful information which pheromone traps can provide is to indicate the absence of a pest species. In general the traps are so sensitive that their failure to catch individuals is a good indication that the pest is not present, at least in the adult stage. This has general application in pheromone monitoring but can be used to great effect when checking consignments of stored food products and also fresh fruit and vegetables for quarantine purposes. In studies conducted in a cargo terminal at Wisconsin, USA, the presence of *Trogoderma* spp. males in pheromone traps provided the first evidence that live insects of this genus were present (Barak and Burkholder, 1976).

3.3.2 Threshold

3.3.2.1 Timing of treatments

This is an approach adopted for both immigrant populations of pests such as the pea moth, *Cydia nigricana* (Fabr.), (Macaulay *et al.*,, 1985) and emergent but resident populations such as those of the codling moth, *Cydia pomonella* (L.) (Riedl *et al.*,, 1976; Glen and Brain, 1982). Normally a minimum catch or catching rate is used to signal the start of temperature-summation calculations to determine the time of occurrence of the susceptible or damaging stage in the life cycle. Minks and de Jong (1975) extended this approach with *Adoxophyes orana* (F. v. R.), using subsequent peaks in numbers of moths caught as indicators of renewed oviposition and the preceding troughs as starting points for temperature-summation calculations.

3.3.2.2 Timing of other sampling methods

Often information from pheromone traps alone is insufficient, but it can pinpoint when to make crop inspections for the pest. In Switzerland, traps for *C. pomonella* are used to determine if and when a second-generation flight occurs; however, traps in one orchard may give a false impression by catching males from neighbouring orchards, so it is necessary to examine the fruit for signs of damage before deciding the need to spray (Charmillot, 1980a).

3.3.2.3 Risk assessment

A single threshold catch is used to distinguish high- and low-risk fields for attack by *C. nigricana*. Evidence from a six-year survey shows that the threshold used relates to the economics of spraying against this pest in combining peas (Wall and Perry, 1987). A similar approach has been adopted in many countries for *C. pomonella*, although the value of the threshold varies considerably, apparently depending on trap density, trap design, the number of generations and which one is being monitored, and the local situation. The thresholds range from a single catch (e.g. first catch or peak catch — Riedl *et al.*, 1976) through a catching rate over a short period (e.g. five moths per trap per week — Alford *et al.*, 1979; Glen and Brain, 1982), to cumulative catches associated with a physiological time-scale as indicated by day-degrees (Charmillot, 1980a). In the last example a range of values is used to determine the need for and the number of insecticide sprays. Weslien *et al.* (1988) have shown a sufficiently good relationship between trap catch of *Ips typographus* (L.) and tree mortality for traps to be used effectively for risk assessment.

3.3.3 Density estimation

3.3.3.1 Population trends

One of the objectives of monitoring *C. fumiferana* is to identify population trends over many generations. Since Sanders (1984) has shown that trap catches can be correlated with previous larval density it seems reasonable to conclude that the trends in trap catches obtained since 1973 do broadly reflect the trends in population density during that period. However, there are examples of this type of use without a relationship between trap catch and density being established; for instance, Bakke (1985) attempted to monitor population trends of *I. typographus*. This has also been done for stored products moths even though catches in control (no pheromone) traps at a number of locations clearly show that there is not a consistent relationship between density and pheromone trap catch (Buchelos and Levinson, 1985).

3.3.3.2 Dispersion and risk assessment

A good example of this application is the monitoring of the boll weevil,

Anthonomus grandis grandis Boheman, in Texas. A trap index is computed from catches of overwintered adults as they move into the crop and this is used to predict likely damage (Rummel *et al.*, 1980; Benedict *et al.*, 1985). Sanders (1984) has shown that catches of *C. fumiferana* can be used to predict larval densities in the following year and similar data are available for the Douglas fir tussock moth, *Orgyia pseudotsugata* (McDunnough) (Daterman and Sower, 1977).

3.3.3.3 Effects of control measures

Many workers have used pheromone traps to monitor the effects of control measures, especially in food stores (Buchelos and Levinson, 1985). However, Henneberry and Clayton (1982) found that traps for the pink bollworm, *Pectinophora gossypiella* (Saunders) did not indicate post-treatment population declines because they attracted the highly mobile moths from elsewhere. This approach has been extended to the evaluation of attempts to control by pheromone disruption, although the use of pheromone-monitoring traps for this purpose is questionable, because large reductions in trap catch are often not accompanied by significant reductions in mating frequency.

3.4 THE VALUE OF PHEROMONE MONITORING TRAPS

The most useful characteristic of pheromone traps is that they catch selectively at low population densities. It has been shown many times that they will catch individuals of the species under investigation earlier than other available sampling methods; in many cases there are no other sampling methods suitable for the adult stage, which enhances the value of pheromone traps but makes verification impossible other than by empirical means. Pheromone traps tend to detect the presence of low-density nocturnal species before light traps (Cranham, 1980), although at moderate to high densities light traps can be superior (Howell, 1981). In the case of diurnal species pheromone traps are unrivalled for early detection of emergent or immigrant adults. However, it is important that traps are deployed before the flight season starts, to avoid misleadingly high initial catches because emerging or immigrating insects have accumulated (Alford *et al.*, 1979; Howell and Quist, 1980). It is possible to inspect cotton crops for the boll weevil, *A. grandis grandis* in the spring, but Rummel *et al.* (1980) demonstrated that pheromone traps are more reliable at detecting the emergence and dispersal of overwintered weevils. A less practical crop inspection procedure for newly laid eggs was adopted for the pea moth, *C. nigricana* in England, until it was demonstrated by Macaulay (1977) that pheromone traps consistently catch immigrating male moths before any eggs can be found.

The selectivity of pheromone traps, and their relative ease of use (e.g. no power required), make them ideal for use by people who are not trained entomologists. There are few associated taxonomic problems and data are obtained rapidly without the need for laborious sorting. This not only means that farmers, foresters and growers can run traps (thus leaving advisers/extension workers to tackle other problems), but also that the traps can be used extensively on a local basis. This is very important for those pests whose timing and density varies significantly on a local scale, such as *C. nigricana*, for which it is necessary to trap in each pea field because infestations can be very different in level and timing in adjacent fields (Perry *et al.*, 1981), or even within a single field (Perry and Wall, 1984).

Pheromone traps for monitoring are cheap to produce, particularly since the lure contains so little chemical, and can therefore be sold to growers or advisory/extension services at a price which is not only reasonable but also represents a tiny fraction of the value of the crop. Unfortunately the relatively small-scale chemical production involved has often made it difficult for research workers to interest the agrochemical industry in producing and marketing pheromone-monitoring traps; thus traps for many pest species can be obtained only because some research laboratories are prepared to provide the chemical lures required, or even the whole trap. However, there are signs that some agrochemical companies are starting to provide traps to growers, at a discount or even free, when they sell pesticide. This 'package' approach is an exciting way of increasing the usage of such monitoring traps provided it does not limit the grower's choice of pesticide too much. The specialist nature of the product, involving detail relevant perhaps to only one pest species (chemical, trap design, retentive material, instructions for use, etc.—see later), makes the commercialization of monitoring traps difficult. Unfortunately there are already far too many examples of unsatisfactory monitoring trap systems being sold; a company may settle on one or two 'standard' trap designs and sell these for a range of pest species, with only the chemical lure being different in each case, and in addition there is often a singular lack of adequately researched instructions on the use of traps and the interpretation of the catches.

The major drawback of pheromone-monitoring traps is the difficulty of relating catches to pest density or damage levels, and there are few examples where this relationship has been defined adequately. This is due in part to the unfortunate characteristic that the sampling efficiency of some traps seems to change with pest density. Several authors (Miller and McDougall, 1973; Riedl and Croft, 1974; Cardé, 1979; Hartstack and Witz, 1981) have stated that trap efficiency declines with increasing population density, and this has been attributed to competition with wild females. Theoretical models have been proposed to describe th'ϡ competition effect (Knipling and McGuire, 1966; Howell, 1974; Nakamuɪ . ʾnd Oyama, 1978). However, work designed to demonstrate such an effect has been limited. Baker *et al.* (1980) found no evidence for such an effect in *Grapholitha molesta* Busck, and Vick *et al.* (1979) showed that the presence of released female stored-product moths did not affect pheromone trap catches, although they did their experiment only at one density. Hartstack and Witz (1981) found that early-season catches of *Heliothis virescens* (F.) could be used to estimate population densities, but the competition effect of high densities caused problems later in the season. Howell (1974) released adult *C. pomonella* into a moth-free apple orchard and observed a 75% reduction in catches in traps containing virgin females when there was a moderate density of free-flying females present to compete with the caged females. Graham (1984, reported in Wall, 1984) showed that trapping efficiency fell off at high densities of *C. nigricana*, suggesting female competition, and Elkinton and Cardé (1984) observed trap-catch reductions of *L. dispar* when 200 laboratory-reared females were placed in the centre of an experimental site. Finally Croft *et al.* (1986) have quantified the reduction effect per female for three densities of female *Argyrotaenia citrana* (Fernald) in an elegant mark and recapture experiment in a semi-enclosed courtyard.

Another reason why so few examples of well-defined relationships between trap catch and pest density or damage exist is that there has been a lack of adequate studies. If pheromone monitoring is ever to become a sophisticated ecological technique, then sufficient good evaluation work must be done to derive some general principles. Without such principles the development of the technique will continue to be very slow and inefficient. Whilst it is acknowledged that traps and their use have to be tailor-made for each species, the current policies of either developing systems by repeating work done elsewhere with slight (unnecessary?) modifications, or not bothering to do all the work required, cannot be cost-effective.

In a number of cases the problem of trap-catch interpretation has been overcome partially by using the traps to determine if and when a threshold number/density of a pest is exceeded. In general, failure to achieve a threshold is taken as a strong indication that control measures are not necessary and this is one of the most valuable applications of pheromone traps. The threshold is usually set to indicate a very low density, at which the traps are very efficient sampling tools, so that much confidence can be placed on subthreshold catches truly indicating little or no subsequent damage (Charmillot, 1980a; Cranham, 1980; Wall and Perry, 1987). This seems to be the main value of using pheromone traps to assess the effects of control measures, as is often done in food stores, whilst the use of the traps to study or monitor population fluctuations may be questionable because the relationship between trap catch and density has not been quantified (see Buchelos and Levinson, 1985). At least periods of low or zero catches ought reliably to indicate very low levels of infestation (Fig. 3.1). A threshold catch

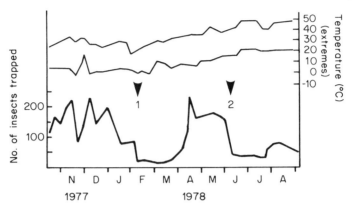

Fig. 3.1 Monthly pheromone trap catches of *Ephestia cautella* (Walker) in a peanut warehouse, showing low population levels after fumigation and when the warehouse was empty. Arrows: 1, warehouse fumigated; 2, warehouse empty (redrawn from Vick *et al.*, 1981)

can be used as an indication of the need for and/or timing of sprays (Madsen and Vakenti, 1973; Alford *et al.*, 1979; Macaulay *et al.*, 1985) or of sampling by other methods, such as crop inspection, which themselves are used to determine the need for insecticidal control (Charmillot, 1980a).

The problems of using pheromone traps for quantitative monitoring of pest species have been addressed by only a few research groups; significant advances could be made with existing trapping systems if more emphasis was placed on this type of field research. Both Rummel *et al.* (1980) and Benedict *et al.* (1985) have demonstrated significant, although differing, relationships between catches of adult boll weevils, *A. grandis grandis*, in pheromone traps placed on the borders of cotton fields and damage to the crop. Both groups operated a trap index system, which is a series of thresholds each predicting damage within certain limits. Both groups have demonstrated that their system works well, and Rummell *et al.* (1980) have shown how much better the traps are than crop inspection, but it is worrying that Benedict *et al.* should have found it necessary to research a trap index system with so many differences from the Rummel *et al.* (1980) system (trap, dose of attractant, spacing of traps, trap catch parameters) when they were working in the same state (Texas), although in a different part of it. The only logical difference is the trap-catch parameter used to calculate the trap index; the high mortality in adult weevils in the semi-arid areas to the west precludes the use of trap catches over a six-week period (Rummel *et al.*, 1980) as is done on the lower Gulf Coast (Benedict *et al.*, 1985).

The variation in the quantitative relationship between pheromone trap catch and pest phenology/damage is perhaps best demonstrated by *C. pomonella* (see Section 3.6.3 for details). It seems that almost every research group has produced different results which are explained by the particular circumstances. Recently McNally and Van Steenwyk (1986) have shown that although a significant relationship existed between trap catch of *C. pomonella* and subsequent damage to walnuts in any one year, other factors as yet unquantified changed the relationship from year to year, making it impossible to predict damage with precision. However, it is clear that the traps can be used successfully in rather simple ways (determination of thresholds in first generation, peaks of oviposition during the second generation, timing of the start of the first generation flight, timing of sprays), especially if the information from the traps is considered along with other trapping/sampling data (Charmillot, 1980a). Perhaps too often pheromone-monitoring traps have been regarded as a panacea rather than additional tools in an integrated pest-sampling programme.

The weather undoubtedly affects pheromone trap catches since they depend on the behavioural responses of the insects. This may be a disadvantage or an advantage depending on the use. As background work to monitoring boll weevils with pheromone traps, Carroll and Rummel (1985) examined the relationship between the timing of emergence of adults from their overwintering habitats and the catches of weevils in pheromone traps. The results indicate a strong correlation between the timing of these events, but the numbers of weevils caught in traps bore little relation to the numbers emerging. The authors suggest this may result from the effects of the weather on weevil response to the pheromone. But this is where it is important to have the objective of the trapping clearly stated: if the traps are intended to monitor numbers of emerging weevils, then they do not perform well; however, if the traps are used to monitor attack on the cotton crop, then the same weather conditions which affect pheromone response may also affect flight into the crop, thus explaining why the trap catches can be used for risk assessment. Certainly in the pea moth, *C. nigricana*, there is evidence that pheromone trap catches of males indicate the rate of immigration of females into the pea crop (Graham, 1984), and may well reflect the level of oviposition of those females once they are in the crop (i.e. on days when low temperatures prevent catches of males, females are probably not flying around in the crop laying eggs). Sanders (1981) found that the catching rate of *C. fumiferana* males in pheromone traps reflected the density of the next generation of larvae more accurately than other measures of population density, such as the density of the previous larval generation.

Finally, the value of pheromone-monitoring traps may be overstated in some cases. Because they are so easy to use there may be a tendency to use them in situations where a more reliable monitoring method, such as crop sampling, is more appropriate. For instance, larvae of *Mamestra brassicae* L. become detectable in brassica crops before they start to cause serious damage; a crop-sampling programme is therefore likely to provide more reliable information than trapping adults with pheromone traps (Theunissen, personal communication). Sanders (1981) states that, where control strategies against C. *fumiferana* depend largely upon the aerial application of insecticides on high-density populations, there is little need for monitoring with pheromone traps because a combination of egg and larval sampling is both efficient and quite easy to do. It is only when attempts are made to reduce the possibility of severe outbreaks that sampling of very low densities becomes necessary by using pheromone traps. Similarly Sower *et al.* (1984) point out that whilst pheromone traps are useful for surveying large areas of ponderosa pine for the Western pine shoot borer, *Eucosma sonomana* Dougl. ex Laws., shoot inspection (for visual damage or larvae) is better for more precise assessment of individual plantations.

3.5 THE COMPONENTS OF A PHEROMONE MONITORING SYSTEM

3.5.1 Attractant

The chemical composition of the attractant will depend on the species to be trapped and is usually a unique mixture of synthetic compounds identical to the natural pheromone, as identified from chemical extracts or collections made during air entrainment. Lists of identified attractants are available in the literature (Inscoe, 1982; Minks, 1984; Arn *et al.*, 1986) and from supply organizations. The required purity of individual components and accuracy of mixtures will vary with the species: in the pea moth, C. *nigricana*, it is possible to use either the single-component synthetic pheromone (E, E)-8,10,dodecadien-1-yl acetate or its analogue (E)-10-dodecen-1-yl acetate (Greenway and Wall, 1981). Even at low doses the pheromone can attract undesirably large numbers of moths, saturating the traps and making interpretation of the catches impossible. The analogue on the other hand attracts far fewer insects (Greenway and Wall, 1981), but sufficient to provide a useful range of catches within the extremes of high and low infestations usually encountered in combining crops. One approach could be to operate traps at a range of doses in any one location, to cater for a range of possible population densities, as shown to be possible by Sanders (1981). In contrast to C. *nigricana* the main problem with monitoring C. *pomonella* seems to have been that insufficient moths are caught in the traps. Most of the thresholds used range from two to five moths per trap per week, with the single-component pheromone (E,E)-8,10-dodecadien-1-ol (Roelofs *et al.*, 1971). Recent findings (Arn *et al.*, 1985; Einhorn *et al.*, 1984; Einhorn *et al.*, 1986) that there are other components in the natural female pheromone raise hopes that more attractive lures can be produced for this species, leading to more sensitive discrimination between low- and high-risk infestations. *A. orana* was one of the first species known to use a sex pheromone of more than one compound (Meijer *et al.*, 1972). It was shown by field trapping that maximum attraction was obtained with a 9 : 1 ratio of the two positional isomers, (Z)-9-tetradecenyl acetate and (Z)-11-tetradecenyl acetate (Minks and Voerman, 1973). This species has been monitored successfully using this mixture (Minks and de Jong, 1975; Alford *et al.*, 1979) but more recent work has shown the blend to be more complex (Den Otter and Klijnstra, 1980; Charmillot, 1981; Guerin *et al.*, 1986). In certain very critical species the chirality of the pheromone molecule is important. Several reports (Vité *et al.*, 1976; Cardé *et al.*, 1977a,b; Miller *et al.*, 1977; Plimmer *et al.*, 1977) have established that lures of disparlure greatly enriched in the (+)-enantiomer produce trap catches of *L. dispar* up to ten times greater than the racemic mixture.

3.5.2 Dispenser

The dose and formulation of attractant used are critical. Lures used in monitoring traps should release attractant at a constant rate throughout the trapping period, which may be several months. Whether or not it is necessary to change lures regularly during the trapping period depends on the formulation (rubber, polyethylene, polyvinyl chloride, hollow fibres) and the response characteristics of the species.

Monitoring traps for both *C. fumiferana* and *C. nigricana* can be operated without the need for the lures to be changed within a flight period. In both cases experiments were performed with pre-aged lures to establish the effects of weathering on the consistency of trap-catch achieved by different formulations. Wall and Greenway (1981) showed that rubber lures containing 3 mg (E)-10-dodecen-1-yl acetate catch *C. nigricana* consistently for at least 3 months (traps are typically in the field for a maximum of 2½ months), and Sanders (1984) demonstrated similar consistency of capture for Bend plastic capsules releasing 2 μg fulure per hour over a 6-week period, which again is sufficient time to cover the flight period of *C. fumiferana*. These examples contrast with monitoring traps for *I. typographus* in Norway which have been shown to decline in attractiveness by about half in 5–6 weeks (Bakke *et al.*, 1983) but which are still used for monitoring with only one change of the lure in mid-summer (Bakke, 1985). There are many other examples where traps are being used for monitoring with lures which clearly decline in effectiveness from the day they are put in the field or for which the response characteristics through time have not been quantified.

One topic which requires more attention is the release rate of chemical required to sample adequately the area (unit) of crop to which specific insecticide sprays will be applied. There are many examples in the literature of monitoring traps catching large numbers of insects associated with low damage levels, possibly as a result of attraction over distances far beyond the boundaries of the crop (Charmillot, 1980a; Henneberry and Clayton, 1982). Baker and Roelofs (1981) raised this point and proposed that, as far as possible, the attractancy of a trap should be tailored to the situation so that the trap samples insects from within the crop and not beyond it (Fig. 3.2). To do this it is necessary to obtain some information on the range of attraction of the trap; this has been attempted in only very few cases (Mitchell and Hardee, 1976; Wall and Perry, 1978, 1980, 1981; Elkinton and Cardé, 1980; Riedl, 1980b), but it may be a key to successful quantitative monitoring. Daterman (1982) expands on this point. Lowering the dose of attractant is one way of reducing the attractive range, but another may be to use a 'less attractive' analogue of the pheromone. This may have been one reason why the monitoring system for *C. nigricana* in combining peas has been successful, although the choice of attractant was made originally for other reasons (Greenway and Wall, 1981).

3.5.3 Trap design

The variety of trap designs used in pheromone trapping is extraordinary. Most designs have been arrived at empirically and in some cases explanations for the varied performance of different designs have been sought (Lewis and Macaulay, 1976). Traps used for insect monitoring need to be practical as well as efficient; they should be cheap and easy to examine and/or empty. Thus most practical pheromone-monitoring systems are based on fairly simple designs, exceptions being only where the behaviour of the insect dictates certain complex features.

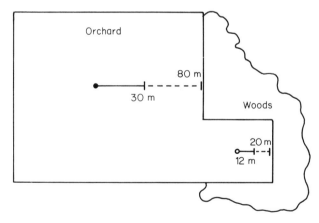

Fig. 3.2 Possible optimal placement of monitoring traps for *G. molesta*. ●, 100 µg; ○, 10 µg; ——— Av. max. drawing range; - - - - max. obs. drawing range (redrawn from Baker and Roelofs, 1981)

Traps range from simple vertical sticky plates (Fig. 3.3a) or paper strips through sticky tubes of various cross-sectional shapes (Fig. 3.3b–d) and other sticky designs like the wing trap (Fig. 3.3e), which is the basis for the various Pherocon models. Traps using a vaporous insecticide to kill include the funnel trap (Fig. 3.3f) and the milk carton trap (Fig. 3.3g). Some traps depend on one-way valves to capture the insects and typical examples are the

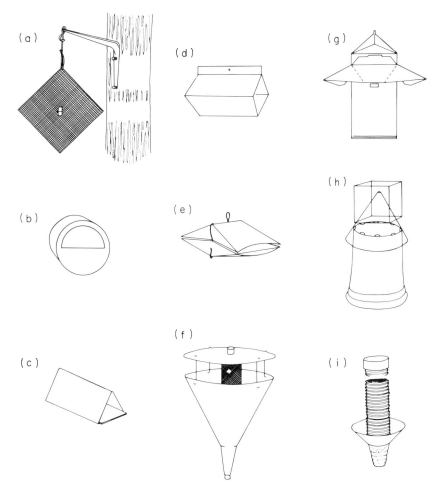

Fig. 3.3 Examples of pheromone trap designs. (a) Vertical sticky plate for *S. multistriatus* (after Peacock and Cuthbert, 1975); (b) IOBC trap for orchard Lepidoptera (after Granges *et al.*, 1969); (c) delta trap for *C. nigricana* (after Lewis and Macaulay, 1976); (d) tent trap; (e) wing trap for many species of Lepidoptera; (f) funnel trap for *C. fumiferana* (after Ramaswamy and Cardé, 1982); (g) milk carton trap for *L. dispar*; (h) Hardee trap for *A. grandis grandis* (after Mitchell and Hardee, 1974); (i) cylinder trap for *I. typographus* (after Bakke and Strand, 1981)

'Hardee' trap for some Coleoptera, including boll weevils (Fig. 3.3h) and the cylinder trap for *I. typographus* (Fig. 3.3i). Some designs, such as the delta trap, can be modified as water or oil traps, but they are not so practical for monitoring except in certain situations where very large catches are expected. In the specialized stored products environment rather novel designs of trap have been used, for example pheromone-impregnated corrugated paper (Barak and Burkholder, 1976) or sacking (Levinson and Levinson, 1977) for crawling insects which are thigmotactic, and tubular traps which can be coupled to probes for catching grain beetles up to several metres below the surface in grain bins (Burkholder, 1985).

The size of the trap may be important, especially with traps relying on a sticky retentive surface. The trap should be large enough for the purpose without being unnecessarily large. There are few examples where this has been investigated adequately (Macaulay and Lewis, 1977).

In sticky traps the retentive material must be efficient for the species being trapped. There are a number of sticky materials on the market which are suitable for use in insect traps, partly because they are odourless, but they are not equally effective (Lewis and Macaulay, 1976), although comparisons may be different for different species. In some cases it may be necessary to apply the sticky material thickly for maximum efficiency (Bradshaw *et al.*, 1983). The efficiency of sticky materials in insect traps declines with exposure, mainly because increasing numbers of insects get caught. It is therefore very important to know both the saturation level and the time taken to achieve such a catch, on average, in order to plan the trap-emptying programme. An effect of trap saturation on subsequent catches is shown in Fig. 3.4 (Sanders, 1981) and discussed in detail by Riedl (1980b), who emphasizes the need for the standardization of trapping procedures partly for this reason.

Trap colour has not generally been regarded as important, although some studies have shown that colour can affect the catch. Sanders (1978) demonstrated that yellow traps are less attractive to males of *C. fumiferana* than white, blue or green ones. The boll weevil, *A. grandis grandis*, responds positively to yellow from a distance of approx. 7 m (Leggett and Cross, 1978) and at 2 m the majority of weevils will orient to this colour rather than pheromone. Most pheromone traps used for this species, such as the 'Hardee' trap (Mitchell and Hardee, 1974) are now yellow. In those insects known to use colour as a critical stimulus, such as fruit flies, non-chemical traps have been carefully designed to create an optimum or even super-stimulus (Prokopy, 1968); this approach has not yet been tried with traps containing specific attractants.

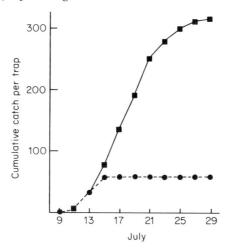

Fig. 3.4 Catches of male *C. fumiferana* moths in Pherocon 1CP traps. ——trap bottoms changed every 2 days; - - - -trap bottoms unchanged throughout (redrawn from Sanders, 1981)

3.5.4 Trap deployment

There are three main considerations in trap placement: trap height and position with respect to vegetation, and trap density. Ideally, traps should be placed at the optimum height for catching insects, but practical considerations may dictate otherwise. For instance Miller and McDougall (1973) showed that maximum catches of *C. fumiferana* occur in the upper canopy of host trees, but Sanders (1978) proposed a more practical monitoring height of 2 m. Wind speed may influence the flight behaviour of the insects and therefore the optimum height. However, in practice a compromise height may need to be chosen, and it may be more meaningful to relate the height of the trap to the height of the vegetation (Riedl *et al.*, 1979) and make adjustments as the latter grows (Lewis and Macaulay, 1976). No matter what height is chosen, standardization is essential. This applies also to the orientation of the trap, if the design is such that the size of the catch is affected by wind direction. The commonly used delta trap is one such design; misalignment with the wind affects the linearity of the short-range instantaneous pheromone plume and therefore the ability of the insects to reach the trap and be caught (Lewis and Macaulay, 1976). With directional designs self-orientating versions may be preferable, but too expensive for normal monitoring purposes. An alternative approach is to use more than one trap at a location, each oriented differently, and to use the largest catch on any one occasion. This is the approach adopted with the pea moth system in which pairs of traps are placed at right angles to each other (Macaulay, 1977).

Trap density is a much more complex problem. Two questions need to be answered: (1) What density of traps is required to sample adequately the pest population? (2) At what density do traps interact, thereby influencing the size of catch in individual traps? Very little work has been done to answer these questions, perhaps because the experimental designs involved may be difficult and time-consuming to implement (Perry *et al.*, 1980) and the results difficult to interpret. All the published studies have been on tortricid moths in either orchards (Riedl and Croft, 1974; Charmillot and Schmid, 1981; McNally and Barnes, 1981; van der Kraan and van Deventer, 1982) or arable crops (Wall and Perry, 1978, 1980, 1981).

The range of action of a trap may be considerable; for instance, Riedl (1980b) established that monitoring traps for *C. pomonella* interact at densities greater than 1 per ha (i.e. range of action of approx. 350 m). If local variations in pest density are important and need to be monitored (Perry and Wall, 1984) there may be a conflict between the density of traps required and the possibility of their interacting, with consequent effects on the individual trap catches (usually reduction). At the very least, therefore, the density of traps ought, where possible, to be standardized. However, there may be a further complication; the range of action of a trap includes both the distance over which insects are attracted and the distance over which they may have travelled prior to attraction (through migration or appetitive behaviour) (Wall and Perry, 1987). If traps are placed within each other's range of attraction, insects may fly predominantly to the upward sources as shown in *C. nigricana* by Wall and Perry (1978). In this type of situation even standardizing trap spacing would not overcome the problem of interactions because individual trap catches would be either increased or reduced depending on the wind direction.

Sanders (1981) showed that survey traps for *C. fumiferana* spaced at 30 km intervals all indicated similar trends in catches over a number of years (Fig. 3.5), from which he concluded that a few trapping locations could be used to indicate general trends over large areas. He also claimed that catches had paralleled population densities, as shown by larval sampling, over a number of years; although this was true in general, there were some years in which high catches of adults were obtained despite low larval densities.

3.5.5 Pest biology

Effective interpretation of trap catches relies on a sound knowledge of the biology of the pest. Indeed, the likelihood of developing a successful monitoring trap can be assessed with this information. For instance, monitoring the emergence or immigration of species with discrete populations (in time) is much more likely to succeed than monitoring

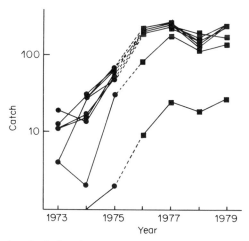

Fig. 3.5 Catches of male *C. fumiferana* at seven locations approximately 30 km apart along a triangular route in northwestern Ontario, and in the centre of a 6000 ha area burned in 1969 (lowest line). All traps containing PVC containing 4.5 mg fulure. 1973–1975 Sectar 1 traps (●); 1975–1979 Pherocon 1CP traps (■) (redrawn from Sanders, 1981)

population fluctuations of multivoltine species with overlapping generations; this is demonstrated in a simple way by the differing relationships between pheromone trap catch and adult phenology in the two generations of *C. pomonella* (Riedl *et al.*, 1976). It is important to know how the dynamics of the two sexes are related; in resident species do the males emerge earlier than females? In migrating species do both sexes migrate? at the same time? in the same numbers? Such questions are often very difficult to answer, particularly for species with low-density adult populations. However, even in such cases it may be possible to devise sampling methods for adults which, although not suitable for practical monitoring, can be used in experimental work. For instance Graham (1984) devised an ingenious modification of a vacuum sampler which enabled him to sample adequately populations of adult *C. nigricana* in pea and wheat crops at densities of the order of 0.1 moths/m².

The responsiveness of the males in relation to emergence, and, where appropriate, dispersal should be known. Schwalbe (1981) and Elkinton and Cardé (1980) showed that less than 5% of released male *L. dispar* disperse more than 800 m, and the latter authors (1981) argue that the pattern of trap catch in a grid of traps as a whole should reveal the location of population centres, despite the problems of larger-scale dispersal by a few individuals. As previously stated (see Section 3.4) Carroll and Rummel (1985) were able to show that the timing of pheromone trap catches of boll weevils did reflect the timing of emergence of overwintering adults. On the other hand

Henneberry and Clayton (1982) were unable to relate trap catches of *P. gossypiella* males to damage of cotton bolls adequately because of the immigration of highly mobile moths from elsewhere. Thus the scheduling of insecticide treatment was possible only for the first treatment after initial arrival of the moths; subsequently high numbers of moths were caught even when insecticides had been successful, making it impossible to determine from the trap catches when further sprays were required.

For timing insecticide applications it may be necessary to calculate the time between recorded catches and the susceptible stage (say, larva). This period will depend on the relationship between male catch and female numbers, the duration of the pre-oviposition and oviposition periods of the females, and the time taken from eclosion to the susceptible stage. If the timing of the male and female flights coincides, pre-oviposition and oviposition are short, and the susceptible stage for control is the egg or newly hatched larva, then there is a good chance of success in timing sprays by means of some temperature-driven model. This has been the case for *C. nigricana* and *C. pomonella* (see Sections 3.6.2 and 3.6.3) and also *A. orana* (Minks and de Jong, 1975). Finally, for quantitative monitoring it is necessary not only to establish a relationship between pest density and trap catch but also between pest density and damage. Sanders (1981, 1984) has made much progress with *C. fumiferana* because trap catches can be related to larval density over the range of larval densities known to be critical for major outbreaks of this pest. Thus the pheromone trap catches are related not to damage but to the likelihood of a major outbreak. In general, the best that has been achieved is a very rough correlation between trap catch and subsequent damage, as in the case of *A. grandis grandis* (see Section 3.6.4). Where more precise relationships have been sought, they have proved to be inconsistent either from locality to locality or from year to year.

3.6 SUMMARY OF SELECTED MONITORING SCHEMES

3.6.1 Eastern spruce budworm, *Choristoneura fumiferana*

Sampling plots are located along accessible routes in stands of susceptible host trees át intervals of up to 50 km. Ideally, the intervals should be less than this, perhaps every 10 km, but in reality the number of locations that can be sampled is limited by practical considerations of manpower and time. The stands selected are principally those which, according to current knowledge, will support the most rapid spruce budworm population growth rates, i.e. mature spruce/fir stands where tree crowns are well exposed to the sun. Two clusters of five double-funnel traps are placed out in each location, the clusters separated by at least 200 m to avoid interference and the traps within each cluster separated by 10 m. Both clusters contain PVC lures, 4 mm diam. × 10 mm long, one cluster containing 0.03% fulure by weight, the other 0.0003%. Traps are deployed at a height of 2 m on marked trees about 10 days before the start of moth flight. Traps are collected after the moth flight. Deployment and collection of traps can be timed to coincide with the sampling of large larvae and egg populations, respectively, if they are being done in the same areas, to save the number of visits to the sampling plots. Trapping is done annually. When catches show significant population increases or exceed the prescribed threshold, then a more intensive sampling programme is conducted in the area using techniques which provide accurate population estimates. This can include sex attractant traps (if catches are found to be well correlated) (Sanders, 1981).

Details of the spruce budworm monitoring scheme are changing continually. The following details are now standard throughout eastern North America: each trap cluster consists of three traps, spaced 40 m apart at a height of 2 m and at least 40 m from the edge of a stand. The number of clusters is largely up to the users, depending on costs and manpower. Traps are Multi-pher (Extermination Sevigny, 2949 Chemin Ste-Foy, Ste-Foy, Quebec, Canada, G1X 1P3) containing PVC lures with a release rate of 100 ng h^{-1} (Sanders, 1986 and personal communication).

3.6.2 Pea moth, *Cydia nigricana*

The traps are used to monitor within individual fields of combining peas; in fields of less than 50 ha one system is used, but in larger fields two systems are recommended, the whole field being sprayed if either system indicates the need. Each system consists of two delta-shaped sticky traps which are placed 100 m apart and 5 m into the crop on adjacent headlands on the side of the prevailing wind. The traps are mounted on stands, and their height is adjusted to two-thirds the height of the crop. Each trap contains a rubber lure impregnated with 3 mg (*E*)-10-dodecen-1-yl acetate. Ideally each system is set up by mid-May (before the flight season starts) and, *unless spraying takes place*, is run until mid-August or harvest, whichever is the sooner. The two traps are examined every other day, and the sticky plate changed if any moths have been caught. Records are kept of the catches. Spraying is necessary only if a 'threshold' catch is achieved—this is ten or more moths in either trap during two consecutive 2-day periods. The date of the 'threshold' is taken to indicate the start of substantial oviposition in the crop. Insecticide sprays are applied to kill newly hatched larvae before they enter the pods. A developmental model using max./min. daily temperatures is used to calculate the date of the first spray, using the date of the 'threshold' as a starting point. These calculations can be done using a simple calculator supplied with the traps and locally recorded temperatures, but more commonly farmers in England make use of the ADAS/PGRO 'phone-in' service, which provides estimated first-spray dates assuming a series of different 'threshold' dates. The 'phone-in' service is based on a computer model using recorded and forecast daily temperatures from a series of local meteorological stations throughout the combining pea area (Macaulay, 1977; Macaulay et al., 1985; Wall and Perry, 1987).

3.6.3 Codling moth, Cydia pomonella

This is the classic example of an insect which has been worked on worldwide, and for which an astonishing variety of pheromone-monitoring systems have been developed. Thus, guidelines for trap operation and catch interpretation also vary. As pointed out by Riedl (1980a) differences in catch interpretation are not surprising in the light of the uniqueness of each fruit area (climate, topography, cultural practices, orchard type, host species and variety, moth voltinism, etc.) and the lack of standardization of trap type, density and use. However, there are certain general features common to most systems, as follows.

Traps are simple sticky designs such as the Pherocon 1C (Alford et al., 1979; Cranham, 1980; Glen and Brain, 1982), the Pherocon 2 (Riedl, 1980a), cylindrical traps (Madsen and Vakenti, 1973; Audemard and Milaire, 1975; Charmillot, 1977, 1980b), the wing trap (Howell, 1972) or the ice-cream carton trap (Butt and Hathaway, 1966). All systems use the single-component pheromone (E,E)-8,10-dodecadien-1-ol (Roelofs et al., 1971), although recent findings indicate that there are minor components which may increase attractiveness (Einhorn et al., 1984, 1986; Arn et al., 1985). Lures usually consist of rubber septa containing 1 mg pheromone, although more recently microfibre dispensers have been used (Riedl et al., 1985).

Traps are placed in trees, usually at a convenient height, although Riedl et al. (1979) showed clearly that trap height should be standardized with respect to canopy height (rather than absolute height), preferably with traps in the upper half of the canopy (McNally and Barnes, 1981). Trap density varies from 1 per ha to 4 per ha despite the fact that Riedl (1980b) showed that catch per trap increased with decreasing trap density at densities greater than 1 trap per 7 ha (see also Charmillot and Schmid, 1981; McNally and Barnes, 1981).

There are large differences in the way in which trap catches are interpreted and this depends, in part, on the generation. In North America, New Zealand, France, Switzerland and Italy, there are two overlapping generations (sometimes three) per annum, whereas in England there is only one, with a partial second generation in some years. In South Africa there are three generations. In general the traps are better phenological indicators for the first generation (Riedl and Croft, 1974; Riedl et al., 1976) and the way in which the information from the catches is used therefore tends to differ in the subsequent generations (Charmillot, 1980a). In England a simple threshold of five moths per trap per week is used to determine the need to spray against the first (and usually only) generation larvae (Alford *et al.*, 1979) and the timing of the first spray should be approx. 140 day-degrees after threshold (Glen and Brain, 1982), with a second spray 100 day-degrees later. A basically similar approach was adopted previously in Canada (Madsen and Vakenti, 1973; Vakenti and Madsen, 1976), France (Audemard and Milaire, 1975), South Africa (Myburgh and Madsen, 1975) and the USA (Riedl and Croft, 1974; Riedl *et al.*, 1976), although with widely differing thresholds and in the last example different day-degree calculations. Probably the most integrated system is that adopted by the Swiss (Charmillot, 1980a). First cardboard band traps are placed on tree trunks and examined during the winter for diapausing larvae. The captures in 40 traps are, together with an estimate of the number of fruits per tree, used to predict a damage level for the first-generation larvae. At the beginning of May pheromone traps are started and the cumulative catch at 250 day-degrees is used to predict the need to spray the first-generation larvae. Less than 20 moths per trap indicates that control is not necessary, 20–50 moths indicates one spray and more than 50 moths two sprays. However, the trap catches alone are not considered completely reliable because males may be attracted from nearby orchards (especially abandoned ones), thereby indicating higher densities of females than actually exist. Therefore growers are encouraged to compare the predictions from the pheromone traps and the larval band traps when deciding on a control strategy. However, zero or very low catches of males in the pheromone traps are taken as a clear indication that control measures are not required. The last step for the first generation is to examine fruits for signs of damage (from 300 day-degrees onwards); on the basis of this inspection a final decision is made about spraying. The procedure is repeated for the second generation; band traps are examined at about the end of July, trap catches are watched closely for any peaks which will indicate the timing of oviposition, and fruit is inspected some 10–15 days later for signs of damage. Once again very low catches in pheromone traps are treated with more confidence than high catches which may simply be the result of the attraction of males from outside the orchard. However, Thwaite and Madsen (1983) have shown that this can be overcome by running traps in the adjacent orchards.

3.6.4 Cotton boll weevil, *Anthonomus grandis grandis*

In one system, developed for semi-arid cotton areas of west Texas, traps are located along the margins of individual cotton fields nearest to favourable overwintering habitats. The number of traps varies between five and ten per field, which are 4–40 ha in area. The minimum distance between traps is 30 m. Leggett traps (Leggett and Cross, 1971) containing 3 mg grandlure in a physical barrier formulation are operated from mid-April, although some crops are not planted until late May/early June. The mean number of overwintered weevils captured per trap per field during a specified time at the appearance of the first small fruit buds is designated the trap index (TI). The TI values are grouped into three distinct categories: (1) TI < 1.0 — do not treat; (2) TI 1.0–2.5 — treatment may or may not be justified, so crop inspected at ⅓-grown square stage for presence of damaged squares or adult weevils; (3) TI > 2.5 — treat just prior to appearance of ⅓-grown squares (Rummell et al., 1980).

Another system (developed later for the lower Gulf Coast of Texas) is similar but uses Hardee traps, placed 76–100 cm above the ground with a minimum spacing of 60 m and containing 24 mg grandlure in Hercon dispensers, replaced every 2 weeks (later changed to 4 weeks). Traps are examined weekly. The TI is calculated from the sum of the mean trap catch for 6 weeks prior to the first ⅓-grown square stage (Benedict et al., 1985, 1986).

3.7 CONCLUSIONS

Pheromone traps are useful additions to the armoury of entomologists involved in monitoring pest insects to ensure the rational application of pesticides. They are sensitive at low population densities, selective, cheap and require no power. Consequently they have great potential as local monitoring tools operated by non-entomologists. However, the need for much good ecological work to be done to overcome the apparent difficulties in interpreting trap catches has been appreciated in only a few quarters. Unless such work is given more prominence we shall continue to see exaggerated claims published and inadequately researched monitoring systems marketed, and pheromone monitoring traps will fail to make their rightful impact on future pest control.

3.8 REFERENCES

Alford, D. V., Carden, P. W., Dennis, E. B., Gould, H. J. and Vernon, J. D. R. (1979). Monitoring codling and tortrix moths in United Kingdom apple orchards using pheromone traps. Ann. Appl. Biol. 91, 165–178.
Arciero, M. F. (1979). Use of multilure-baited traps in the California Dutch Elm disease program for survey and detection of Scolytus multistriatus. Bull. Ent. Soc. Amer., 25, 119–121.

Arn, H., Guerin, P. M., Buser, P. R., Rauscher, S. and Mani, E. (1985). Sex pheromone blend of the codling moth, *Cydia pomonella*: Evidence for a behavioural role of dodecan-1-ol. *Experientia*, **41**, 1482–1484.

Arn, H., Toth, M. and Priesner, E. (1986). List of sex pheromones of Lepidoptera and related attractants. *IOBC/WPRS*, Paris, 124 pp.

Audemard, H. and Milaire, H. G. (1975). Le piégeage du carpocapse (*Laspeyresia pomonella* L.) avec une phéromone sexuelle de synthèse; premiers resultats utilisables pour l'estimation des populations et la conduite de la lutte. *Ann. Zool. Écol. Anim.*, **7**, 61–80.

Baker, T. C., Cardé, R. T. and Croft, B. A. (1980). Relationship between pheromone trap capture and emergence of adult Oriental fruit moths, *Grapholitha molesta* (Lepidoptera: Tortricidae). *Can. Entomol.*, **112**, 11–15.

Baker, T. C. and Roelofs, W. L. (1981). Initiation and termination of Oriental fruit moth male response to pheromone concentrations in the field. *Environ. Entomol.*, **10**, 211–218.

Bakke, A. (1985). Deploying pheromone-baited traps for monitoring *Ips typographus* populations. *Z. Ang. Ent.*, **99**, 33–39.

Bakke, A., Saether, T. and Kvamme, T. (1983). Mass trapping of the spruce bark beetle, *Ips typographus*. Pheromone and trap technology. *Medd. Nor. Inst. Skogforsk.*, **38** (3), 1–35.

Bakke, A. and Strand, L. (1981). Feromoner og feller som ledd i integrert bekjempelse av granbarkbillen. Noen resultater fra barkbilleaksjonen i Norge 1979 og 1980. [Pheromones and traps as part of an integrated control of the spruce bark beetle. Some results from a control program in Norway in 1979 and 1980]. *Rapp. Nor. Inst. Skogforsk.*, **5/81**, 1–39.

Barak, A. and Burkholder, W. E. (1976). Trapping studies with dermestid sex pheromones. *Environ. Entomol.*, **5**, 111–114.

Benedict, J. H., Urban, T. C. George, D. M., Severs, J. C., Anderson, D. J., McWhorter, G. M. and Zummo, G. R. (1985). Pheromone trap thresholds for management of overwintering boll weevils (Coleoptera:Curculionidae). *J. Econ. Entomol.*, **78**, 169–171.

Benedict, J. H., Segers, J. C., Anderson, D. A., Parker, R. D., Walmsley, M. R. and Hopkins, S. W. (1986). Use of pheromone traps in the management of overwintered boll weevil on the lower gulf coast of Texas. *Tex. Agric. Exp. Stn. Misc. Publ. 1576*, 11 pp.

Blight, M. M., Pickett, J. A., Smith, M. C. and Wadhams, L. J. (1984). An aggregation pheromone of *Sitona lineatus*. Identification and initial field studies. *Naturwissenschaften*, **71**, 480–481.

Bogenschutz, H. (1982). Die Uberwachung schadlicher Lepidopteren im Wald. *Zeit. Pflanzenkrank. Pflanzenschutz*, **89**, 586–594.

Bradshaw, J. W. S., Baker, R., Longhurst, C., Edwards, J. C. and Lisk, J. C. (1983). Optimization of a monitoring system for the pine beauty moth, *Panolis flammea* (Denis & Schiffermuller), using sex attractants. *Crop Protection*, **2**, 63–73.

Buchelos, C. Th. and Levinson, A. R. (1985). Population dynamics of *Ephestia elutella* (Huebner) in tobacco stores with and without insecticidal treatments: a survey by pheromone and unbaited traps. *Z. Angew. Entomol.*, **100**, 68–78.

Burkholder, W. E. (1985). Pheromones for monitoring and control of stored-product pests. *Ann. Rev. Entomol.*, **30**, 257–272.

Butt, B. A. and Hathaway, D. O. (1966). Female sex pheromone as attractant for male codling moths. *J. Econ. Entomol.*, **59**, 476–477.

Campion, D. G., Bettany, B. W., McGinnigle, J. B. and Taylor, L. R. (1977). The distribution and migration of *Spodoptera littoralis* (Boisduval) (Lepidoptera: Noctuidae), in relation to meteorology in Cyprus, interpreted from maps of pheromone trap samples. *Bull. Entomol. Res.*, **67**, 501–522.

Cardé, R. T. (1979). Behavioural responses of moths to female-produced pheromone and the utilization of attractant-baited traps for population monitoring. In *Movement of Highly Mobile Insects: Concepts and Methodology in Research*, R. L. Rabb and G. G. Kennedy (Eds), North Carolina State University, Raleigh, pp. 286–315.

Cardé, R. T., Doane, C. C., Baker, T. C., Iwaki, S. and Marumo, S. (1977a). Attractancy of optically active pheromone for male gypsy moths. *Environ. Entomol.*, **6**, 768–772.

Cardé, R. T., Doane, C. C., Granett, J., Hill, A. S., Kochansky, J. and Roelofs, W. L. (1977b). Attractancy of racemic disparlure and certain analogues to male gypsy moths and the effect of trap placement. *Environ. Entomol.*, **6**, 765–767.

Carroll, S. C. and Rummel, D. R. (1985). Relationship between time of boll weevil (Coleoptera, Curculionidae) emergence from winter habitat and response to grandlure-baited pheromone traps. *Environ. Entomol.*, **14**, 447–451.

Chambers, D. L. (1977). Attractants for fruit fly survey and control. In *Chemical Control of Insect Behaviour: Theory and Application*, H. H. Shorey and J. J. McKelvey (Eds), Wiley, New York, pp. 327–344.

Charmillot, P. J. (1977). Carpocapse des pommes (*Laspeyresia pomonela* L.): Contribution à l'étude de l'efficacité des pièges a attractif sexuel synthétique. *Bull. Soc. Ent. Suisse*, **50**, 37–45.

Charmillot, P. J. (1980a). Developpement d'un système de prévision et de lutte contre le carpocapse (*Laspeyresia pomonella* L.) en Suisse romande: rôle du service régional d'avertissement et de l'arboriculture. *Bull. OEPP*, **10**, 231–239.

Charmillot, P. J. (1980b). Le piègeage sexuel du carpocapse (*Laspeyresia pomonella* L.) en tant que moyen de prévision. *Acta Oecologica*, **1**, 111–122.

Charmillot, P. J. (1981). Technique de confusion contre la tordeuse de la pelure *Adoxophyes orana* F. v. R. (Lep., Tortricidae): II. Deux ans d'essais de lutte en vergers. *Mitt. Schweiz. Ent. Ges.*, **54**, 191–204.

Charmillot, P. J. and Schmid, A. (1981). Influence de la densité des pièges sexuels sur les captures de capua, la tordeuse de la pelure (*Adoxophyes orana* F. v. R.) *Rev. Suisse Vitic. Arboric. Hortic.*, **13**, 93–97.

Cogan, P. M. and Hartley, D. (1984). The effective monitoring of stored product moths using a funnel pheromone trap. *Proc. 3rd Int. Working Conf. Stored Product Ent., Kansas*, 631–639.

Cranham, J. E. (1980). Monitoring codling moth (*Cydia pomonella* L.) with pheromone traps. *Bull. OEPP*, **10**, 105–108.

Croft, B. A., Knight, A. L., Flexner, J. L. and Miller, R. W. (1986). Competition between caged virgin female *Argyrotaenia citrana* (Lepidoptera: Tortricidae) and pheromone traps for capture of released males in a semi-enclosed courtyard. *Environ. Entomol.*, **15**, 232–239.

Daterman, G. E. (1982). Monitoring insects with pheromones: Trapping objectives and bait formulations. In *Insect Suppression with Controlled Release Pheromone Systems* (Vol. 1), A. F. Kydonieus and M. Beroza (Eds), CRC Press, Boca Raton, Florida, pp. 195–212.

Daterman, G. E. and Sower, L. L. (1977). Douglas-fir tussock moth pheromone research using controlled-release systems. *Proc. Internat. Controlled Release Pesticide Symp.*, **4**, 68–77.

Den Otter, C. J. and Klijnstra, J. W. (1980). Behaviour of male summer fruit tortrix moths. *Adoxophyes orana* (Lepidoptera: Tortricidae), to synthetic and natural female sex pheromone. *Ent. Exp. Appl.*, **28**, 15–21.

Einhorn, J., Beauvais, F., Gallois, M., Descoins, C. and Causse, R. (1984). Constituants secondaires de la phéromone sexuelle du Carpocapse des Pommes, *Cydia pomonella* L. (Lepidoptera, Tortricidae). *CR Acad. Sci. Paris*, **299**, serie III, 773–778.

Einhorn, J., Witzgall, P., Audemard, H., Boniface, B. and Causse, R. (1986). Constituants secondaires de la phéromone sexuelle du Carpocapse des Pommes, *Cydia pomonella* L. (Lepidoptera, Tortricidae). II. Première approche des effets comportementaux. *CR Acad. Sci. Paris*, **302**, serie III, 263–266.

Elkinton, J. S. and Cardé, R. T. (1980). Distribution, dispersal, and apparent survival of male gypsy moths as determined by capture in pheromone-baited traps. *Environ. Entomol.*, **9**, 729–737.

Elkinton, J. S. and Cardé, R. T. (1981). The use of pheromone traps to monitor distribution and population trends of the gypsy moth. In *Management of Insect Pests with Semiochemicals: Concepts and Practice*, E. R. Mitchell (Ed.), Plenum, New York, pp. 41–55.

Elkinton, J. S. and Cardé, R. T. (1984). Effect of wild and laboratory-reared female gypsy moth (Lepidoptera: Lymantriidae) on the capture of males in pheromone-baited traps. *Environ. Entomol.*, **13**, 1377–1385.

Glen, D. M. and Brain, P. (1982). Pheromone-trap catch in relation to the phenology of codling moth (*Cydia pomonella*). *Ann. Appl. Biol.*, **101**, 429–440.

Graham, J. C. (1984). Emergence, dispersal and reproductive biology of *Cydia nigricana* (F.) (Lepidoptera: Tortricidae). Unpublished PhD Thesis, University of London, 376 pp.

Granges, J., Stahl, J., Baggiolini, M. and Murbach, R. (1969). Essais préliminaires sur le piègeage du carpocapse. *Comptes Rendues IVth IOBC Symp. on Integrated Control in Orchards, Avignon, France*, 9.

Greenway, A. R. and Wall, C. (1981). Attractant lures for males of the pea moth, *Cydia nigricana* (F.), containing (*E*)-10-dodecen-1-yl acetate and (*E,E*)-8,10-dodecadien-1-yl acetate. *J. Chem. Ecol.*, **7**, 563–573.

Guerin, P. M., Arn, H., Buser, H. R. and Charmillot, P. J. (1986). Sex pheromone of *Adoxophyes orana*: additional components and variability of (*Z*)-9- and (*Z*)-11-tetradecenyl acetate. *J. Chem. Ecol.*, **12**, 763–772.

Hardee, D. D., Cross, W. H. and Mitchell, E. B. (1969). Male boll weevils are more attractive than cotton plants to boll weevils. *J. Econ. Entomol.*, **62**, 165–169.

Hartstack, A. W. and Witz, J. A. (1981). Estimating field populations of tobacco budworm moths from pheromone trap catches. *Environ. Entomol.*, **10**, 908–914.

Henneberry, T. J. and Clayton, T. E. (1982). Pink bollworm of cotton (*Pectinophora gossypiella* (Saunders)): male moth catches in gossyplure-baited traps and relationships to oviposition, boll infestation and moth emergence. *Crop Protection*, **1**, 497–504.

Howell, J. F. (1972). An improved sex attractant trap for codling moths. *J. Econ. Entomol.*, **65**, 609–611.

Howell, J. F. (1974). The competitive effect of field populations of codling moth on sex attractant trap efficiency. *Environ. Entomol.*, **3**, 803–807.

Howell, J. F. (1981). Codling moth: blacklight trapping and comparisons with fermenting molasses bait and sex pheromone traps. *USDA, ARS, ARR-W-22*, 14 pp.

Howell, J. F. and Quist, J. A. (1980). Codling moth: effect of postemergence placement and location on the predictive value of pheromone traps. *USDA, ARR-W-13*, 8 pp.

Inscoe, M. N. (1982). Insect attractants, attractant pheromones, and related compounds. In *Insect Suppression with Controlled Release Pheromone Systems* (Vol. II), A. F. Kydonieus and M. Beroza (Eds), CRC Press, Boca Raton, Florida, pp. 201–298.

Knipling, E. P. and McGuire, J. U. Jr (1966). Population models to test theoretical effects of sex attractants used for insect control. *Inf. Bull—US Dep. Agric.* **308**.

Leggett, J. E. and Cross, W. H. (1971). A new trap for capturing boll weevils. *USDA Coop. Econ. Insect Rep.*, **21**, 773–774.

Leggett, J. E. and Cross, W. H. (1978). Boll weevils: the relative importance of color and pheromone in orientation and attraction to traps. *Environ. Entomol.*, **7**, 4–6.

Levinson, H. Z. and Levinson, A. R. (1973). The dual function of the assembling scent of the female Khapra Beetle *Trogoderma granarium*. *Naturwissenschaften*, **60**, 352–353.

Levinson, H. Z. and Levinson, A. R. (1977). Integrated manipulation of storage insects by pheromones and food attractants—a proposal. *Z. Ang. Entomol.*, **84**, 337–343.

Lewis, T. and Macaulay, E. D. M. (1976). Design and elevation of sex-attractant traps for the pea moth, *Cydia nigricana* (Steph.), and the effect of plume shape on catches. *Ecol. Entomol.*, **1**, 175–187.

Livingston, R. L. and Daterman, G. E. (1977). Surveying for Douglas-fir tussock moth with pheromone. *Bull. Ent. Soc. Am.*, **23**, 172–174.

Macaulay, E. D. M. (1977). Field trials with attractant traps for timing sprays to control pea moth. *Plant Path.*, **26**, 179–188.

Macaulay, E. D. M., Etheridge, P., Garthwaite, D. G., Greenway, A. R., Wall, C. and Goodchild, R. E. (1985). Prediction of optimum spraying dates against pea moth, (*Cydia nigricana* (F.)), using pheromone traps and temperature measurements. *Crop Protection*, **4**, 85–98.

Macaulay, E. D. M. and Lewis, T. (1977). Pheromone trap design. *Rothamsted Rep. for 1976, Part 1*, 125–126.

McNally, P. S. and Barnes, M. M. (1981). Effects of codling moth pheromone trap placement, orientation and density on trap catches. *Environ. Entomol.*, **10**, 22–26.

McNally, P. S. and Van Steenwyk, R. (1986). Relationship between pheromone-trap catches and sunset temperatures during the spring flight to codling moth (Lepidoptera: Olethreutidae) infestations in walnuts. *J. Econ. Entomol.*, **79**, 444–446.

Madsen, H. F. and Vakenti, J. M. (1973). Codling moth: use of codlemone baited traps and visual detection of entries to determine need of sprays. *Environ. Entomol.*, **2**, 677–679.

Meijer, G. M., Ritter, F. J., Persoons, C. H., Minks, A. K. and Voerman, S. (1972). Sex pheromones of summer fruit tortrix moth *Adoxophyes orana*: two synergistic isomers. *Science*, **175**, 1469–1470.

Miller, C. A. and McDougall, G. A. (1973). Spruce budworm moth trapping using virgin females. *Can. J. Zool.*, **51**, 853–858.

Miller, J. R., Mori, K. and Roelofs, W. L. (1977). Gypsy moth field trapping and electro-antennogram studies with pheromone enantiomers. *J. Insect Physiol.*, **23**, 1447–1453.

Minks, A. K. (1984). Attractants and pheromones of noxious insects (selected references). *IOBC/WPRS Bulletin VII (1)*, 175 pp.

Minks, A. K. and de Jong D. J. (1975). Determination of spraying dates for *Adoxophyes orana* by sex pheromone traps and temperature recordings. *J. Econ. Entomol*, **68**, 729–732.

Minks, A. K. and Voerman, S. (1973). Sex pheromones of the summerfruit tortrix moth, *Adoxophyes orana*: Trapping performance in the field. *Ent. Exp. Appl.*, **16**, 541–549.

Mitchell, E. B. and Hardee, D. D. (1974). In field traps: a new concept in survey and suppression of low populations of boll weevils. *J. Econ. Entomol.*, **67**, 506–508.

Mitchell, E. B. and Hardee, D. D. (1976). Boll weevils: attractancy to pheromone in relation to distance and wind direction. *J. Ga. Entomol. Soc.*, **11**, 113–115.

Myburgh, A. C. and Madsen, H. F. (1975). Interpretation of sex trap data in codling moth control. *Deciduous Fruit Grower*, **25**, 272–275.

Nakamura, K. and Oyama, M. (1978). An equation for the competition between pheromone traps and adult females for adult males. *Appl. Entomol. Zool.*, **13**, 176–184.

Peacock, J. W. and Cuthbert, R. A. (1975). Pheromone-baited traps for detecting the smaller European elm bark beetle. *USDA Coop. Econ. Ins. Rep.*, **25**, 497–500.

Perry, J. N., Macaulay, E. D. M. and Emmett, B. (1981). Phenological and geographical relationships between catches of pea moth. *Ann. Appl. Biol.*, **97**, 17–26.

Perry, J. N. and Wall, C. (1984). Local variation between catches of pea moth, *Cydia nigricana* (F.) (Lepidoptera: Tortricidae), in sex-attractant traps, with reference to the monitoring of field populations. *Prot. Ecol.*, **6**, 43–49.

Perry, J. N., Wall, C. and Greenway, A. R. (1980). Latin-square designs in field experiments involving insect sex-attractants. *Ecol. Entomol.*, **5**, 385–396.

Plimmer, J. R., Schwalbe, C. P., Paszek, E. C., Bierl, B. A., Webb, R. E., Maruno, S. and Iwaki, S. (1977). Contrasting effectiveness of (+) and (−) enantiomers of disparlure for trapping native populations of the gypsy moth in Massachusetts. *Environ. Entomol.*, **6**, 518–522.

Prokopy, R. J. (1968). Sticky spheres for estimating apple maggot adult abundance. *J. Econ. Entomol.*, **61**, 1082–1085.

Ramaswamy, S. B. and Cardé, R. T. (1982). Nonsaturating traps and long-life attractant lures for monitoring spruce budworm males. *J. Econ. Entomol.*, **75**, 126–129.

Riedl, H. (1980a). Monitoring and forecasting methods for codling moth management in the United States and Canada. *Bull. OEPP*, **10**, 241–252.

Riedl, H. (1980b). The importance of pheromone trap density and trap maintenance for the development of standardized monitoring procedures for the codling moth (Lepidoptera: Tortricidae). *Can. Entomol.*, **112**, 655–663.

Riedl, H. and Croft, B. A. (1974). A study of pheromone trap catches in relation to codling moth (Lepidoptera: Olethreutidae) damage. *Can. Entomol.*, **106**, 525–537.

Riedl, H., Croft, B. A. and Howitt, A. J. (1976). Forecasting codling moth phenology based on pheromone trap catches and physiological-time models. *Can. Entomol.*, **108**, 449–460.

Riedl, H., Hoying, S. A., Barnett, W. W. and Detar, J. E. (1979). Relationship of within-tree placement of the pheromone trap to codling moth catches. *Environ. Entomol.*, **8**, 765–769.

Roelofs, W. L., Comeau, A., Hill, A. and Milicevic, G. (1971). Sex attractant of the codling moth: characterisation with electroantennogram technique. *Science*, **174**, 297–299.

Rummel, D. R., White, J. R., Carroll, S. C. and Pruitt, C. R. (1980). Pheromone trap index system for predicting need for overwintered boll weevil control. *J. Econ. Entomol.*, **73**, 806–810.

Sanders, C. J. (1978). Evaluation of sex attractant traps for monitoring spruce budworm populations (Lepidoptera: Tortricidae). *Can. Entomol.*, **110**, 43–50.

Sanders, C. J. (1981). Sex attractant traps: Their role in management of spruce budworm. In *Management of Insect Pests with Semiochemicals*, E. R. Mitchell (Ed.), Plenum, New York, pp. 75–91.

Sanders, C. J. (1984). Sex pheromone traps and lures for monitoring spruce budworm populations—the Ontario experience (*Choristoneura fumiferana* forest pest monitoring). *USDA Forest Service General Tech. Rep.*, **88**, 17–22.

Sanders, C. J. (1986). Evaluation of high-capacity, nonsaturating sex pheromone traps for monitoring population densities of spruce budworm (Lepidoptera: Tortricidae). *Can. Entomol.* **118**, 611–619.

Sartwell, C., Daterman, G. E., Koerber, T. W., Stevens, R. E. and Sower, L. L. (1980). Distribution and hosts of *Eucosma sonomana* in the Western United States as determined by trapping with synthetic sex attractants. *Ann. Ent. Soc. Amer.*, 73, 254–256.

Schwalbe, C. P. (1981). Disparlure-baited traps for survey and detection. In *The Gypsy Moth: Research Toward Integrated Pest Management*, C. C. Doane and M. L. McManus (Eds), *USDA Tech. Bull.*, No. 1584, 542–548.

Sower, L. L., Daterman, G. E. and Sartwell, C. (1984). Surveying populations of western pine shoot borers (Lepidoptera: Olethreutidae). *J. Econ. Entomol.*, 77, 715–719.

Thwaite, W. G. and Madsen, H. F. (1983). The influence of trap density, trap height, outside traps and trap design on *Cydia pomonella* (L.) captures with sex pheromone traps in New South Wales apple orchards. *J. Aust. Ent. Soc.*, 222, 97–99.

Vakenti, J. M. and Madsen, H. F. (1976). Codling moth (Lepidoptera: Olethreutidae): monitoring populations in apple orchards with sex pheromone traps. *Can. Entomol.*, 108, 433–438.

van der Kraan, C. and Van Deventer, P. (1982). Range of action and interaction of pheromone traps for the summer fruit tortrix moth, *Adoxophyes orana* (F. v. R.) *J. Chem. Ecol.*, 8, 1251–1262.

Vick, K. W., Coffelt, J. A. and Mankin, R. W. (1981). Recent developments in the use of pheromones to monitor *Plodia interpunctella* and *Ephestia cautella*. In *Management of Insect Pests with Semiochemicals: Concepts and Practice*, E. R. Mitchell (Ed.), Plenum, New York, pp. 19–28.

Vick, K. W., Kvenberg, J., Coffelt, J. A. and Steward, C. (1979). Investigation of sex pheromone traps for simultaneous detection of Indian meal moths and Angoumois grain moths. *J. Econ. Entomol.*, 72, 245–249.

Vité, J. P., Klimetzek, D., Loskant, G., Hedden, R. and Mori, K. (1976). Chirality of insect pheromones: response interruption by inactive antipodes. *Naturwissenschaften*, 68, 582–583.

Wall, C. (1984). The exploitation of insect communication—fact or fantasy? In *Insect Communication*, T. Lewis (Ed.), *Symp. R. Ent. Soc. Lond.*, 12, 379–400.

Wall, C., Garthwaite, D. G., Blood Smyth, J. A. and Sherwood, A. (1987). The efficacy of sex-attractant monitoring for the pea moth, *Cydia nigricana* (F.), in England, 1980–85. *Ann. Appl. Biol.* 110, 223–229.

Wall, C. and Greenway, A. R. (1981). An effective lure for use in pheromone monitoring traps for the pea moth, *Cydia nigricana* (F.) *Plant Path.*, 30, 73–76.

Wall, C. and Perry, J. N. (1978). Interactions between pheromone traps for the pea moth, *Cydia nigricana* (F.). *Entomol. Exp. Appl.* 24, 155–162.

Wall, C. and Perry, J. N. (1980). Effects of spacing and trap number on interactions between pea moth pheromone traps. *Entomol. Exp. Appl.* 28, 313–321.

Wall, C. and Perry, J. N. (1981). Effects of dose and attractant on interactions between pheromone traps for the pea moth, *Cydia nigricana* (F.) *Entomol. Exp. Appl.* 30, 26–30.

Wall, C. and Perry, J. N. (1987). Range of action of moth sex-attractant sources. *Entomol. Exp. Appl.* 44, 5–14.

Wanabe, T., Kato, J., Arai, K., Kohno, M., Chuman, T., Mochizuki, K. and Kato, K. (1982). Integrated control of cigarette beetle (*Lasioderma serricorne* F.) with a synthetic sex pheromone and food attractant in the cigarette manufacturing factory. *Sci. Pap. Cent. Res. Inst. Jpn. Tobacco Salt Public Corp.* 124, 23–32.

Weslien, J., Annila, E., Bakke, A., Bejer, B., Eidmann, H. H., Narvestad, K., Nikula, A. and Ravn, H. P. (1988). Estimating risks for spruce bark beetle (*Ips typographus* (L.)) damage using pheromone-baited traps and trees. *Scand. J. For. Res.* 3, (in press).

Insect Pheromones in Plant Protection
Edited by A. R. Jutsum and R. F. S. Gordon
©1989 John Wiley & Sons Ltd

4

Mass Trapping

A. Bakke
Norwegian Forest Research Institute, Box 61, 1432 As-NLH, Norway
R. Lie
Borregaard Industries Ltd, N-1701 Sarpsborg, Norway

4.1 INTRODUCTION

Mass trapping of pest insects is an alternative to mass killing by using insecticides. The control measure is directed against the adult insect. The intention is to catch a pest species selectively and thereby suppress the population to a level below the threshold of damage.

This method of control will be limited to insects with certain behavioural patterns and to those with a highly developed ability to respond to attractants. The range of application is mostly restricted to the protection of crops where the use of insecticides is difficult or environmentally undesirable, and where less than 100% control can be accepted.

World-wide a great variety of insects has been mass-trapped in pheromone-baited traps. The results vary from some success to no effect at all on the insect population.

4.2 COTTON PESTS

Cotton is an important cash crop in many countries; one-third of the pesticide consumption in the world is used for the management of cotton pests. Even with this use of insecticides, many insects cause extensive damage to crops.

The boll weevil, *Anthonomus grandis* Boheman, is a most serious cotton pest in the USA. Cotton with a value of $200 to $300 million is lost through boll weevil attacks every year (Hardee, 1982). In early trap experiments cottonseed meal, molasses and cotton plant extracts were used as baits with little success, but the discovery of the male-released pheromone (grandlure) provided an effective attractant (Tumlinson et al., 1969). The male pheromone also acts as an aggregation pheromone for both sexes in the early and late season. Grandlure is a mixture of four components. Maximum attraction is achieved by using a mixture of two alcohols and two aldehydes. Numerous field evaluations of boll weevil mass trapping have been conducted in the USA between 1970 and 1980 with different pheromone formulations and traps (Hardee, 1982). It was possible to demonstrate an effect on the insect population, but the suppression was not sufficient to avoid insecticide use. For this reason, work with mass trapping of boll weevil with pheromone-baited traps is not believed to be a suitable way to manage the insect (Hardee, 1984).

The pink bollworm, *Pectinophora gossypiella* (Saunders), is an important pest causing great damage in many cotton growing areas. Hummel *et al.* (1973) identified the female released sex pheromone to be a 1:1 mixture of the Z,Z and Z,E isomers of 7,11-hexadecadienyl acetate, generally known under the name of gossyplure. Mass-trapping experiments have been conducted with different pheromone dispensers and traps of various designs. For many of the cotton fields methods like mass trapping are too labour-intensive to be used on a large scale, but some promising results are reported from Israel (Shani, 1982). Mass trapping of one insect species in a field with competing pests will not necessarily avoid the use of insecticides. For this reason many programmes of pink bollworm mass trapping have not been continued.

The polyphagous Egyptian cotton leaf worm, *Spodoptera littoralis* (Boisduval), is a serious pest to many cash crops in Mediterranean countries and Africa. Labour-intensive egg-mass collection is still used in Egyptian cotton fields as a management technique (Campion and Nesbitt, 1981), and intensive spraying with insecticides does not always prevent or eliminate damage (Neumark *et al.*, 1974).

The female sex pheromone of *S. littoralis* was identified by Nesbitt *et al.* (1973). Field evaluation of the four components of the sex pheromone demonstrated (Z,E)-9,11-tetradecadienyl acetate to be a long-range attractant (Campion *et al.*, 1974). In practical use this diene is sufficient as a bait. The three other components have little attractant effect or inhibit the attraction of male *S. littoralis*. When the diene is exposed to high temperature, it will polymerize. For this reason it is advantageous to mix it with an antioxidant. Various designs of traps have been tested. A funnel trap connected to a plastic bag is the most commonly used.

Large-scale mass trapping started in Israel in 1975 (Teich *et al.*, 1979). With one trap per 1.7 ha it was possible to reduce the egg clusters by 40–50% and to reduce insecticide treatment by 20–25% (Shani, 1982). The mass-trapped area was 2500 ha in 1975, 12 000 ha in 1980 and 20 000 ha in 1981. With a trap density of 1 trap per 0.5–0.6 ha the results from Israel are promising. The mass trapping of Egyptian cotton leaf worm is an alternative to classical insecticide spraying, and is now in large-scale use throughout Israel.

Campion and Nesbitt (1981) worked on trap technology in Crete and Egypt. Trap densities ranging from 3 to 27 traps per hectare were studied. With more than 9 traps per hectare the total number of insects caught did not significantly increase. With 3–9 traps per hectare it was possible to reduce the number of egg masses on some areas while on others it was not possible to detect any effect of the mass trapping. Campion and Nesbitt concluded from their work on the mass trapping of *S. littoralis* that the method is not suitable for pest management in Egypt and Crete. This is in total contrast to the promising results in Israel.

4.3 STORED-PRODUCT PESTS

Stored food is an excellent breeding and living place for a great variety of insects. Insect damage on stored human food accounts for great economic losses. The majority of insect species infesting stored food belong to the orders Coleoptera and Lepidoptera. The great variety of species complicates the use of species-specific attractants like sex pheromones in pest management. The use of food baits is also complicated since large amounts of food are available in the store. It is also of great interest to reduce the use of insecticides and avoid fumigation of stored food. The use of the common fumigant methyl bromide will probably become more restricted in many countries.

The structure of sex pheromones is established for all the important stored-product insects (Burkholder, 1981; Inscoe, 1982). Different designs of sticky traps are effective for most of the stored-product insects (Levinson and Levinson, 1979), but other designs are also in use. Levinson and Levinson in 1979 reported mass trapping as part of an integrated pest management programme for stored-product pests. Trematerra and Battaini (1987) reported success in controlling the Mediterranean flour moth, *Ephestia kuehniella* (Zeller), by mass trapping with pheromone-baited funnel traps at a density of one trap per 260–270 cu. m.

In a moderate insect population it is possible over some time to suppress the population below the economic damage level. In this way it is possible to reduce the use of insecticides and number of fumigations. This will also reduce the probability of development of insecticide-resistant strains. In many of the programmes of mass trapping for stored-product insects it was not possible to correlate trap catch with reduction in population. It is therefore probable that mass trapping of stored-product insects will be used only to a limited extent in pest management.

4.4 HORTICULTURAL PESTS

It is necessary to apply insecticides in virtually all fruit-growing areas of the world. Fruit is a cash crop with a high export value in many countries. For these reasons much research and applied work on pheromones has been directed to insects damaging fruit. Interesting results have been achieved by disruption, but only a few experiments on mass trapping were successful.

Of the great number of insects using citrus crops as a host the citrus flower moth, *Prays citri* (Milliere), is the main pest of lemon in parts of the Mediterranean. Nesbitt and co-workers isolated and identified the female pheromone in 1977 (Nesbitt *et al.*, 1977). The female sex pheromone of the citrus flower moth, (Z)-7-tetradecenal, is identical to the chemical released from female olive moth, *Prays oleae* (Bernard) (Campion *et al.*, 1979). With only one main pest on the monocultured citrus the necessary condition exists for insect suppression by mass trapping. This is the situation in Israel, where the necessary tools have been developed for mass trapping of the citrus flower moth (Sternlicht *et al.*, 1981). Field evaluation has demonstrated low-density polyethylene fibres and rubber septa to have a suitable release rate over 4 months for (Z)-7-tetradecenal (Sternlicht, 1984). Various types of sticky traps and funnel traps have been tested for practical use in a mass-trapping campaign (Sternlicht *et al.*, 1981). Both types of traps can be used, and a trap with a renewable sticky floor is very effective. By using 120–140 traps per hectare it was possible in most areas to control the citrus flower moth (Shani, 1982). In 1978, 800 hectares were managed and there was no need for insecticide application on 80% of the area. Today almost all lemon growers in Israel use pheromone for male mass trapping to control the citrus flower moth. Insecticides are generally not used against this pest.

The olive fruit fly, *Dacus oleae* (Gmelin), is a major olive pest in the Mediterranean. In several overlapping generations it attacks olives from early summer up to harvest. The eggs are laid singly into maturing olives. Attack by olive fly may result in fruit drop as well as reduced quality and quantity of olive oil. In many olive-growing areas the olive fly is the most damaging insect on this important cash crop. Control practices involve the use of insecticides over large areas.

For many years various traps baited with ammonium salts or protein hydrolysates have been used for trapping of the olive fruit fly (Jones *et al.*, 1983). The main component of the olive fly pheromone has been identified as a spiroacetal and synthesized (Baker *et al.*, 1980).

Later works have also characterized minor constituents in the pheromone blend (Garibaldi *et al.*, 1983; Baker *et al.*, 1982). A great variety of traps have been tested and used for trapping the olive fly. For practical use the yellow vertical sticky trap is very effective, but the catch of non-target insects is a disadvantage

(Jones, 1983). Rubber septa and polyethylene vials are in use as dispensers for the spiroacetal.

In Greece, Haniotakis (1981) and Broumas *et al.* (1983) have achieved promising results with mass trapping of olive fly. In 1981, a 100 ha olive area was treated with pheromone-baited traps. In a total of 10 000 trees a yellow sticky trap was placed on every ninth tree. With this approach a single aerial spray of insecticide was required to give equivalent control to that in the area without traps, which was sprayed three times. Even with a high olive fly population in the mass-trapping area there was no difference in the final gross yield per tree between the two areas.

Insecticide spraying of olives will also protect the crop from damage by the olive moth, *Prays oleae*, and olive scale as well as other insects using olives as a host. Jones (1983) has employed mass trapping for olive fly in Spain. Even with high catches it was not possible to prevent insecticide spraying.

The smaller tea tortrix, *Adoxophyes* sp., is a pest of tea plants in Japan. The female pheromone is a four-component mixture of acetates which is effective as a bait in a mass-trapping programme (Negishi *et al.*, 1980). Field evaluation demonstrated pheromone mass trapping to be superior to conventional insecticide spraying on a control field. In this way it was possible to reduce the number of insecticide sprays from six to three. The work was carried out on a small test area. Further experiments must be conducted to see whether or not mass trapping is viable on large areas.

4.5 POLYPHAGOUS INSECTS

Of the many polyphagous insects some, like Japanese beetle and gypsy moth, have been a focus for applied pheromone research.

The Japanese beetle, *Popillia japonica* Newman, is native to Japan and north-eastern China. It was imported to the USA in 1916 and is now a serious pest in all eastern states. This polyphagous insect attacks more than 300 different kinds of plant. The larvae feed on the roots of a wide variety of garden, truck crop and ornamental plants. Adult beetles attack the aerial parts of fruit trees and flowers.

The Japanese beetle is destructive to many plants but it is not a primary pest of any major crop. It has been the focus of research for its attacks on ornamental and garden plants.

Since 1928 many chemicals have been tested and used as an attractant for Japanese beetle (McGovern and Ladd, 1984). The female sex pheromone (japonilure) was identified as (R,Z)-5-(1-decenyl)dihydro-2(3H)-furanone (Tumlinson *et al.*, 1977).

Field experiments have demonstrated the need for japonilure to be optically pure for maximum biological effect. Admixture of as little as 1% of the S,Z isomer significantly reduced attraction to males (Tumlinson, 1979). Comparison of japonilure and a chemical attractant like phenethyl propionate/eugenol demonstrated the former to attract more male beetles, while females dominated in the latter (Ladd and Klein, 1982). Combination of japonilure and the chemical attractant was shown to be more attractive to both sexes than either lure alone. Commercial companies have been interested in developing a trap suitable for mass trapping of Japanese beetle (Klein, 1981). The most commonly used trap has plastic fins and a funnel with a disposable plastic bag to collect the trapped insects. Both hollow fibre and laminate flake systems have been used as a dispensing device for the dual lure. The system with trap and bait has been available to the consumer for some years. The traps have been deployed in mass trapping mainly by garden owners, but they are also recommended by the USDA for population reduction of Japanese beetle (Schwartz, 1975).

By using dual lures and traps it has been possible to trap high numbers of Japanese beetle. However, the effect of mass trapping on insect population and crop damage is questionable. So far, no reports have documented the direct effect of mass trapping on preventing crop losses.

Over the years the gypsy moth, *Lymantria dispar* L., has been a target for mass trapping. Gypsy moths are found all over Europe and east Asia, but very serious outbreaks occur on the east coast of the USA. In 1980, 2 million hectares of forest were defoliated, and in 1981 more than 5 million hectares were attacked (Plimmer *et al.*, 1982). This polyphagous insect feeds on about 500 species of vegetation, but oak is the preferred host. Even with only one generation per year the gypsy moth population varies enormously from year to year. The insect can be controlled by conventional insecticide spraying or the use of bacteria, parasites and predators. The application of pesticides is restricted in the USA, and the large extent of the area is a problem for effective pest management.

The female pheromone is *cis*-7,8-epoxy-2-methyl-octadecane (Bierl *et al.*, 1970). The (+)-disparlure enantiomer is more attractive than the racemic mixture (Plimmer *et al.*, 1982).

For practical mass-trapping of gypsy moth a great variety of dispensers and trap designs have been tested (Plimmer *et al.*, 1982; Webb, 1982). The first mass-trapping tests started in 1971 with virgin females as bait (Cameron, 1973). In this and later experiments using racemic disparlure it was not possible to demonstrate that the effect of trapping was a reduction in population (Webb, 1982). With a high population of gypsy moth over as many as 20 million hectares, pest management based only on mass trapping is probably not possible. Extensive use of traps in areas of low and medium infestation may be effective. Different traps and pheromone dispensers have been marketed for garden owners in the USA (Kydonieus, 1981). During the outbreak in 1980–1982 these products had a wide application. However, no scientific documentation of the results is available.

4.6 BARK BEETLES

Many bark beetles have developed a semi-social behaviour pattern. Both sexes participate in gallery construction and many individuals cooperate when living trees are attacked and made accessible as breeding sites.

Chemical communication is fundamental in such behaviour. Attractants, which are released partly from the host tree, but mainly from pioneer beetles, lead the beetles in their search for breeding material. Chemical components also regulate the spacing of egg galleries.

Host odours and pheromones may act differently during the process of host colonization. In the first phase, when pioneer beetles are searching for breeding sites, they depend solely on host stimuli. Later, when the pioneering beetles have found a suitable host and started pheromone production, the importance of the host odour diminishes.

This is true for some tree-killing bark beetles species and ambrosia beetles with effective aggregation pheromones. However, the addition of host components will usually enhance the response to the pheromone in most species.

Aggregation pheromones have been identified in several species of bark beetles (Borden, 1982). The sex which initiates gallery construction is the main pheromone producer. In monogamous bark beetle species this is the female; in polygamous beetles it is the male. In some species, however, both sexes contribute to the bouquet of pheromone components. The pheromone generally consists of more than one component, which are either bicyclic ketals or alcohols and their corresponding ketones (Vité and Francke, 1976). Terpenes of the host tree are often the precursors. The pheromones are released with the frass after successful penetration by the beetles into the host tree.

The aggregation behaviour of bark beetles was exploited in control measures long before the pheromones were demonstrated and identified. Mass trapping of beetles by means of trap trees has been practised in Europe for 200 years (see Gmelin, 1787). Pheromones produced by boring beetles were the main source of attraction. Living, mature spruce trees were felled in early spring and left in the forest during the main flight period. After the trunks had been colonized by the beetles, they were removed from the forest while the beetles were still in the bark. The logs were debarked, the bark being burned or destroyed in some other way.

Trap trees have also been used in the north-western part of the USA and Canada to control the spruce beetle, *Dendroctonus rufipennis* (Kirby)

(Nagel *et al.*, 1957). The method has been modified in recent years (Schmid and Frye, 1977). Arsenic components are injected into the living tree and translocated throughout the bole. When the tree is attacked after felling, the biocide kills the brood and further treatment is not required.

The trap tree method is expensive, time-consuming and has many other disadvantages. When synthetic pheromones became available, experiments were initiated to replace the trap trees with artificial traps baited with pheromones. For some species a successful mass-trapping technology has been developed.

Large-scale experimental control programmes, where mass trapping has been integrated, are conducted in cases where bark beetles are responsible for three types of damage: tree killing; transmission of a phytotoxic fungus; and boring in saw timber.

4.6.1 The tree killers

Bark beetles of the genus *Dendroctonus* are responsible for tree killing over large areas, particularly in western and southern North America and Central America. In central and northern Europe and in north-eastern Asia the spruce bark beetle, *Ips typographus* (L.), accounts for great timber losses. These tree killers have much in common in their behaviour and attack strategy, and their epidemics are often initiated by the same factors.

All are inhabitants of the subcortical tissues of living trees. They manage to overcome the resistance of the host tree, colonize the bole and utilize the inner bark for brood production. There are two main reasons for the extraordinary capabilities of these aggressive bark beetles. First, they have developed an effective chemical signal system to coordinate the attack and to aggregate in masses on selected trees. This cooperation is necessary to overcome the resistance of the tree. Second, they introduce pathogenic blue-stain fungi that invade the sapwood of the tree, induce water stress and render the tree susceptible to colonization by the beetles. Without the action of the fungus the tree would stay alive and produce resin even when the beetles made tunnels in the inner bark.

There is a strong correlation between host tree condition and success of bark beetle attack. A given forest stand can tolerate a bark beetle population of a certain size because of the trees' ability to repel attacking beetles. The critical population level is considered a continuous function of stand resistance. Berryman (1978) has suggested a simple threshold model relating stand resistance and bark beetle population level (Fig. 4.1). The model shows that control measures to suppress the beetle population may bring the population below the critical threshold. The chance of successfully controlling the beetles is best when stand resistance is high.

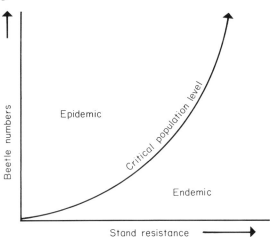

Fig. 4.1 The critical beetle population level which separates endemic from epidemic is considered a continuous function of stand resistance (after Berryman, 1978)

Large-scale outbreaks are usually the result of special weather conditions that have caused an increase in the beetle population by supplying large amounts of breeding material. This may include wind-thrown trees, or trees suffering from drought. An increasing beetle population will overcome the resistance of increasingly healthier trees and will thus gradually spread to more vigorous stands.

There are variations in the behaviour of different tree-killing species. Furthermore, their pheromones may be of different potency. Some species only breed in standing trees, while others can also utilize wind-thrown trees. The latter may reach an epidemic level after a heavy windstorm, while both groups may profit from a decrease in host resistance, which is often caused by drought. These differences, as well as differences in silvicultural practices throughout the world, will require different control strategies.

The western pine beetle, *Dendroctonus brevicornis* Le Conte, attacks the ponderosa pine, *Pinus ponderosa*, which is one of the most important forest trees in the western United States and Canada. Research to develop a mass-trapping suppression tactic over large, forested areas started as early as 1970 (Bedard and Wood, 1981).

Flight and attacks start in late spring and continue until stopped by cold weather. There are one to four generations and more sister broods annually, depending on latitude. Initial bores into the standing trees are made by the female beetles. The attacking beetles transfer blue-stain fungi, which invade the sapwood. *Ceratocystis minor* (Hedge.) plays a pathogenic role. The beetle offspring overwinter in the tree, in all stages except for pupae.

In the 1920s and 1930s the beetle was epidemic during an extended drought in the Pacific coast states. Infested trees were felled and the bark burnt in order to protect commercially valuable stands for later harvest. Fire and insecticide treatment have been used for destroying beetles in trees that cannot be logged (Miller and Keen, 1960).

Silverstein *et al.* (1968) identified *exo*-brevicomin from female-produced frass and Kinzer *et al.* (1969) identified frontalin from male hindguts. Wood *et al.* (1976) showed that a combination of (+)-*exo*-brevicomin, (−)-frontalin, and myrcene from the host tree is a highly attractive mixture to both males and females. Thus an easily portable sticky trap with a large trapping surface was developed (Browne, 1978).

In 1970, a large-scale mass-trapping experiment was conducted at Bass Lake, California (Bedard and Wood, 1974). In one area, traps were deployed in two configurations over 65 km² for one season. Nearly one million bark beetles were trapped. Tree mortality declined throughout the area (DeMars *et al.*, 1980) and remained low for about 4 years. There were no control areas without traps. The sticky traps also caught a large number of predators which responded to the pheromone components.

In a large-scale experiment in a heavily infested area (McCloud Flats, California) the results were less promising. Tree mortality was apparently not affected despite the trapping of 7 million beetles (Silverstein, 1981). For western pine beetles it is concluded that, despite the successful demonstration at Bass Lake, the economics of forest management are unlikely to justify the use of pheromone for control.

The spruce bark beetle, *Ips typographus* (L.), is the main pest of mature Norway spruce *Picea abies* in Europe and north Asia. Several severe outbreaks are described during the last century in central Europe, Scandinavia and Japan.

The flight takes place mainly in spring, from April through May. Sister brood flights may occur throughout the summer, particularly during epidemics. There is one annual generation in northern areas, where the beetles hibernate as adults. In southern areas there may be two and even three generations a year. The beetles breed in trees thrown by wind, in slash after logging operations, and during outbreaks mainly in healthy trees. The male beetle initiates the boring and attracts one to four females by releasing a pheromone. The beetles transfer several species of blue-stain fungi, one of which, *Ceratocystis polonica* (Siem.), is able to kill healthy trees (Horntvedt *et al.*, 1983). *cis*-Verbenol and methylbutenol are the major components of the aggregation pheromone. Ipsdienol also is present and plays a minor role (Bakke *et al.*, 1977).

Dispensers containing three-component mixtures have been developed. Exposed in the field in early May, the dispenser continues to release pheromones during a 2–3-month period. The dispensers were approved by the Pesticides Board of the Ministry of Agriculture in Norway and Sweden for practical use in forestry and are used as bait in drainpipe traps made of polyethylene tubes with about 900 holes (Bakke, 1981; Bakke *et al.*, 1983) (Fig. 4.2).

The pheromone trap technology was developed to be used as part of an integrated control programme against *I. typographus* in Norway and Sweden (Bakke and Strand, 1981; Eidmann, 1983). Storm fellings and drought precipitated severe outbreaks during the 1970s (Worrell, 1983). In Norway, trees equivalent to 5 million cubic metres of timber were attacked and killed in an area of 140 000 km². About half of this quantity was killed before organized control measures were undertaken.

Mass trapping constituted a major part of the integrated control programme, but was combined with other measures. The major long-term

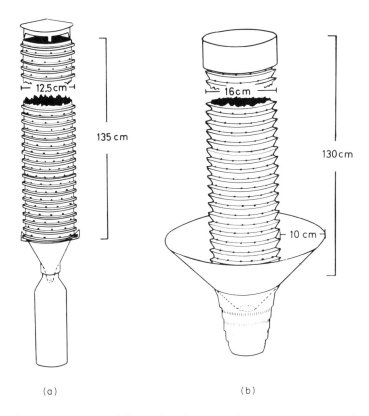

(a)   (b)

Fig. 4.2 Drainpipe trap models used in the control programme against the spruce bark beetle in Norway. (a) 1979 model; (b) 1980 model

objective was to stimulate increased harvesting of overmature stands in areas threatened by the beetles. A forest practice law was amended to prohibit storage of unbarked logs in the forest during summer, and to require clean-up after storm damage and logging. In addition to mass trapping, felling and removal of beetle-infested trees (sanitation cutting) was the main short-term measure. It was strongly recommended that infested trees should be removed from the forest in June–July, before the adults of the next generation had emerged.

Pheromone-baited drainpipe traps were deployed in clear-cuts throughout the infested areas (Table 4.1). Test traps were selected in counties with the most severe beetle attacks. The average catches per trap varied widely, the highest number being recorded in counties with the most severe beetle attack (Fig. 4.3). The catches were highest in 1980. Traps located in areas cut during the previous winter had the highest catches. Many traps had more than 50 000 beetles and some caught close to 100 000.

TABLE 4.1 Number of traps deployed in Norway during the bark beetle control programme and estimated number of beetles caught in traps

Year	No. of traps	Estimated total capture of beetles (millions)
1979	600 000	2900
1980	590 000	4500
1981	530 000	2100

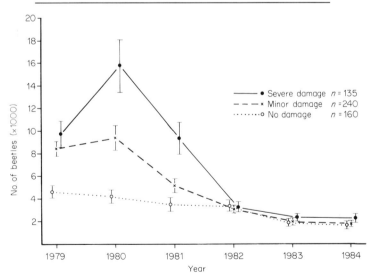

Fig. 4.3 Average catches per trap in areas of southern Norway with different degrees of bark beetle damage during the period 1979–1981. After 1981 there was only insignificant damage in all areas

More than 99% of the insects trapped were beetles of the genus *Ips*. *I. typographus* predominated, but in certain localities small percentages of *I. duplicatus* (Sahlberg) were found. Predatory clerids of the genus *Thanasimus* are attracted by *cis*-verbenol and ipsdienol (Bakke and Kvamme, 1981) and a few were found in the traps. However, the holes in the trap tube were made too small to admit most of the attracted predators. The best trap model had only 0.5 clerids caught per thousand *Ips*. This indicates that the drainpipe trap is very selective, and unlike sticky traps or window traps prevents the killing of useful predators. This may increase the ratio of natural enemies to bark beetle and result in a higher bark beetle mortality.

Attempts were made to examine the local effect of the control measures. Counts were made of newly infested trees and trees killed the previous year in 958 selected trap sites throughout the epidemic area (Bakke and Strand, 1981). Whereas the total damage throughout Norway was almost unchanged from 1979 to 1980, the results of the trap sites were as shown in Table 4.2. There was a distinct reduction in number of killed trees compared to the previous year on sites with all grades of damage. Sites where salvage logging was combined with trapping had one-third less attacked trees compared to sites where the infested trees remained in the forest.

The epidemic started to decline after 1980 and in 1982 it was difficult to find trees killed by the beetles (Fig. 4.3). It was supposed that the decline in beetle populations was a combined effect of several factors. Cool and wet summer weather limited the flight activity of the beetles and restored the resistance of the trees to beetle attack, and the extensive control programme suppressed the beetle population. The effect of mass trapping alone is not easy to evaluate isolated from the other measures of the control programme and from the influences of weather.

A mass-trapping programme started in central Europe (West Germany, Poland) in 1984 to suppress the populations of *I. typographus* and thereby prevent the beetle from utilizing the spruce trees weakened by air pollution (Vité, 1984). Gritsch (1987) reviewed the integrated control of *I. typographus* and other bark beetles with pheromones and insecticide-treated trap trees.

TABLE 4.2 Effect of mass trapping on sites with different levels of damage

Damage level in 1979 (No. of trees killed)	No new infestations in 1980 (% of sites)
1–5	70
6–30	51
More than 30	30

4.6.2 Transmitters of phytotoxic fungus

The fungus *Ceratocystis ulmi* (Buisman) C. Moreau, which causes Dutch elm disease, is transmitted primarily by various bark beetles. The smaller European elm bark beetle, *Scolytus multistriatus* (Marsham), is a principal vector in North America and Europe. This phytotoxic fungus has virtually eliminated the elm as a shade tree throughout most of the United States and Western Europe. It has become clear that there is no single measure or method capable of controlling Dutch elm disease. In North America, the aggregation pheromone of *S. multistriatus* was identified and technology was developed to include mass trapping of the flying beetles as part of the control measures (Peacock *et al.*, 1981).

The aggregation attractant of *S. multistriatus* is a combination of (−)-*α*-multistriatin and (−)-4-methyl-3-heptanol produced by virgin female beetles, and (−)-*α*-cubebene, a volatile compound from the host tree (Lanier *et al.*, 1976, 1977). The release of the elm tree component is enhanced by the beetle boring activity and the presence of *C. ulmi* (Peacock *et al.*, 1981). For field trials synthetic components were dispersed from laminated plastic flakes or hollow fibres. Both sexes were affected. The traps used for beetle suppression are 45×66 cm sheets of white, double poly-coated cardboard coated with sticky material.

Mass-trapping experiments for elm bark beetle suppression have been conducted in several parts of the United States. Two basic trap deployment strategies were used: deployment in a grid within areas; or deployment in one or more 'barrier' rows around areas to be protected. In a study in Detroit, Michigan, 1 million beetles were captured in 400 traps 50 m apart on the boles of healthy elms in a 1 km^2 plot. Only 20% of the beetles estimated to have emerged within the plot were captured and the number of diseased trees increased during the trapping period. It was concluded that the trapping had little appreciable effect in suppressing the beetle population (Cuthbert *et al.*, 1977). Several possible reasons are given for the failure of these experiments: the placement of traps directly on the bole; the extremely high local beetle population; and beetle immigration from surrounding areas. Another study was conducted in Independence, California, which had a relatively small population and no Dutch elm disease (Birch, 1979). The results support the conclusion drawn from Detroit that trapping does not directly affect population size within or between years.

Experiments with barrier deployments of traps have been conducted under various conditions, in a number of areas in the United States. Analysing the effectiveness is complicated by the effects of weather on infection, but results from the studies indicated that mass trapping can contribute to the control of Dutch elm disease when conditions are appropriate, e.g. where elms within the groves to be protected are free from diseased or weakened trees that may serve as competitive pheromone sources if infested by beetles. An intensive sanitation programme is also essential to the success of grove trapping.

Peacock *et al.* (1981) summarized the results after several years of study in the United States. They concluded that there is a potent attractant trap system for killing the beetle, but the results do not give a clear indication of its ability to reduce the incidence of Dutch elm disease. The existing system for reducing Dutch elm disease in large-area, city-wide barrier trapping efforts has so far been unsuccessful or inconclusive. The trapping programmes have had a significant impact on flying beetle populations but Dutch elm disease rates were not significantly altered when the system was applied over large areas.

4.6.3 Ambrosia beetles boring in saw timber

Logs stored in the forest or in timber-processing areas during the summer season may be heavily infested by ambrosia beetles. *Trypodendron lineatum* (Olive) is a serious pest in coniferous saw-logs throughout the Northern Hemisphere. In the north-western forests of North America species of *Gnathotrichus* also cause great damage. The beetles tunnel in the sapwood and inoculate the wood with species-specific ambrosia fungi. Their small dark-stained galleries devalued the timber considerably. Research in Canada led to the isolation, identification and synthesis of the pheromones for these species of ambrosia beetles (Borden and McLean, 1981).

T. lineatum is the most common and widespread species. In late summer emergent beetles fly from the brood logs to their main overwintering site in the litter within forests surrounding a sawmill. Large populations may build up around such industrial areas or other areas where timber of conifers is stored. Early the following spring the beetles fly to logs, which they attack, preferentially those felled before mid-winter. Metabolism in these logs results in the production of ethanol, which is a primary attractant of *T. lineatum* (Moeck, 1970). There is only one generation a year, the main flight takes place in April–May, but flight occurs throughout the summer because some parent beetles leave the galleries and attack new logs.

The pheromone of *T. lineatum*, lineatin, has been identified (MacConnell *et al.*, 1977) and response has been demonstrated in field tests in Canada (Borden *et al.*, 1979) and in Europe. The two host components α-pinene and ethanol attract the beetle (Bauer and Vité, 1975) and enhance the response to dispensers holding lineatin as baits in traps (Vité and Bakke, 1979; Lindgren and Borden, 1983).

The *Gnathotrichus* spp. attack logs within 2 weeks after felling. The two species *G. sulcatus* (LeConte) and *G. retusus* (LeConte) overwinter in their brood logs and attack fresh logs in May–June. *G. sulcatus* can attack also late into autumn in British Columbia and is a constant economic problem there. The pheromone, sulcatol, has been identified from both species (Byrne *et al.*, 1974). Both sexes of *G. sulcatus* produce and respond to a mixture of both enantiomers of the same compound (Borden *et al.*, 1980). Ethanol is an effective synergist for the response of both species to their pheromones.

A. Bakke and R. Lie

In a sawmill in British Columbia, pheromone-baited traps were applied as early as 1975 to protect timber from being attacked by *G. sulcatus* (McLean and Borden, 1977). With encouraging results from 1975, a suppression programme was started on a larger scale in 1976 (McLean and Borden, 1979). Traps made of large fibreglass screen vanes coated with sticky materials were deployed in the area. About 43 000 *G. sulcatus* were captured. There was a significant reduction of damage in yards with traps compared to control yards. Although complete suppression was not achieved, the mill personnel evaluated the attack rate on lumber as exceedingly low based on past experience (Borden and McLean, 1981).

A programme for survey and mass trapping of all three species of ambrosia beetles was carried out in a timber-processing area on Vancouver Island in the years of 1979–1981 (Lindgren and Borden, 1983). The total catch of all species in 1981 was 2 800 000, of which approximately 79% were *T. lineatum*. When compared with high and low population estimates of overwintering beetles the catches represented 44–77% of the total population. A mass-trapping programme, utilizing synthetic pheromones, multiple funnel traps and sticky vane traps, was recommended in conjunction with improved log inventory management.

An extended mass-trapping programme was started in West Germany in 1984 to suppress the increasing population of *T. lineatum* (Vité, 1984).

4.7 CONCLUSIONS

Suppression by mass trapping has been demonstrated in a number of field experiments, but only very few are reported to be successful in providing adequate control of the pest species. The most promising results are obtained in cases of low-density populations. Attempts to apply trapping as the only measure in high-density populations have often failed. Trapping of the flying population is becoming an important component of integrated pest management, particularly for some bark beetle species. Several conditions must be fulfilled if successful mass-trapping is to be achieved:

(1) The insect must respond effectively to an attractive pheromone.

(2) This pheromone must be fully identified, the synthetic components commercially available and formulated in appropriate dispensers.

(3) An economically feasible mass-trapping technology must have been developed.

(4) A certain level of damage must be tolerated.

(5) The insect must be the main pest, and not part of a pest complex which will require insecticidal treatment to be controlled to the necessary extent.

(6) The trapping devices must protect predators and parasites from being captured.

Mass-trapping has been applied with limited success in horticultural and agricultural crops, while in forestry the results have been more promising.

There are, however, very few experiments which demonstrate the direct effect of mass trapping isolated from other measures of the management programme or from the effect of changing weather conditions. In this area there is a need for further accelerated research.

4.8 REFERENCES

Baker, R., Herbert, R. H. and Parton, A. H. (1982). Isolation and synthesis of 3- and 4-hydroxy-1,7-dioxaspiro (5,5) undecanes from the olive fly (*Dacus oleae*). *J. Chem. Soc. Chem. Commun.*, **1982**, 601–603.

Baker, R., Herbert, R., Howse, P. E. and Jones, O. T. (1980). Identification and synthesis of the major sex pheromone of the olive fly (*Dacus oleae*). *J. Chem. Soc. Chem. Commun.*, **1980**, 52–53.

Bakke, A. (1981). The utilization of aggregation pheromone for the control of the spruce bark beetle. In *Insect Pheromone Technology; Chemistry and Applications*, B. A. Leonhardt and M. Beroza (Eds), ACS Symposium Series, no. 190, Washington DC, pp 219–229.

Bakke, A. and Kvamme, T. (1981). Kairomone response in *Thanasimus* predators in pheromone components of *Ips typographus*. *J. Chem. Ecol.*, **7**, 305–312.

Bakke, A. and Strand, L. (1981). Pheromones and traps as part of an integrated control of the spruce bark beetle. Some results from a control programme in Norway in 1979 and 1980. (In Norwegian with English summary.) *Rapp. Nor. Inst. Skogforsk.*, 5/81, 1–39.

Bakke, A., Froyen, P. and Skattebol, L (1977). Field response to a new pheromonal compound isolated from *Ips typographus*. *Naturwissenschaften*, **64**, 98.

Bakke, A., Sether, T. and Kvamme, T. (1983). Mass-trapping of the spruce bark beetle *Ips typographus*. Pheromone and trap technology. *Medd. Nor. Inst. Skogforsk.*, **38**, 1–35.

Bauer, J. and Vité, J. P. (1975). Host selection by *Trypondendron lineatum*. *Naturwissenschaften*, **62**, 539.

Bedard, W. D. and Wood, D. L. (1974). Programme utilizing pheromones in survey or control. Bark beetles—the western pine beetle. In *Pheromones*, M. C. Birch (Ed.), North-Holland, Amsterdam, pp. 441–449.

Bedard, W. D. and Wood, D. L. (1981). Suppression of *Dendroctonus brevicornis* by using a mass-trapping tactic. In *Management of Insect Pest with Semiochemicals*, E. R. Mitchell (Ed.), Plenum Press, New York, pp. 103–114.

Berryman, A. A. (1978). A synoptic model of the lodgepole pine/mountain pine beetle interaction and its potential application in forest management. In *Theory and Practice of Mountain Pine Beetle Management in Lodgepole Pine Forests. Symposium Proceedings*, A. A. Berryman, G. D. Amman, R. W. Stark and D. L. Kibbee (Eds), University of Idaho, Moscow, pp. 98–105.

Bierl, B. A., Beroza, M. and Collier, C. W. (1970). Potent sex attractant of the gypsy moth: Its isolation, identification, and synthesis. *Science*, **170**, 87–89.

Birch, M. C. (1979). Use of pheromone traps to suppress populations of *Scolytus multistriatus* in small, isolated Californian communities. *Bull. Ent. Soc. Am.*, **25**, 112–115.

Borden, J. H. (1982). Aggregation pheromones. In *Bark Beetles in North American Conifers*, J. B. Millou and K. B. Sturgeon (Eds), University of Texas Press, Austin, pp. 74–139.

Borden, J. H. and McLean, J. A. (1981). Pheromone-based suppression of ambrosia beetles in industrial timber processing areas. In *Management of Insect Pests with Semiochemicals*, E. R. Mitchell (Ed.), Plenum Press, New York, pp. 133–154.

Borden, J. H., Hadley, J. R., Johnston, B. D., McConnel, J. G., Silverstein, R. H., Slissor, K. N., Swigar, A. A. and Wong, D. T. W. (1979). Synthesis and field testing of lineatin, the aggregation pheromone of *Trypodendron lineatum* (Coleoptera: Scolytidae). *J. Chem. Ecol.*, **5**, 681–689.

Borden, J. H., Hadley, J. R., McLean, J. A., Silverstein, R. M., Chong, L., Slissor, K. N., Johnston, B. D. and Schuler, H. R. (1980). Enantiomer-based specificity in pheromone communication by two sympatric *Gnathotrichus* species (Coleoptera: Scolytidae). *J. Chem. Ecol.*, **6**, 445–456.

Broumas, T., Katasoyannos, P., Yamvrias, C., Liaropoulos, C., Haniotakis, G. and Strong, F. (1983). Control of the olive fruit fly in a pest management trial in olive culture. In *Fruit Flies of Economic Importance*, R. Cavalloro (Ed.), A. A. Balkema, Rotterdam, pp. 384–592.

Browne, L. E. (1978). A trapping system for the western pine beetle using attractive pheromones. *J. Chem. Ecol.*, **4**, 261–275.

Burkholder, W. E. (1981). Biomonitoring for stored-product insects. In *Management of Insect Pests with Semiochemicals*, E. R. Mitchell (Ed.), Plenum Press, New York, pp. 29–40.

Byrne, K. J., Swigar, A. A., Silverstein, R. M., Borden, J. H. and Stokkink, E. (1974). Sulcatol: population aggregation pheromone in the Scolytid beetle, *Gnathotrichus sulcatus*. *J. Insect. Physiol.*, **20**, 1895–1900.

Cameron, E. A. (1973). Disparlure: A potential tool for gypsy moth population manipulation. *Bull. Entomol. Soc. Am.*, **19**, 15–19.

Campion, D. G. and Nesbitt, F. (1981). Recent advances in the use of pheromones in developing countries with particular reference to mass-trapping for the control of the Egyptian cotton leaf worm *Spodoptera littoralis* and mating disruption for the control of pink bollworm *Pectinophora gossypiella*. In *Les Médiateurs Chimiques Agissant sur le Comportement des Insectes*, Institut National de la Recherche Agronomique, Paris, pp. 335–342.

Campion, D. G., Bettany, B. W., Nesbitt, B. F., Beevor, P. S., Lester, R. and Poppi, R. G. (1974). Field studies of the female sex pheromone of the cotton leafworm *Spodoptera littoralis* (Boisd.) in Cyprus. *Bull. Ent. Res.*, **64**, 89–96.

Campion, D. G., McVeigh, L. J., Polyarkis, J., Michaelakis, S., Stravrakis, G. N., Beevor, P. S., Hall, D. R. and Nesbitt, B. F. (1979). Laboratory and field studies of the female sex pheromone of the olive moth, *Prays oleae*. *Experientia*, **35**, 1146–1147.

Cuthbert, R. A., Peacock, J. W. and Cannon, W. N., Jr (1977). An estimate of the effectiveness of pheromone-baited traps for the suppression of *Scolytus multistriatus* (Coleoptera: Scolytidae). *J. Chem. Ecol.*, **3**, 527–537.

DeMars, C. J., Slaughter, G. W., Bedard, W. D., Norick, N. X. and Roettgering, B. (1980). Estimating western pine beetle-caused tree mortality for evaluating an attractive pheromone treatment. *J. Chem. Ecol.*, **6**, 853–866.

Eidmann, H. H. (1983). Management of the spruce bark beetle *Ips typographus* in Scandinavia using pheromones. *Proc. 10th Int. Congress Plant Protection*, **3**, 1042–1050.

Garibaldi, P., Verotta, L. and Fanelli, R. (1983). Studies on the sex pheromone of *Dacus oleae*. Analysis of the substances contained in the rectal glands. *Experientia*, **39**, 502–505.

Gmelin, J. F. (1787). *Abhandlung uber die Wurmtrocknis*, Verlag d. Crusiusschen buchhandlung, Leipzig, 176 pp.

Gritsch, I. W. (1987). Control of bark beetles using baits. *Pflanzenarzt*, **40**, 3–4.

Haniotakis, G. E. (1981). Field evaluation of the natural female pheromone of *Dacus oleae* Gmelin. *Environ. Entomol.*, **10**, 832–834.

Hardee, D. D. (1982). Mass trapping and trap cropping of the boll weevil *Anthonomus grandis* Boheman. In *Insect Suppression with Controlled Release Pheromone System* (Vol. II), A. F. Kydonieus and M. Beroza (Eds), CRC Press, Boca Raton, Florida, pp. 65–72.

Hardee, D. D. (1984). Personal communication.

Horntvedt, R., Christiansen, E., Solheim, H. and Wang, S. (1983). Artificial inoculation with *Ips typographus* associated blue-stain fungi can kill healthy Norway spruce. *Medd. Nor. Inst. Skogforsk.*, **38**, 1–20.

Hummel, H. E., Gaston, L. K., Shorey, H. H., Kaae, R. S., Byrne, K. J. and Silverstein, R. M. (1973). Clarification of the chemical status of the pink bollworm sex pheromone. *Science*, **181**, 873–875.

Inscoe, M. H. (1982). Insect attractants, attractant pheromones, and related compounds. In *Insect Suppression with Controlled Release Pheromone Systems* (Vol. II), A. F. Kydonieus and M. Beroza (Eds), CRC Press, Boca Raton, Florida, pp. 201–295.

Jones, O. T. (1983). Personal communication.

Jones, O. T., Lisk, J. C., Longhurst, C., Howse, P. E., Ramos, P. and Campos, M. (1983). Development of a monitoring trap for the olive fly, *Dacus oleae* (Gmelin) (Diptera: Tephritidae), using a component of its sex pheromone as a lure. *Bull. Ent. Res.*, **73**, 97–106.

Kinzer, G. W., Gentiman, A. F., Jr, Page, T. F., Jr, Foltz, R. L., Vité, J. P. and Pitman, G. B. (1969). Bark beetle attractants: identification, synthesis and field bioassay of a new compound isolated from *Dentroctonus*. *Nature* (Lond.), **221**, 477–478.

Klein, M. G. (1981). Mass trapping for suppression of Japanese beetles. In *Management of Insect Pests with Semiochemicals*, E. R. Mitchell (Ed.), Plenum Press, New York, pp. 183–190.

Kydonieus, A. F. (1981). Personal communication.

Ladd, R. L., Jr and Klein, M. G. (1982). Trapping Japanese beetles with synthetic female sex pheromone and food-type lures. In *Insect Suppression with Controlled Release Pheromone Systems* (Vol. II), A. F. Kydonieus and M. Beroza (Eds), CRC Press, Boca Raton, Florida, pp. 57–64.

Lanier, G. N., Silverstein, R. M. and Peacock, J. W. (1976). Attractant pheromone of the European elm bark beetle (*Scolytus multistriatus*): Isolation, identification, synthesis and utilization studies. In *Perspectives in Forest Entomology*, J. H. Anderson and H. K. Koya (Eds), Academic Press, New York, pp. 149–175.

Lanier, G. N., Gore, W. E., Pearce, G. T., Peacock, J. W. and Silverstein, R. M. (1977). Response of the European elm bark beetle (*Scolytus multistriatus*) (Coleoptera, Scolytidae), to isomers and components of its pheromone. *J. Chem. Ecol.*, **3**, 1–8.

Levinson, H. Z. and Levinson, A. R. (1979). Trapping of storage insects by sex and food attractants as a tool of integrated control. In *Chemical Ecology: Odour Communication in Animals*, F. J. Ritter (Ed.), Elsevier, Amsterdam, pp. 327–341.

Lindgren, B. S. and Borden, J. H. (1983). Survey and mass trapping of ambrosia beetles (Coleoptera: Scolytidae) in timber processing areas on Vancouver Island. *Can. J. For. Res.*, **13**, 481–493.

MacConnell, J. G., Borden, J. H., Silverstein, R. M. and Stokkink, E. (1977). Isolation and tentative identification of lineatin, a pheromone from the frass of *Trypodendron lineatum* (Coleoptera: Scolytidae). *J. Chem. Ecol.*, **5**, 549–561.

McGovern, T. P. and Ladd, T. L., Jr (1984). Japanese beetle (Coleoptera: Scarabaeidae) attractant: Test with eugenol substitutes and phenetyl propionate. *J. Econ. Entomol.*, **77**, 370–373.

McLean, J. A. and Borden, J. H. (1977). Suppression of *Gnathotricus sulcatus* with sulcatol-baited traps in a commercial sawmill and notes on the occurrence of *G. retusus* and *Trypodendron lineatum. Can. J. For. Res.*, **7**, 348–356.

McLean, J. A. and Borden, H. H. (1979). An operational pheromone based suppression program for an ambrosia beetle, *Gnathotrichus sulcatus*, in a commercial sawmill. *J. Econ. Entomol.*, **12**, 165–172.

Miller, J. M. and Keen, F. P. (1960). Biology and control of the western pine beetle. *US Dept. Agric.*, Misc. Publ. 800, 381 pp.

Moeck, H. A. (1970). Ethanol as the primary attractant for the ambrosia beetle *Trypodendron lineatum* (Coleoptera: Scolytidae). *Can. Entomol.*, **102**, 985–995.

Nagel, R. H., McComb, D. and Knight, F. B (1957). Trap tree method for controlling the Englemann spruce beetle in Colorado. *J. For.*, **55**, 894–898.

Negishi, T., Ishiwatari, T., Asano, S. and Fugikawa, H. (1980). Mass-trapping for the smaller tea tortrix control. *App. Ent. Zool.*, **15**, 894–1114.

Nesbitt, B. F., Beevor, P. S., Cole, R. A., Lester, R. and Poppi, R. G. (1973). Sex pheromones of two noctuid moths. *Nature New Biol.*, **244**, 208–209.

Nesbitt, B. F., Beevor, P. S., Hall, D. R., Lester, R., Sternlicht, M. and Goldenberg, S. (1977). Identification and synthesis of the female sex pheromone of the citrus flower moth, *Prays citri. Insect Biochem.*, **7**, 355–359.

Neumark, S., Jacobsen, M. and Teich, I. (1974). Field evaluation of the four synthetic components of the sex pheromone of *Spodoptera littoralis. Environ. Letters*, **6**, 219–230.

Peacock, J. W., Cutherbert, R. A. and Lanier, G. N. (1981). Deployment of traps in a barrier strategy to reduce populations of the European elm beetle, and the incidence of Dutch elm disease. In *Management of Insect Pests with Semiochemicals*, E. R. Mitchell (Ed.), Plenum Press, New York, pp. 155–174.

Plimmer, J. R., Leonhardt, B. A., Webb, R. E. and Schwalbe, C. P. (1982). Management of the gypsy moth with its sex attractant pheromone. In *Insect Pheromone Technology: Chemistry and Applications*, B. A. Leonhardt and M. Beroza (Eds), ACS Symposium series no. 190, Washington, pp. 231–242.

Schmid, J. M. and Frye, R. H. (1977). Spruce beetle in the Rockies. *US Dept. Agric. Forest Service. Gen. Tech. Rep. RM-49*, 38 pp.

Schwartz, P. H., Jr (1975). Control of insects in deciduous fruits and tree nuts in the home orchard without insecticides. *US Dept. Agric., Home Garden Bull.*, **211**, 36.

Shani, A. (1982). Field studies and pheromone application in Israel, presented at the 3rd Israeli meeting on pheromone research, May 4, 1982, pp. 18–22.

Silverstein, R. M. (1981). Pheromones: Background and potential for use in insect pest control. *Science*, **213**, 1326–1332.

Silverstein, R. M., Brownlee, R. G., Bellas, T. E., Wood, D. L. and Browne, L. E. (1968). Brevicomin: Principal sex attractant in the frass of the female western pine beetle. *Science*, **159**, 889–890.

Sternlicht, M. (1984). Personal communication.

Sternlicht, M., Goldenberg, S., Nesbitt, B. F. and Hall, D. R. (1981). Further field trials of pheromone dispensers and traps for males of *Prays citri* (Millière) (Lepidoptera: Yponomeutidae). *Bull. Ent. Res.*, **71**, 267–274.

Teich, I., Neumark, S., Jacobson, M., Klug, J., Shani, A. and Waters, R. M. (1979). Mass trapping of males of Egyptian cotton leafworm (*Spodoptera littoralis*) and large-scale synthesis of prodlure. In *Chemical Ecology: Odour Communication in Animals*, F. J. Ritter (Ed.) Elsevier, Amsterdam, pp. 343–350.

Trematerra, P. and Battaini, F. (1987). Control of *Ephestia kuehniella* (Zeller) by mass-trapping. *J. Appl. Entomol.* **104**, 336–340.

Tumlinson, J. H. (1979). The need for biological information in developing strategies for applying semiochemicals. In *Chemical Ecology: Odour Communication in Animals*, F. J. Ritter (Ed.), Elsevier, Amsterdam, pp. 301–311.

Tumlinson, J. H., Klein, M. G., Doolittle, R. E., Ladd, T. L. and Proveaux, A. T. (1977). Identification of the female Japanese beetle sex pheromone: Inhibition of male response by an enantiomer. *Science*, **197**, 789–792.

Tumlinson, H. H., Hardee, D. D., Gueldner, R. C., Thompson, A. C., Hedin, P. A. and Minyard, J. P. (1969). Sex pheromones produced by male boll weevil: Isolation, identification, and synthesis. *Science*, **166**, 1010–1012.

Vité, J. P. (1984). Erfahrugen und Erkentnisse zur akuten Gefahrdung des mitteleuropaischen Fichtenwaldes durch Kaferbefall. *Allq. Forst. Zeitschr.*, **39**, 249–252.

Vité, J. P. and Bakke, A. (1979). Synergism between chemical and physical stimuli in host selection by an ambrosia beetle. *Naturwissenschaften*, **66**, 528–529.

Vité, J. P. and Francke, W. (1976). The aggregation pheromones of bark beetles: Progress and problems. *Naturwissenschaften*, **63**, 550–555.

Webb, R. E. (1982). Mass trapping of the gypsy moth. In *Insect Suppression with Controlled Release Pheromone Systems* (Vol. II), A. F. Kydonieus and M. Beroza (Eds), CRC Press, Boca Raton, Florida, pp. 27–56.

Wood, D. L., Browne, L. E., Ewing, B., Lindahl, K., Bedard, W. D., Tilden, P. E., Mori, K., Pitman, G. B. and Hughes, P. R. (1976). Western pine beetle: Specificity among enantiomers of male and female components of an attractant pheromone. *Science*, **192**, 896–898.

Worrell, R. (1983). Damage by the spruce bark beetle in South Norway 1970–80: A survey, and factors affecting its occurrence. *Medd. Nor. Inst. Skogforsk.*, **38**, 1–34.

Insect Pheromones in Plant Protection
Edited by A. R. Jutsum and R. F. S. Gordon
Published 1989 by John Wiley & Sons Ltd

5

Mating Disruption

D. G. Campion, B. R. Critchley and L. J. McVeigh

Overseas Development Natural Resources Institute,
Central Avenue, Chatham Maritime, Chatham, Kent ME4 4TB, UK

5.1 INTRODUCTION

Control of insect pests by the mating disruption technique is achieved by the widespread application of synthetic pheromone over the target crop. The insects are then unable to locate their mates when using their own pheromone system and mating is therefore reduced or eliminated.

The mechanisms by which the technique may succeed are still not fully understood. They may include:

(a) Confusion—adaptation of the antennal receptors and habituation of the central nervous system following constant exposure to relatively high levels (or a 'fog') of the pheromone, such that the insect cannot respond to signals from a potential mate.

(b) Trail masking—an obliteration of the natural pheromone plume making trail following impossible.

(c) False-trail following—the diversion of insects from the naturally occurring pheromone sources by the application of a relatively small number of synthetic sources.

Various slow-release pheromone formulations have been developed which either permeate the air with a fog of pheromone so as to achieve sensory adaptation and habituation or provide numerous discrete point sources so as to mask trail following or to create false trails. However, mating disruption can be achieved by any of these mechanisms depending on the rates of pheromone application or the density of the pheromone point sources applied.

The pheromones of Lepidoptera are typically long-chain, unsaturated esters, alcohols or aldehydes. They generally consist of more than one component and these may be isomers with respect to the geometry and position of the unsaturation, or they may be structurally related compounds differing in chain length or the nature of the functional group. The precise role of the separate components in eliciting trail following, masking or total habituation is not known.

An early example of a multicomponent pheromone is provided by the sex attractant of the red bollworm, *Diparopsis castanea* (Hamps.), which was shown to consist of up to five components (Moorhouse *et al.*, 1969; Nesbitt *et al.*, 1973), while as many as seven components have since been identified in the pheromone blends of the cotton bollworms belonging to the genus *Heliothis* (Klun *et al.*, 1979; Tumlinson *et al.*, 1982).

The number of components present in any one species may also vary with geographical distribution. For example, the Egyptian cotton leafworm, *Spodoptera littoralis* (Boisd.), had two components in Crete but four or five in Israel and Egypt, respectively (Campion *et al.*, 1980).

Many pheromones are inherently unstable and in some instances isomerize rapidly in the presence of ultraviolet light. The resulting geometric isomers may inhibit attraction, for example in the case of the pheromone of the pink bollworm, *Pectinophora gossypiella* Saunders (Jacobson and Beroza, 1963; Bierl *et al.*, 1974), and the spotted bollworm, *Earias vittella* (F.) (Cork *et al.*, 1985).

Instability of the pheromone molecule may also be related to the level of unsaturation and type of functional group. Short-chain aldehydes are as a rule more reactive and therefore more difficult to stabilize than longer-chained acetates. However, stabilized pheromone formulations containing ultraviolet screeners and antioxidants have been devised (see Chapter 9).

5.2 SYNTHETIC PHEROMONE MIMICS

Numerous synthetic compounds with molecular structures similar to the naturally occurring pheromones have been examined and tested for their ability to elicit similar behavioural responses. Successful compounds have been termed 'parapheromones' or 'antipheromones' depending on whether they attract the insects or interfere with the way in which the insects perceive the natural pheromone (Gaston *et al.*, 1972). Although none of these compounds has been found to be more attractive or inhibitory than the naturally occurring pheromone components it is not inconceivable that they could achieve similar effects to those of habituation if they were to mask the pheromone emitted by calling insects.

5.3 DEVELOPMENT OF SLOW-RELEASE FORMULATIONS

Slow-release formulation is essential for pheromones used in mating disruption in order to prolong the release and efficacy of compounds which are otherwise highly volatile and to provide stabilization of remaining material under field conditions.

Four formulations have so far been used commercially:

(a) Hollow fibres developed by Albany International (subsequently Scentry, now Yellowstone International) in the USA. The fibres are made of polyacetyl resin measuring 1.5 cm in length, 200 μm internal diameter, and are sealed at one end. Each fibre contains about 250–275 μg pheromone. A special glue is mixed with the fibres which makes them adhere to the foliage of the target crop. The fibre and glue mixture is applied by hand or using specially designed applicators attached to aircraft (see Chapter 8 for further details).

(b) Laminate flakes produced by the Hercon Division of the Health-Chem Corporation in the USA. In this formulation the pheromone is contained in a plastic reservoir layer which is sandwiched between two layers of semi-porous vinyl plastic. The size of individual flakes can be varied to achieve the required release rate, number of point sources and active ingredient per hectare. This system also requires a strong adhesive to ensure adherence to the foliage, as well as specially designed applicators or hand application (see Chapter 7).

(c) Microcapsules developed by ICI Agrochemicals in the UK. In this formulation the pheromone is enclosed within tiny polyurea spheres of 3 μm median diameter. The formulation is water-based, may be sprayed with conventional equipment and shows excellent rain-fastness on the crop without the need for adhesives (see Chapter 9).

(d) Polyethylene tubes (or twist-tie dispensers) made by the Shin-Etsu Co. in Japan are marketed by Shin-Etsu, the Mitsubishi Corporation of Japan and by BioControl Ltd. This formulation consists of a polyethylene tube, 10–20 cm long, containing the pheromone and a soft wire stiffener. The tube must be applied by hand, being tied around the stems of the plant.

5.4 PROBLEMS IN CONDUCTING FIELD TRIALS AND IN THE INTERPRETATION OF DATA

Pheromones are not biocides. It is therefore possible that unmated insects in a pheromone treatment area may disperse to untreated areas where mating can take place, and that gravid females from surrounding areas may migrate into the treatment area, lay their eggs and so reduce the treatment effects. Initial field trials, therefore, have been large or in areas which are well removed or isolated from likely sources of infestation.

However, for many pests it is not known over what distances adult insects can fly, whether unmated females actively search for males, or whether mating occurs before or after dispersal flights. Mating disruption is therefore likely to be most effective with insects having specialized or a limited range of food plants among which the host crops are grown as large-scale monocultures.

The need to treat large areas with pheromones leads to difficulties in operating experimental layouts for statistical analyses of comparisons with conventional treatments and untreated controls. The logistics of sampling more than a very few replicates may prove to be limiting and the cost of the pheromone formulations may not permit applications on the scale required.

These problems are particularly noticeable in developing countries in the tropics where large-area crop monocultures normally receive regular prophylactic conventional pesticide treatments to prevent devastating increases in pest numbers. In such circumstances it is rarely possible to maintain untreated control plots and the only comparisons possible are those between pheromone and conventional treatments. Thus it becomes impossible to decide whether in the absence of the prophylactic treatments pests would have reached economically damaging levels. In such circumstances the best that can be said is that at least pheromone applications do no harm. Where the commercial cost is similar to conventional insecticides their use is preferable, both from the point of view possibly of user safety and the protection of the environment from the loss of beneficial insects and in some cases from pollution by residues.

The prophylactic use of pheromones is unlikely to create 'resistance' problems. Resistance and cross-resistance to insecticides may eliminate whole ranges of compounds from use, whereas with pheromones it will be a matter of adjusting the balance of the components in the blend. Insects must have their mate-finding pheromones; adaptation to different chemicals can be quickly overcome by modern analytical means, once the basic pattern of use and the right formulation have been established.

5.5 MATING DISRUPTION IN PRACTICE

Although many pheromones have now been isolated, identified and synthesized, few successful mating disruption trials have yet been reported. Many such trials were undertaken before the pheromones were properly characterized, with inadequate slow-release formulations or poor basic biological knowledge of the pest species (see Campion, 1985). One exception and major success has been the control of the pink bollworm, P. gossypiella, by use of synthetic pheromone. It should be illustrative to consider the factors which led to that success as a paradigm for projects in the future.

P. gossypiella is one of the most important cotton pests occurring in many parts of the world (Fig. 5.1). Control by conventional agents has sometimes

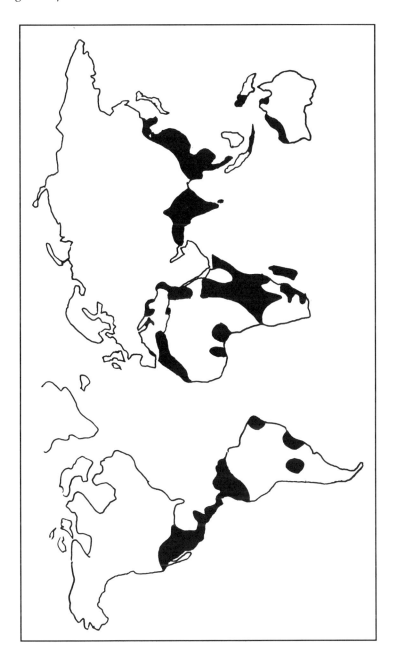

Fig. 5.1 Global distribution of pink bollworm, *Pectinophora gossypiella*

proved difficult, because the newly emerged larvae quickly enter the fruiting parts of the plant and are protected from conventional pesticides. In contrast, the target of the sex pheromone is the adult moth, which is exposed throughout its 1–2-week existence.

The sex pheromone of P. gossypiella is a 1 : 1 mixture of (Z,Z)- and (Z,E)-7,11-hexadecadienyl acetate (Hummel et al., 1973; Bierl et al., 1974). Unlike many pheromones, it contains only two components and is relatively stable, which makes it particularly suitable as a selective control agent with a potentially large worldwide market.

5.6 PHEROMONAL CONTROL OF
PECTINOPHORA GOSSYPIELLA IN THE USA

The parapheromone (Z)-7-hexadecenyl acetate was first shown to be effective in reducing mating frequency and thus in controlling P. gossypiella by Shorey et al. (1974). As well as a lower larval infestation in green bolls in comparison with that obtained by commercial insecticide applications, a 90% reduction in resident male and female moths was also achieved in the pheromone-treated areas. This dispelled fears that large numbers of moths might be attracted to the pheromone-treated fields from surrounding areas. The parapheromone was hand-applied using pheromone dispensers and the areas treated (4.8 ha) were relatively small.

With the advent of the hollow-fibre formulation in 1974, larger-scale application of a pheromone became possible. Greater biological activity was demonstrated using the synthetic pheromone of P. gossypiella, gossyplure, formulated in the hollow fibres than had been achieved with the para-pheromone (Shorey et al., 1976; Gaston et al., 1977). These trials also showed that reductions in conventional pesticide usage were attainable.

In 1978, the United States Environmental Protection Agency (EPA) granted the first registration of a sex pheromone product to gossyplure formulated in hollow fibres (Brooks et al., 1979). Following this, approximately 20 000 sprayed hectares of cotton were treated with gossyplure in the south-western USA, particularly in Arizona and Southern California.

In 1979, the EPA granted an experimental use permit (EUP) to the laminate flake formulation and a total of 192 ha of cotton was treated with that formulation (Kydonieus and Beroza, 1981). Both of these formulations were tested in larger-scale trials in 1980 and a combined total of 50 000 sprayed hectares was treated in 1981. Following fears that whitefly, Bemisia tabaci (Gennadius), and Heliothis spp. outbreaks were associated with insecticide treatments, farmers in California's Imperial Valley voted in 1982 to use pheromones in the entire valley, on about 14 700 ha of cotton, to try to reduce the amount of pesticides they were using. A Cotton Pest Abatement District

Fig. 5.2 A twist-tie dispenser being attached to a cotton plant

(CPAD) was established in which a farmer who did not use pheromones
to control *P. gossypiella* on his cotton could be fined. The programme was
considered a success, with fewer applications of conventional pesticides
being made, yields higher and only 5% of the crop damaged, compared with
over 30% in conventionally treated neighbouring fields (Doane *et al.*, 1983).
The mandatory programme was discontinued in 1984 with over 80% of the
farmers expected to continue using pheromones without the need for a
CPAD. Unfortunately this did not prove to be the case and many farmers
reverted to using regular schedules of conventional pesticides.

More recently the twist-tie formulation of *P. gossypiella* pheromone has
been successfully used in Arizona in small-scale trials totalling 10 ha (Flint
et al., 1985). Trials were conducted again, on a larger scale, in California in
1985 (Staten *et al.*, 1987) (Fig. 5.2). Twist-ties were applied at a rate of 1000
per ha (78 g a.i. ha^{-1}) with an expected half-life of 58 days. Mean larval
counts in the 23 pheromone treated fields were lower than in the
conventional insecticide treated fields (significantly so in the Mexicali valley).
The average number of insecticide treatments was reduced by 40% as a result
of the pheromone treatments. A limiting factor may be the necessity for hand
application in a country where labour costs are high.

5.7 PHEROMONAL CONTROL OF
PECTINOPHORA GOSSYPIELLA IN EGYPT

There has been for some years a collaboration between the Overseas
Development Natural Resources Institute (ODNRI) and ICI Agrochemicals
to develop microencapsulated formulations of lepidopteran sex pheromones.
These formulations were favoured because they are easily manufactured on
a large scale using existing technology; because, unlike hollow fibres and
laminated flakes, they can be applied using conventional spray equipment;
and because they adhere to the foliage without the use of special adhesives
(Campion, 1983).

During 1979, small plot trials were conducted on cotton in Egypt using
microencapsulated *P. gossypiella* pheromone formulations applied by
knapsack sprayer and compared with the hollow-fibre formulation applied
by hand. The results indicated that both formulations induced high levels
of disruption when applied at rates of 5–10 g ha^{-1} (Hall *et al.*, 1982).

During 1981, enlarged field trials of the microencapsulated formulation
of *P. gossypiella* pheromone were undertaken in Egypt. Two 50 ha blocks
of cotton located in central Egypt were sprayed with the formulation from
the air using fixed-wing aircraft. Five applications were made at a rate of
10 g pheromone per hectare and at intervals of between 2 and 3 weeks. Two
areas of similar size were subjected to the conventional insecticide practice

to control pink bollworm and other pests. The level of infestation of *P. gossypiella* in all the areas was assessed from samples of bolls collected at 10-day intervals. The results showed no significant differences between insecticide- and pheromone-treated areas, either by comparing the number of damaged bolls or in the level of damaged cotton seed and yields of seed cotton (Critchley *et al.*, 1983).

The efficiencies of the three formulations then available—the hollow fibres, laminate flakes and microcapsules—were compared for the first time in similar large-scale trials in Egypt in 1982 (Critchley *et al.*, 1985) (Fig. 5.3). All three gave adequate levels of control in terms of final yield and in comparison with conventional insecticide treatments. The laminate flake formulation applied at a rate of 800 point sources per hectare gave a poorer reduction of boll infestation than the hollow-fibre formulation applied at a rate of 12 000 point sources per hectare. Despite the lower numbers of flakes applied, the level of active ingredient (a.i.) applied per hectare was the same for both formulations. In subsequent trials the distribution rate of the flake formulation was increased to 3000 point sources per hectare.

The microencapsulated formulation was applied by air in 1982 using helicopters (Fig. 5.4), but it was necessary to apply the hollow-fibre and flake formulations by hand (Fig. 5.5) because the special aerial applicators were

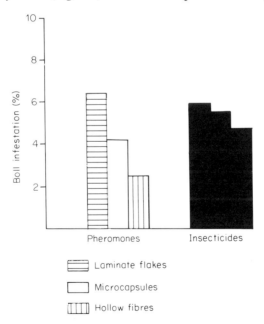

Fig. 5.3 Mean percentage boll infestation by *P. gossypiella* larvae in pheromone- and insecticide-treated areas of approximately 50 ha in size; Fayoum Governorate, Egypt, 1982

unavailable. This demonstrated the feasibility of hand-treating large areas with these formulations and also the relevance of the technique to developing countries where labour costs are relatively low and where the need for savings in application costs may be particularly important.

During 1983, large-scale trials were undertaken for the three *P. gossypiella* pheromone formulations, covering totals of 500 ha per formulation comprising discrete or contiguous cotton fields of 5–20 ha.

One such large-scale trial with the microcapsules covered 250 ha of cotton. Application was by helicopter using boom and nozzle spray apparatus. Five applications of pheromone were made throughout the season, each of $10 \, g \, ha^{-1}$ and at intervals of 2–3 weeks, commencing at the beginning of June. Effectiveness of the treatments was again measured by comparing the infestation levels of *P. gossypiella* in the pheromone-treated area with infestations in 160 ha of cotton situated in the same locality where a conventional regime of pesticides was applied. This consisted of successive sprays, at recommended rates, of:

(a) a mixture of chlorpyrifos and diflubenzuron;
(b) synthetic pyrethroid;
(c) chlorpyrifos;
(d) profenofos.

The interval between each spray was 2 weeks, with the first application at the beginning of July.

The levels of infestation in the sampled bolls are summarized in Fig. 5.6. In only one of the eight sampling periods (around 23 July) did the mean percentage boll infestation in the pheromone-treated area exceed that in the insecticide-treated area. The level of infestation in the pheromone-treated area never exceeded the recommended economic threshold of 10%, although this was exceeded in three sampling periods where insecticides had been used.

There was no indication of unacceptable increases in the populations of other insect pests in the pheromone-treated areas, which were feared in the absence of insecticide treatments. This may be related to the presence of greater numbers of beneficial insects which were present in the pheromone-treated area (Table 5.1).

In the following year, 1984, as a result of the successful earlier trials, ICI and Sandoz (under licence from Scentry) sold formulations of *P. gossypiella* pheromone in Egypt at prices similar to those of conventional pesticides. This was the first such purchase of a pheromone formulation for use in controlling an insect outside the USA. 2000 ha of cotton were treated season-long with each formulation (four applications) using helicopters and fixed-wing aircraft. In terms of boll infestation the results achieved using both formulations compared favourably with the conventional pesticide programme and similar yields of seed cotton were obtained from the pheromone and insecticide-treated areas.

Fig. 5.4 High-volume water-based applications of microencapsulated *Pectinophora* pheromone formulation in Egypt by a Bell-47 helicopter equipped with boom and nozzle sprayers

Fig. 5.5(a)

Fig. 5.5 (a) Hand application of hollow fibres in Egypt (inset: fibres on a cotton leaf). (b) Hand application of laminate flakes in Egypt (inset: flake on a cotton leaf). (c) Gangs of field workers applying pheromone formulations in a large cotton field in Egypt. Each worker can cover 1 ha per day

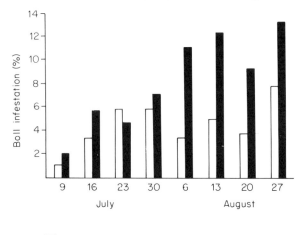

□ Microencapsulated pheromone only
■ Conventional insecticide programme

Fig. 5.6 Weekly boll infestation by *P. gossypiella* larvae in a 250 ha area of cotton treated season-long with microencapsulated pheromone, compared with cotton treated with conventional insecticides; Fayoum Governorate, Egypt, 1983

TABLE 5.1 Mean numbers of predatory insects per hectare in cotton sampled by D-Vac suction apparatus following the first application of insecticide in pheromone- and insecticide-treated areas, Central Egypt, 7th July, 1983

		Insecticide treated		Pheromone treated	Ratio (insecticides : pheromone)	
Genus		Area 1	Area 2	Area 3	1 : 3	2 : 3
Coccinellida	adults	33	250	3717	1 : 112.6	1 : 14.9
Paederus	adults	417	17	1717	1 : 4.1	1 : 101
Scymnus	adults	33	0	1184	1 : 35.9	—
Chrysoperla	adults	17	0	200	1 : 11.8	—
Orius	adults	0	0	583	—	—
Orius	nymphs	0	0	183	—	—
Total		500	267	7584	1 : 15.2	1 : 28.4

For the first time in many years bee-keepers were able to put hives out in cotton fields where pheromone applications were being undertaken, without fear of losing their bees from pesticide applications.

In 1985 the total area treated throughout the season with *P. gossypiella* pheromone increased still further to 15 000 ha, with the microencapsulated and hollow-fibre formulations from ICI and Sandoz being used, and also the laminate flake formulation from BASF of West Germany, under licence from Hercon. Satisfactory results were again achieved (Fig. 5.7).

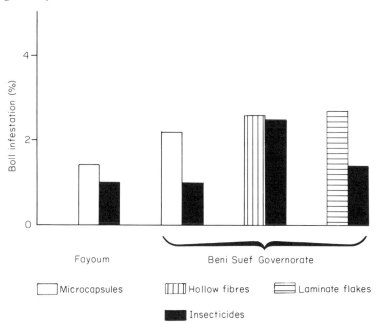

Fig. 5.7 Mean percentage boll infestation by *P. gossypiella* larvae in pheromone- and insecticide-treated areas of cotton 4000 ha in size; Fayoum & Beni Suef Governorates, Egypt, 1985

The success of the pheromone programme between 1981 and 1985 encouraged the Egyptian Ministry of Agriculture to increase the area treated in 1986 to 20 000 ha. Up until 1986 all the trials had been conducted in the Fayoum and Beni Suef regions of middle Egypt. 1986 saw the first large-scale pheromone applications in the Nile Delta region, with further work in middle Egypt. The total area was divided between the three formulations: the fibres, flakes and microcapsules. An additional area of 400 ha was treated with the Shin-Etsu/Mitsubishi twist-tie or 'rope' formulation.

In general, *P. gossypiella* infestations were significantly greater in 1986 than in the preceding three years. This applied to both pheromone and insecticide treated areas. The rise in boll infestation above the 'economic threshold' was seen first in some of the pheromone areas and these were subsequently over-sprayed with insecticide after two or three scheduled applications of pheromone.

Although reported yields in the conventionally (insecticide) treated areas were lower on average in 1986 than in 1985, yield reductions in the pheromone areas (Fig. 5.8) were seen as disappointing and the Egyptian Ministry of Agriculture was to take a cautious approach to further pheromone use in 1987.

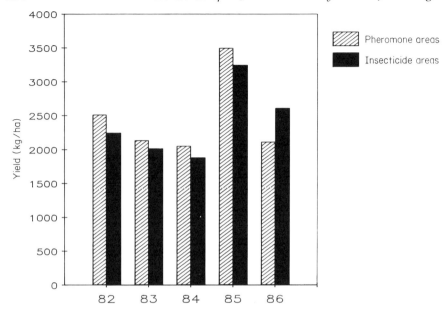

Fig. 5.8 Seed cotton yield from areas treated with pheromones followed by insecticides or a conventional insecticide programme. Egypt, 1982–86

A total area of 2 400 ha of cotton in the Delta and middle Egypt was treated with the microcapsule or the hollow fibre formulation in 1987. As in 1984–86, the need for and the timing of a switch from pheromone to insecticide applications in the 'pheromone' areas was at the discretion of the local chief pest control officers. In all pheromone areas, three pheromone applications were made, followed by one or two insecticide applications.

Compared to 1986, general infestation levels of *P. gossypiella* were higher in 1987 in the Fayoum (middle Egypt) and lower in 1987 in Dakahlia (Delta). Boll infestations in the Fayoum were marginally higher in the pheromone area than in the insecticide area (Fig. 5.9), while in Dakahlia infestations were lower in the pheromone area. Similarly, seed cotton yields were marginally lower in middle Egypt pheromone areas compared to the insecticide areas. However, yields in Dakahlia pheromone areas were significantly higher than those from local insecticide treated areas.

Dakahlia honey collectors benefitted greatly from the pheromone applications in 1987. Total honey production rose from zero in 1986 when only conventional insecticides were used on the cotton, to a total for the district of over 10 700 kg with the pheromones in 1987. Average honey production rose from zero to 4.5 kg per hive between 1986 and 1987. This encouraging 'side benefit' was achieved with a program of three pheromone applications, starting at the first flower stage, followed by two applications of insecticide.

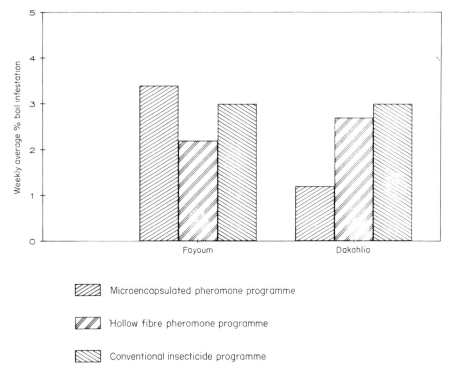

Fig. 5.9 Cotton boll infestation by *P. gossypiella* and *E. insulana* in areas treated with pheromone followed by insecticide and areas treated with a conventional insecticide programme. Fayoum (middle Egypt) and Dakahlia (Nile Delta), 1987

In 1988, a total of 12 500 ha of cotton will receive one or more applications of pink bollworm pheromone. Three formulations (hollow fibres, micro-capsules and twist-tie ropes) will have equal shares of the total area. The strategy will again be pursued of early-season pheromone application followed by insecticide application as required.

The success of the mating disruption technique against pink bollworm in Egypt can be attributed to the following factors, additional to the stability of its pheromone and the economic and convenient methods developed to apply it:

(a) In most years *P. gossypiella* is the only major pest of cotton requiring control using pesticides and these applications are entirely under the control of government agencies. Early-season sucking pests are not normally a problem on Egyptian cotton of the *Gossypium barbadense* varieties because of the hairy nature of the plants. A potentially important early-season pest, the cotton leafworm, *S. littoralis*, is kept in check by hand picking of the egg masses or judicious small-scale application of selective pesticides.

Late-season attacks of the spiny bollworm, *Earias insulana* (Boisd.), can be controlled by conventional practices at a time when numbers of beneficial insects are naturally declining (Hosny, 1980).

(b) The willingness of the Egyptian Government to experiment with an innovative control measure when faced with the possible alternatives of pesticide-resistant strains of cotton pests or the almost insuperable, man-made problem of whitefly which has occurred in neighbouring Sudan (Eveleens, 1983) and in parts of Turkey (Anon., 1983).

(c) Collaborative development and sharing of costs between governments (i.e. the United Kingdom and Egypt) and the scientific research community (ODNRI in Chatham, Ain Shams University and the Plant Protection Research Institute of the Egyptian Ministry of Agriculture in Cairo) and industry (ICI, Sandoz, BASF and Shin-Etsu/Mitsubishi).

5.8 PHEROMONAL CONTROL OF *PECTINOPHORA GOSSYPIELLA* IN PAKISTAN

Trials using *P. gossypiella* pheromone formulations have also been conducted by ODNRI at Multan in the Punjab Province of Pakistan in cooperation with the Central Cotton Research Committee and with financial support from the FAO.

In contrast to the situation in Egypt, there is no governmental control of application of insecticide use against cotton pests in Pakistan. The selection of insecticides by the farmers therefore depends to a greater extent on the promotional activity of the agrochemical companies. The situation is therefore more similar to that prevailing in the USA and with individual farmers as potential customers the unit areas of treatment are very much smaller than in Egypt.

The cotton pest complex also differs from that in Egypt since early-season sucking pests such as jassids often occur, while two species of *Earias — E. insulana*, the spiny bollworm, and *E. vittella*, the spotted bollworm — are sometimes found in damaging numbers at the same time as *P. gossypiella*. Late-season attacks of tetranychid mites have also increased in importance in recent years.

Aerial application of pesticides is not at present used for cotton pest control in Pakistan. Trials conducted in 1984 at Multan therefore used ground-based techniques. Of the three formulations tested in a replicated series of 10 ha blocks of cotton, the microencapsulated formulation was applied by motorized knapsack sprayer while the fibre and flake formulations were applied by hand using teams of village women. All three formulations were applied three to five times during the season and gave as good control of boll damage attributable to *P. gossypiella* as did conventional pesticide treatments (Fig. 5.10).

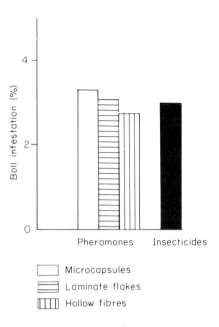

Fig. 5.10 Mean % green boll infestation due to *P. gossypiella* larvae in replicated 10 ha blocks of cotton; Multan, Pakistan, 1984

Insecticides were needed for the control of other pest species in the pheromone treatment areas, but the total area so treated during the season was less than that required by the conventional programme (Table 5.2).

Given the nature of the cotton pest complex in the Punjab it would be unrealistic for an agrochemical company to promote the sale of a pheromone formulation for the control of *P. gossypiella* without due regard for the other pest species, in particular *E. insulana* and *E. vittella*. The commercial promotion of *P. gossypiella* pheromone formulation as a selective pesticide is likely to be possible only when it is offered as part of a pest control package for the cotton season as a whole. Such a package should have the advantage of prolonging the duration of the use of associated conventional pesticides by delaying the onset of insecticide resistance and pest resurgence problems, while the inclusion of the pheromone formulation would enhance the survival of the beneficial insects.

A programme along these lines has been conducted at Multan since 1985 with financial support from the Overseas Development Administration (ODA) in the UK and with commercial support from ICI, Sandoz and Mitsubishi. A systemic insecticide, disulfoton, was used early in the season for the control of sucking pests, and followed by two or three applications of *P. gossypiella* pheromone formulation during mid-season to delay the need

TABLE 5.2 Relative number of insecticide treatments in pheromone- and insecticide-treated blocks based on the area of each block treated relative to its size in large farm mating disruption trials conducted at Multan, Pakistan, 1984

Treatment	Size of block (ha)	Total area treated with insecticides (ha)	Insecticide-treated area as a multiple of block size
Microcapsules	13.8	16.0	1.2×
Fibres	11.7	11.7	1.0×
Flakes	11.3	14.5	1.3×
Insecticides Control 1	10.1	34.2	3.4×
Insecticides Control 2	13.4	54.7	4.1×

for insecticidal control of the *Earias* complex. The conventional insecticide programme comprised disulfoton and pyrethroid insecticides (fluvalinate and lambda cyhalothrin).

Comparisons of the levels of boll infestation by *P. gossypiella* in the two programmes are shown in Fig. 5.11. Both strategies gave adequate bollworm control. However, the 'insecticides only' programme suffered an increase in mite population which required further treatment with acaricides.

Some Pakistani critics have argued that pheromonal control strategies for *P. gossypiella* are only possible provided that a similar technique is developed for control of the *Earias* complex (Attique, 1985). These insect species have a restricted range of host plants and in contrast to *P. gossypiella* show no diapause.

Pheromones for *E. insulana* and *E. vitella* have been identified (Hall *et al.*, 1980; Cork *et al.*, 1985) and the main component, (*E,E*)-10,12-hexdecadienal, is common to both species. It was anticipated that the aldehyde functional group and the conjugated double bond would lead to problems of instability in the field. However, preliminary experiments conducted in 1984 using a black hollow-fibre formulation of the principal *Earias* pheromone component indicated that it could be used as a mating disruptant for both species.

During the 1985 season three applications of *Pectinophora* pheromone, three of *Earias* pheromone and one of a pyrethroid insecticide maintained low levels of boll infestation by all three species in a single 12 ha area of cotton.

A combination formulation containing 1:1 Z,Z:Z,E-7,11-hexadecadienal acetate and a 10:1 mixture of E,E-10,12-hexadecadienal and Z-11-hexadecenal gave almost complete pheromone trap catch suppression for *P. gossypiella*, *E. insulana* and *E. vitella* for up to eight weeks following a single application at the pin-square stage (Critchley *et al.*, 1987).

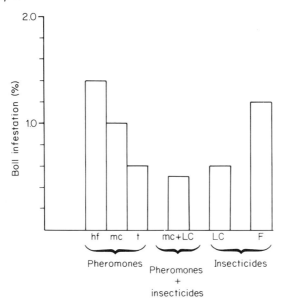

Fig. 5.11 Mean % green boll infestation due to *P. gossypiella* larvae in replicated 10 ha blocks of cotton receiving pheromones, pheromones plus insecticides, or insecticides only as the principal treatment in an integrated pest management programme of cotton pests; Pakistan 1985 (hf = hollow fibres, mc = microcapsules, t = twist-ties, mc + LC = microcapsules + lambda cyhalothrin, LC = lambda cyhalothrin, F = fluvalinate)

Further trials are being undertaken to develop improved *Earias* pheromone formulations on a larger scale, since a pheromonal control strategy for the bollworm complex in Pakistan would also be relevant to the pest situation in India, where a similar cotton pest complex prevails. As in Pakistan, there is also a lack of governmental control over the sale or application of insecticides and this has led to their indiscriminate use, so that as many as thirteen insecticide treatments are common.

As a means of overcoming the problem, integrated pest management (IPM) programmes have been introduced by government agricultural institutions and this has reduced pesticide applications to five or six per season (Sundaramurthy and Basu, 1985). Having established a basic IPM programme it is considered important to introduce, wherever appropriate, selective pest control agents which would suppress the target pest species but leave unharmed the beneficial fauna. A possibility therefore exists of introducing pheromones for the control of *Earias* spp. and pink bollworm. Further inputs could include the use of microbial insecticides for the control of *Heliothis armigera* (Hbn.) and insect growth regulators for the control of *Spodoptera litura* (Fabricius).

5.9 PHEROMONAL CONTROL OF
PECTINOPHORA GOSSYPIELLA IN PERU

Peru is a country where IPM strategies for the control of cotton pests are practised throughout the total cotton-growing area of 150 000 ha, much of which is farmed in cooperatives. There is a maximal utilization of cultural practices and the minimal use of conventional pesticides (Gonzalez, 1979).

P. *gossypiella* was first detected in northern Peru in 1983. It was assumed to have originated from Ecuador to the north following unusually torrential rains which caused widespread flooding. The pest species has since spread to most other cotton-growing areas of the country.

Early-season attacks of the cotton leafworm *Alabama argillaceae* (Hbn.) and the cotton leafminer *Bucculatrix thurberiella* (Busck.) are controlled by applications of lead arsenate and insect growth regulators (IGRs). Attacks of *Heliothis virescens* (Fabricius) are reduced by the hand collection of infested flower buds. These measures have been shown to have a minimal effect on the beneficial insects. Late-season applications of pyrethroid insecticides are directed against the cotton stainer, *Dysdercus peruvianus* (Guerin). The aim of this programme is to delay application of broad-spectrum insecticides until at least 100 days post-sowing so as to avoid the resurgence of what are otherwise unimportant insects (Gonzalez, 1982). Susceptibility to attack by P. *gossypiella* commences with flowering, at 50 days post-sowing. Protection of the plants at this critical stage by one or two P. *gossypiella* pheromone treatments should provide an ideal selective treatment. Trials along these lines were conducted in 1986.

5.10 INTEGRATED USE OF PHEROMONES WITH
CONVENTIONAL AND SELECTIVE PESTICIDES

To summarize the earlier sections, the control of P. *gossypiella* by the mating disruption technique is possible in three agricultural systems:
(a) in Egypt, where Governmental control of insecticide usage occurs;
(b) in Pakistan, under a free enterprise system;
(c) in Peru, as part of an IPM approach organized in cooperatives.
The proposed programmes for the integrated use of conventional and selective pesticides including pheromones in the three agricultural systems are summarized in Table 5.3.

One development coinciding with the introduction of pheromone for the control of P. *gossypiella* in Egypt and Pakistan has been a trend towards selective chemicals for the control of other key members of the cotton pest complex. Thus in Egypt, egg-mass collecting for control of *S. littoralis*

TABLE 5.3 Proposed programmes for the integrated use of conventional and selective pesticides, including pheromones, for control of cotton pests in three differing agro-ecosystems

Country	Programme	Target pests
Egypt	1. PBW pheromone and egg mass picking	*P. gossypiella, S. littoralis*
	2. PBW pheromone and egg mass picking	*P. gossypiella, S. littoralis*
	3. PBW pheromone	*P. gossypiella*
	4. PBW pheromone	*P. gossypiella*
	5. Synthetic pyrethroid	*E. insulana*
Pakistan	1. Systemic organophosphate	Sucking pest complex
	2. PBW pheromone	*P. gossypiella*
	3. PBW pheromone	*P. gossypiella*
	4. Synthetic pyrethroid }	Bollworm complex incl.
	5. Synthetic pyrethroid }	*E. vitella* and *E. insulana*
Peru	1. Lead arsenate	*Alabama argillacea*
	2. Insect growth regulator (IGR)	*Bucculatrix thurberiella*
	3. PBW pheromone	*P. gossypiella*
	4. PBW pheromone	*P. gossypiella*
	5. Synthetic pyrethroid	*Dysdercus peruvianus*

PBW = Pink bollworm

is supplemented by application of IGRs. A request has been made by the Egyptian Ministry of Agriculture for a pheromone formulation to control *E. insulana*. A similar request for pheromonal control of both *Earias* species in Pakistan was discussed earlier, while in Peru a juvenile hormone analogue has been requested by the Cotton Growing Association FUNDEAL (Fundación Para El Desarollo Algodonero) for control of late-season attacks of the cotton stainer, *Dysdercus peruvianus*.

5.11 MATING DISRUPTION FOR CONTROL OF OTHER INSECT PEST SPECIES

If it is assumed that the commercial development of pheromones is of primary importance for their adoption as pest control agents, then, because of their selectivity, only the pheromones of those pest species of widespread occurrence and at the same time of major economic importance will be considered. The choice will then be restricted, at least initially, to relatively few target pests. These include major pests of forests, orchards and vineyards together with the *Heliothis* and *Spodoptera* complexes which attack both cotton and food crops.

A general discussion on the status of mating disruption for the control of other groups of pests is given by Campion (1985).

5.11.1 Forest pests

Forest pests usually have a narrow range of host plants and large plantations make area wide application possible. Successful control of the western pine shoot borer *Eucosoma sonomana* (Kearfott) in the USA by mating disruption has been achieved in areas of up to 600 ha, using a variety of slow-release formulations (Sower *et al.*, 1982). The pheromone, a mixture of (Z)- and (E)-9-dodecenyl acetate, is stable in the field and the components are relatively cheap to synthesize. Rates of application of pheromone per season have not exceeded $20 \, g \, ha^{-1}$. The insect, which attacks ponderosa pine, *Pinus ponderosa*, has only one generation each year and population levels are generally low. The eggs are laid under the scales of young shoots and the larvae burrow straight into the growing point, causing stunted growth; insect damage can therefore be accurately assessed. Critical assessment will ascertain whether the damage caused by the insect is of sufficient economic importance to support a commercially viable control agent.

The gypsy moth, *Lymantria dispar* (L.), is an important forest pest in the United States. Since the female moths are wingless, the spread of the insect is achieved by windborne movements of the young larvae on their silk threads. Initial trials directed primarily against low-density populations using a variety of slow-release formulations of the pheromone (Z)-7,8-epoxy-2-methyloctadecane (disparlure) were conducted in relatively large areas before either effective dose levels were established or adequately stabilized slow-release formulations were available (Cameron, 1979). A further problem was the estimation of low-density larval populations and therefore accurate determination of the level of control achieved. More recently it has been suggested that the pheromone may have use against high-density populations, presumably using higher rates of pheromone application (Plimmer, 1982).

Another important pest of the North American forests is the spruce budworm *Choristoneura fumiferana* (Clem.). The pheromone of this species consists of a blend of (E)- and (Z)-11-tetradecenal (Sanders and Weatherston, 1976) and hence problems exist in stabilizing the aldehyde functional groups. Early trials using slow-release pheromone formulations directed at low-density populations produced equivocal results (Sanders, 1981). More recent large-scale field cage experiments have been aimed at achieving mating disruption at high moth population densities using laminate flakes and microencapsulated pheromone formulations. Results suggest that this can be achieved but a cost-effective application rate has yet to be determined (W. D. Seabrook, personal communication).

5.11.2 Orchard and vineyard pests

Pheromonal control of the oriental fruit moth, *Cydia molesta* (L.), a worldwide pest of peaches and nectarines, was achieved on a commercial scale in Australia during 1985 (Davidson, 1985), following earlier trials conducted by Vickers *et al.* (1985). An area of 160 ha was treated season long with the pheromone blend of (Z)-8-dodecenol and (Z)-8-dodecenyl acetate dispensed in twist-tie dispensers. Levels of control were comparable to those achieved with a conventional programme. However, an upsurge of two-spotted mites which followed four applications of azinphosmethyl and malathion did not occur in the pheromone-treated plots, probably because the beneficial insects were preserved—a situation comparable to that described earlier for cotton in Pakistan. A similar programme was conducted in California peach and nectarine orchards in 1985 (Weakley *et al.*, 1987). The twist-tie or rope pheromone formulation, isomate-M, successfully disrupted mating of *C. molesta* but fruit in orchards not treated with insecticide was subjected to higher risk of damage from other lepidopteran pests.

Successful control of *C. molesta* was also reported from France on orchards of up to 19 ha in size, using a laminate flake formulation at a maximum application rate for the season of 210 g ha^{-1} (Audemard, 1988). Mean application rates of 75 g ha^{-1} were used for the control both of the summer fruit tortrix moth, *Adoxophyes orana* (F.v.R.), and the codling moth, *C. pomonella* (L.), in orchards of similar size, again with laminate flake formulations of the main pheromone components (Charmillot, 1982, 1986). Registration for the use of these latter pheromone formulations has recently been achieved in Switzerland and small-scale commercial applications were undertaken during 1986 (H. Arn, personal communication).

Early attempts to control the European grape berry moth *Eupoecilia ambiguella* (Hubner), an important pest of vines, using a hollow-fibre formulation of the main pheromone component, (Z)-9-dodecenyl acetate, at the relatively low application rate for the season of 10 g a.i. ha^{-1}, were unsuccessful (Schruft, 1982). However, effective mating disruption was achieved using a laminate flake pheromone formulation deployed at a rate of 180 flakes ha^{-1} but with a rate of a.i. per season of 100 g ha^{-1} (Vogt and Schrop, 1985). Commercial registration of the product has recently been granted in Germany and an estimated 2000 ha of cultivation were treated by this pheromone formulation during 1986 (H. Wolgast, personal communication).

Control of the grape moth, *Lobesia botrana* (Schiff.), by mating disruption using the major pheromone component (E,Z)-7,9-dodecenyl acetate, dispensed in large numbers of rubber tubes at the relatively low application rate of 30 g a.i. ha^{-1} per season, gave control in some areas but not others (Roehrich and Carles, 1981).

5.11.3 *Heliothis* complex

Because of their great economic importance much research has been devoted to establishing the feasibility of the mating disruption technique for the control of *Heliothis* spp. Pheromones have been identified for *H. armigera*, *H. punctigera* (Wallengren), *H. subflexa* (Guenée), *H. virescens* and *H. zea* (Boddie). The major pheromone component for all five species is (Z)-11-hexadecenal, which with an aldehyde functional group is chemically more reactive than the acetate functional groups found for example in the pheromones of *E. ambiguella*, *C. molesta* or *P. gossypiella*.

All the *Heliothis* species are polyphagous, the adult moths are known to migrate for considerable distances, and larval infestations are often unpredictable, characteristics which at present are considered unfavourable for the successful development of control strategies using mating disruption.

Small-scale field trials conducted in the United States with hollow-fibre and laminate flake formulations of the main components or analogues of the *H. virescens* and *H. zea* pheromones were either unsuccessful or gave equivocal results (McLaughlin and Mitchell, 1982). Application rates did not, however, exceed $10 \, \text{g a.i. ha}^{-1}$ and therefore, apart from anticipated problems of moth migration, inadequate concentrations of pheromone may also have been a contributing factor. This view receives some support from results of small-scale trials conducted in Egypt using a microencapsulated formulation of the main pheromone components of *H. armigera*. As measured by trap-catch suppression in the treated areas, complete mating disruption was achieved for a period of 22 days when applied at a rate of $100 \, \text{g ha}^{-1}$, whereas at rates of 10 and $50 \, \text{g ha}^{-1}$ such activity persisted for only 4–5 days (Critchley *et al.*, 1986).

The problem of utilizing an inherently unstable pheromone on a large scale at such a high application rate remains to be solved. One possible approach is discussed in the following section.

5.12 LURE AND KILL

Much work has been undertaken with the pheromones of *Spodoptera* spp., several of which are of great economic importance. The utilization of the mating disruption technique has proved difficult, partly perhaps because, like *Heliothis* spp., they are of a polyphagous habit. In addition, while *S. littoralis*, for example, has been shown to have limited migratory potential (Campion *et al.*, 1977; Nasr *et al.*, 1984), it possesses a multicomponent pheromone which is relatively unstable. Work conducted in Egypt showed that mating disruption of *S. littoralis* using microencapsulated pheromone formulations is possible but only at levels of between 40 and $80 \, \text{g a.i. ha}^{-1}$ per application, which at the present cost of active ingredient is too high to be economically viable (Critchley *et al.*, 1986).

Field observations using night vision equipment have shown that *S. littoralis* moths attracted to point sources of pheromone at densities of 500–1000 per hectare will remain in contact with the source for an average of three seconds before flying off (B. W. Bettany, personal communication). At higher rates of distribution of pheromone sources, trail masking or confusion occurs so that the frequency of moth arrivals is reduced. It is therefore possible that a lure and kill technique might be developed using low densities of insecticide-dosed pheromone sources. The level of pheromone required for such regimes is not likely to exceed 2 g ha^{-1} per application and is therefore economically acceptable. In laboratory studies *S. littoralis* male moths were killed after short contact with insecticide-treated surfaces (L. J. McVeigh, unpublished data). Furthermore, sublethal doses of insecticide have been shown to adversely affect mating behaviour in *P. gossypiella* (Haynes and Baker, 1985; Haynes *et al.*, 1986).

To ensure the most effective use of the insecticide in the field, it is essential to determine the optimum distribution of the pheromone sources and small-scale trials with *S. littoralis* are now in progress to this end in Egypt in 1988.

A successful outcome to these trials could provide a model for control not only of other *Spodoptera* species but also the *Heliothis* complex.

5.13 CONCLUSIONS

Although most emphasis in this chapter has been given to control of a cotton pest, *P. gossypiella*, by the mating disruption technique, this is in part due to the experience of the authors. Other examples have, however, been presented for control of forestry, orchard and vineyard pests; hence the control technique is applicable to a variety of insect pests.

One of the earliest attempts at mating disruption for the control of gypsy moth utilized a pheromone application rate of 0.05 g ha^{-1} (Stevens and Beroza, 1972). Successes using this technique for a variety of pest species are now utilizing more realistic rates of application ranging from 40 to 200 g ha^{-1}. There is evidence from laboratory studies that at such high application rates close-range as well as long-range orientation between the sexes is inhibited, which would prevent mating at high population densities, while for certain species adverse effects on female moth behaviour have also been noted. Further behavioural studies are necessary to verify these observations under field conditions. However, if it is confirmed that high application rates are generally necessary for successful control by mating disruption then, in many cases, ways will have to be found to reduce the cost of these chemicals to make the technique commercially competitive with conventional pesticides. More work also needs to be done to integrate insect control by mating disruption with other control strategies in well-constructed integrated pest management programmes.

5.14 ACKNOWLEDGEMENTS

The authors are indebted to the Ministry of Agriculture, Cairo, Egypt, the Pakistan Central Cotton Committee, Karachi and Multan, Pakistan and Fundeal, Lima, Peru for assistance in the projects described in this chapter and for the use of data.

5.15 REFERENCES

Anon. (1983). FAO, UNEP and Turkish Government Regional Symposium on Integrated Pest Control in the Near East, September 5–9 1983, Adana, Turkey.

Attique, M. R. (1985). Pheromones for the control of cotton pests in Pakistan. *The Pakistan Cottons* **29**, 1–6.

Audemard, H. (1988). Les applications practiques des phéromones sexuelles des Lépidoptères. Situation en France. In *Médiateurs Chimiques: Comportement et Systematique des Lépidoptères, Applications en Agronomie*. Proceedings of a Colloquium, Valence, France, 13–14 December, 1985.

Bierl, B. A., Beroza, M., Staten, R. T., Sonnet, P. E. and Adler, V. E. (1974). The pink bollworm sex attractant. *J. Econ. Ent.*, **67**, 211–216.

Brooks, T. W., Doane, C. C. and Staten, R. T. (1979). Experience with the first commercial pheromone communication disruptive for suppression of an agricultural pest. In *Chemical Ecology: Odour Communication in Animals*, F. J. Ritter (Ed.), Elsevier, Amsterdam, pp. 375–388.

Cameron, E. A. (1979). Disparlure and its role in gypsy moth manipulation. *Mitt. Schweiz. Ent. Ges.*, **52**, 333–342.

Campion, D. G. (1983). Pheromones for the control of insect pests in Mediterranean countries. *Crop Protection*, **2**, 3–16.

Campion, D. G. (1985). Survey of pheromone uses in pest control. In *Techniques in Pheromone Research*, H. E. Hummel and T. A. Miller (Eds), Springer-Verlag, New York, pp. 405–469.

Campion, D. G., Bettany, B. W., McGinnigle, J. B. and Taylor, E. R. (1977). The distribution and migration of *Spodoptera littoralis* (Boisduval) (Lepidoptera, Noctuidae) in relation to meteorology in Cyprus interpreted from maps of pheromone trap samples. *Bull. Ent. Res.*, **67**, 501–522.

Campion, D. G., Hunter-Jones, P., McVeigh, L. J., Hall, D. R., Lester, R. and Nesbitt, B. F. (1980). Modification of the attractiveness of the primary pheromone component of the Egyptian cotton leafworm, *Spodoptera littoralis* (Boisduval) (Lepidoptera: Noctuidae), by secondary pheromone components and related chemicals. *Bull. Ent. Res.*, **70**, 417–434.

Charmillot, P. J. (1982). Expérimentation de la technique de confusion contre Capua (*Adoxophyes orana* F.v.R.) en Suisse Romande. In *Médiateurs Chimiques Agissant sur la Comportement des Insectes, Versailles 1981*, (Les colloques d'INRA 7), Institut National de la Recherche Agronomique, Paris, pp. 357–363.

Charmillot, P. J. (1986). Technique de confusion sexuelle contre *Adoxophyes orana* F.v.R. Utilisation des composantes phéromonales principales et mineures au laboratoire et en verger. In *Médiateurs Chimiques: Comportement et Systematique des Lépidoptères, Applications en Agronomie*, Proceedings of a Colloquium, Valence, France, 13–14 December, 1985.

Cork, A., Beevor, P. S., Hall, D. R., Nesbitt, B. F. and Campion, D. G. (1985). A sex attractant for the spotted bollworm, *Earias vitella*. *Tropical Pest Management*, **31**, 158.

Critchley, B. R., Campion, D. G., McVeigh, L. J., Hunter-Jones, P., Hall, D. R., Cork, A., Nesbitt, B. F., Marrs, G. J., Jutsum, A. R., Hosny, M. M. and Nasr, El Sayed A. (1983). Control of the pink bollworm, *Pectinophora gossypiella* (Saunders) (Lepidoptera: Gelechiidae), in Egypt by mating disruption using an aerially applied microencapsulated pheromone formulation. *Bull. Ent. Res.*, **73**, 289–299.

Critchley, B. R., Campion, D. G., McVeigh, L. J., McVeigh, E. M., Cavanagh, G. G., Hosny, M. M., Nasr, El Sayed A., Khidr, A. A. and Naguib, A. A. (1985). Control of the pink bollworm, *Pectinophora gossypiella* (Saunders) (Lepidoptera: Gelechiidae), in Egypt by mating disruption using hollow-fibre, laminate-flake and micro-encapsulated formulations of synthetic pheromone. *Bull. Ent. Res.*, **54**, 329–345.

Critchley, B. R., McVeigh, L. J., McVeigh, E. M., Cavanagh, G. G. and Campion, D. G. (1986). Integrated control of cotton pests in Egypt using pheromones and viruses. Pheromone studies 1979–1983. *Tropical Development and Research Institute Overseas Assignment Report* R1310(R), 59pp.

Critchley, B. R., Campion, D. G., Cavanagh, G. G., Chamberlain, D. J. and Attique, M. R. (1987). Control of three major pests of cotton in Pakistan by a single application of their combined sex pheromones. *Trop. Pest Manage.* **33**, 374.

Davidson, S. (1985). Confusion control of the oriental fruit moth. *Rural Research*, **126**, 9–12.

Doane, C. C., Haworth, J. K. and Dougherty, D. G. (1983). Nomate PBW, a synthetic pheromone formulation for wide area control of the pink bollworm. In *Proceedings of 10th International Congress of Plant Protection*, Brighton, England, 20–25 November, 1983.

Eveleens, K. G. (1983). Cotton insect control in the Sudan Gezira: analysis of a crisis. *Crop Protection*, **2**, 273–287.

Flint, H. M., Merkle, J. R. and Yamamoto, A. (1985). Pink bollworm (Lepidoptera: Gelechiidae): Field testing a new polyethylene tube dispenser for gossyplure. *J. Econ. Ent.*, **78**, 1431–1436.

Gaston, L. K., Payne, T. L., Takahashi, S. and Shorey, H. H. (1972). Correlation of chemical structure and sex pheromone activity in *Trichoplusia ni* (Noctuidae). In *Olfaction and Taste*, D. Schrieden (Ed.), Wissenschaftliche, Stuttgart, pp. 167–173.

Gaston, L. K., Kaae, R. S., Shorey, H. H. and Sellers, D. (1977). Controlling the pink bollworm by disrupting sex pheromone communication between the adults. *Science*, **196**, 904–905.

Gonzalez, J. E. (1979). Cotton pests in South America. *Outlook on Agriculture*, **10**, 197–201.

Gonzalez, J. E. (1982). Manual de evaluacion y control de insectos y acaros del algodonero. FUNDEAL, Bol. Tec. No. 2, Lima, Peru, pp. 79.

Hall, D. R., Beevor, P. S., Lester, R. and Nesbitt, B. F. (1980). (*E*,*E*)-10,12-Hexadecadienal: a component of the sex pheromone of the spiny bollworm, *Earias insulana* (Boisd.) (Lepidoptera, Noctuidae). *Experientia*, **36**, 152–153.

Hall, D. R., Nesbitt, B. F., Marrs, G. J., Green, A. St J., Campion, G. and Critchley, B. R. (1982). Development of microencapsulated pheromone formulations. In *Insect Pheromone Technology: Chemistry and Applications*, B. A. Leonhardt and M. Beroza (Eds), American Chemical Society Symposium Series No. 190, Washington, DC, August 1981, pp. 131–143.

Haynes, K. F. and Baker, T. C. (1985). Sublethal effects of permethrin on the chemical communication system of the pink bollworm moth, *Pectinophora gossypiella*. *Arch. Insect Biochem. Physiol.*, **2**, 283–293.

Haynes, K. F., Li, W.-G. and Baker, T. C. (1986). Control of pink bollworm moth (Lepidoptera: Gelechiidae) with insecticides and pheromones (attracticide): lethal and sub-lethal effects. *J. Econ. Entomol.* **79**, 1466–1471.

Hosny, M. M. (1980). The control of insect pests in Egypt. *Outlook on Agriculture*, **10**, 204–205.

Hummel, H. E., Gaston, L. K., Shorey, H. H., Kaae, R. S., Byrne, K. J. and Silverstein, R. M. (1973). Clarification of the chemical status of the pink bollworm sex pheromone. *Science*, **181**, 873–875.

Jacobson, M. and Beroza, M. (1963). Chemical insect attractants. *Science*, **140**, 1367–1373.

Klun, J. A., Plimmer, J. R., Bierl-Leonhardt, B. A., Sparks, A. N. and Chapman, O. L. (1979). Trace chemicals: the essence of chemical communications in *Heliothis* species. *Science*, **204**, 1328–1329.

Kydonieus, A. F. and Beroza, M. (1981). The Hercon dispenser formulation and recent test results. In *Management of Insect Pests with Semiochemicals: Concepts and Practice*, E. R. Mitchell (Ed.), Plenum Press, New York, pp. 445–453.

McLaughlin, J. R. and Mitchell, E. R. (1982). Practical development of pheromones in *Heliothis* management. In *Proceedings of the International Workshop on Heliothis Management*, W. Read and V. Kumble (Ed.) International Crops Research Institute for the Semi-Arid Tropics (1982) pp. 309–318.

Moorhouse, J. E., Yeadon, R., Beevor, P. S. and Nesbitt, B. F. (1969). Method for use in studies of insect chemical communication. *Nature* (Lond.), **223**, 1174–1175.

Nasr, El-Sayed A., Tucker, M. R. and Campion, D. G. (1984). Distribution of moths of the Egyptian cotton leafworm, *Spodoptera littoralis* (Boisduval) (Lepidoptera: Noctuidae), in the Nile Delta interpreted from catches in a pheromone trap network in relation to meteorological factors. *Bull. Ent. Res.*, **74**, 487–494.

Nesbitt, B. F., Beevor, P. S., Cole, R. A., Lester, R. and Poppi, R. G. (1973). Sex pheromones of two noctuid moths. *Nature New Biol.*, **244**, 208–209.

Plimmer, J. R. (Ed.) (1982). Pesticide Residues and Exposure. Washington DC, American Chemical Society (ACS Symposium Series 182), 213 pp.

Roehrich, R. and Carles, J. P. (1981). Mating disruption in vineyards to control the grape vine moth *Lobesia botrana* (Schiff.). In *Les Médiateures Chimiques Agissant sur le Comportement des Insectes*, Versailles 1981 (Les Colloques d' INRA 7), pp. 365–371.

Sanders, C. J. (1981). Disruption of spruce budworm mating—state of the art. In *Management of Insect Pests with Semiochemicals*, E. R. Mitchell (Ed.), Plenum Press, New York, pp. 339–345.

Sanders, C. J. and Weatherston, J. (1976). Sex pheromone of the eastern spruce budworm: optimum blend of *trans* and *cis*-11-tetradecenal. *Can. Ent.*, **108**, 1285–1290.

Schruft, G. (1982). A study of control of the European grape berry moth *Eupoecilia ambiguella* by the mating disruption technique. *Unpublished internal IOBC report presented at IOBC/WPRS working group on pheromones, special meeting on mating disruption*, Changins/Nyon, Switzerland, 28–29 September, 1982.

Shorey, H. H., Kaae, R. S. and Gaston, L. K. (1974). Sex pheromones of Lepidoptera, development of a method for pheromonal control of *Pectinophora gossypiella* in cotton. *J. Econ. Ent.*, **67**, 347–350.

Shorey, H. H., Gaston, L. K. and Kaae, R. S. (1976). Air permeation with gossyplure for control of the pink bollworm. In *Pest Management with Insect Sex Attractants*, M. Beroza (Ed.), ACS Symp. Ser. 23, American Chemical Society, Washington, DC, pp. 67–74.

Sower, L. L., Overhulser, D. L., Daterman, G. E., Sartwell, C., Laws, D. E. and Koerber, T. W. (1982). Control of *Eucosoma sonomana* by mating disruption with synthetic sex attractant. *J. Econ. Ent.*, **75**, 315–318.

Staten, R. T., Flint, H. M., Weddle, R. C., Quintero, E., Zarate, R. E., Finnell, C. M., Hernandes, M. and Yamamoto, A. (1987). Pink bollworm (Lepidoptera: Gelechiidae): large-scale field trials with a high-rate gossyplure formulation. *J. Econ. Entomol.* **80**, 1267–1271.

Stevens, L. J. and Beroza, M. (1972). Mating inhibition field tests using disparlure, the synthetic gypsy moth sex pheromone. *J. Econ. Ent.*, **65**, 1090–1095.

Sundaramurthy, V. T. and Basu, A. K. (1985). Management of cotton insect pests in polycrop systems in India. *Outlook on Agriculture*, **14**, 79–82.

Tumlinson, J. H., Heath, R. R. and Teal, P. E. A. (1982). Analysis of chemical communication systems of Lepidoptera. In *Insect Pheromone Technology: Chemistry and Applications*, B. A. Leonhardt and M. Beroza (Eds), ACS Symposium Series 190, American Chemical Society, Washington, DC, pp. 1–25.

Vickers, R. A., Rothschild, G. H. L. and Jones, E. L. (1985). Control of the oriental fruit moth, *Cydia molesta* (Busck) (Lepidoptera: Tortricidae), at a district level by mating disruption with synthetic female pheromone. *Bull. Ent. Res.*, **75**, 625–634.

Vogt, H. and Schrop, A. (1985). Field trials on the control of the grape moth *Eupoecilia ambiguella* Hbn. by using the mating disruption technique. *Gesunde Pflanzen*, **37**, 439.

Weakley, C. V., Kirsch, P. and Rice, R. E. (1987). Control of oriental fruit moth by mating disruption. *Calif. Agric.* **41**, 7–8.

Production, Formulation and Application

Insect Pheromones in Plant Protection
Edited by A. R. Jutsum and R. F. S. Gordon
©1989 John Wiley & Sons Ltd

6

Chemistry and Commercial Production of Pheromones and other Behaviour-modifying Chemicals

A. Yamamoto and K. Ogawa

Shin-Etsu Chemical Co. Ltd,
Tokyo, Japan

6.1 INTRODUCTION

In recent years sex pheromones have been identified and synthesized, their functions have been understood, and practical uses have been found for them. In studies of pheromone functions and the prediction of their occurrence, the quantities required were sufficiently small that laboratory preparation methods sufficed, but the development of communication disruption techniques has led to an increasing need for pheromones to be supplied in large quantities at low prices. The industrial production of pheromones is being studied in many countries, and many patents have been applied for, but it is not possible to say which of the methods proposed is most economical. At present, the most economical method is the one that makes the most efficient use of the manufacturer's existing equipment.

The largest number of pheromones of currently known chemical structure belong to the Lepidoptera. This chapter presents some patented methods of synthesizing these and other pheromones. There have been extensive reviews of pheromone synthesis (Henrick, 1977; Rossi, 1978; Mori and Uchida, 1981), but by focusing here on patented procedures it is hoped to draw attention to those compounds which have been considered for commercialization.

6.2 LEPIDOPTERA

Studies of sex pheromones and their use in pest control are most advanced in Lepidoptera, whose pheromones are used both for monitoring and for communication disruption.

Most of these sex pheromones have structures of C_{10} to C_{21} unsaturated aliphatic straight-chain alcohols, acetates, aldehydes and ketone derivatives. Formate derivatives, known to *mimic* aldehyde pheromones, have been applied in communication disruption. It is on lepidopteran pheromones, therefore, that most work on commercial synthesis has been done, and it is in this field that the following discussion centres.

The key points in pheromone structures are:
(1) the double-bond position and number;
(2) the configuration (Z or E) of the double bonds;
(3) the functional group (alcohol, acetate, aldehyde, ketone, etc.).

Once these three points have been clarified, the problem is how to synthesize the correct molecule. A survey of the patents shows that four methods have been employed to introduce double bonds in the desired position:
(1) use of acetylene compounds;
(2) use of the Wittig reaction;
(3) use of compounds with naturally stereoselective double bonds;
(4) disproportionation reaction between a mono-olefin having a functional group and a hydrocarbyl mono-olefin.

Some typical methods of pheromone synthesis are described below.

6.2.1 Alkenyl acetates

These are the simplest and most representative of the sex pheromones of Lepidoptera. Three sex pheromones whose synthesis has been described in patents are typical.

(Z)-8-Dodecen-1-yl acetate is the sex pheromone of the oriental fruit moth (*Cydia molesta*, L.). It has been synthesized from 1-pentyne in two ways (Holan, 1975; Yamamoto *et al.*, 1981). The method employed by Holan is described in Scheme 6.1.

(Z)-9-Tetradecen-1-yl acetate is the sex pheromone of the summer fruit tortrix (*Adoxophyes orana* Fischer von Roslerstamm). Its synthesis by Taguchi *et al.* (1981) using Z-6-undecenyl chloride as the key material is described in Scheme 6.2.

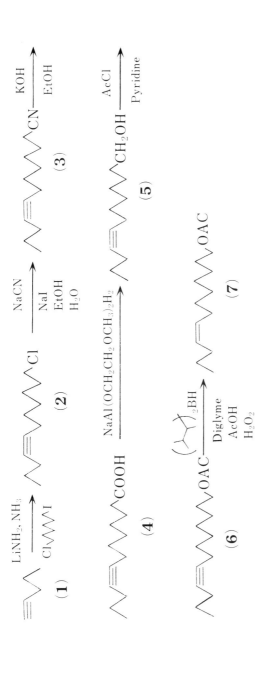

Scheme 6.1 (Z)-8-Dodecen-1-yl acetate ((Z)-8-DDA). 1-Pentyne (**1**) and iodochlorohexane are coupled with LiNH₂ in ammonia to prepare 7-undecynyl chloride (**2**), which is converted by NaCN to the nitrile (**3**). This is hydrolysed to form the corresponding carboxylic acid (**4**), which is reduced to an alcohol and acetylated by the usual method. (Z)-8-DDA (**7**) is synthesized by *cis* partial hydrogenation using disiamyl borane.

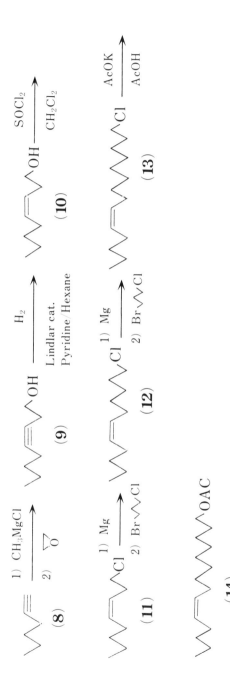

Scheme 6.2 (Z)-9-Tetradecen-1-yl acetate ((Z)-9-TDA). 1-Hexyne (8) is reacted with a Grignard reagent and ethylene oxide to prepare 3-octynol (9). Next (Z)-3-octenol (10) is synthesized from 9 by catalytic hydrogenation, using the Lindlar catalyst in hexane. Compound 10 is chlorinated to form (Z)-3-octenyl chloride (11), the Grignard reagent of which is coupled with bromochloropropane to yield (Z)-6-undecenyl chloride (12). The Grignard reagent of 12 is then in turn coupled with bromochloropropane to make (Z)-9-tetradecenyl chloride (13), which is acetylated to synthesize (Z)-9-TDA (14).

The above examples take the acetylene compound route, but many methods using alkenyl acetates and the Wittig reaction are also known (Szantay *et al.*, 1979; Hoechst AG, 1975). A method employed by Szantay *et al.* (1979) in Hungary to synthesize (Z)-11-hexadecen-1-yl acetate, the sex pheromone of the cabbage army worm (*Mamestra brassicae*, L.) is shown in Scheme 6.3.

6.2.2 Conjugated alkadienyl acetates

The difficulty in synthesizing these conjugated diene compounds is that if an intermediate that contains the conjugated diene is subjected to inappropriate reaction conditions, it rapidly isomerizes into stereo-isomers. The various strategies for overcoming this difficulty fall into two main groups: one is to apply the Wittig reaction as close to the end of the synthesis as possible (Roelofs *et al.*, 1974; Goto, 1975); the other is to synthesize a compound with a conjugated *trans* double bond and triple bond (conjugated ene-yne), and employ partial hydrogenation near the end of the process (Labovitz and Henrick, 1976; Descoins and Samain, 1977; Mori and Uchida, 1981; Passardo *et al.*, 1979; Yushima, 1976). Of the many methods that have been studied, two typical ones are described below.

(E,Z)-7,9-Dodecadien-1-yl acetate, the sex pheromone of the European grapeberry moth, *Lobesia botrana* Schiff, can be synthesized using 1-bromo-(E)-4-nonen-6-yne, a conjugated ene-yne, as the key material. Nearly identical methods have been given by Labovitz and Henrick (1976) (Scheme 6.4) and by Descoins and Samain (1977). Roelofs *et al.* (1974) have described a method using the Wittig reaction.

Japanese patents show two methods of synthesizing (Z,E)-9,11-tetra-decadien-1-yl acetate, the sex pheromone of *Spodoptera* species: one via a conjugated ene-yne, using *cis* partial hydrogenation, due to Mori and Uchida (1981) and Yushima (1976), and one employing *trans* reduction with $LiAlH_4$, due to Passardo *et al.* (1979). Scheme 6.5 presents a method using the Wittig reaction, due to Goto (1975).

6.2.3 Non-conjugated alkadienyl acetates

The key point in the synthesis of these compounds is how to obtain double bonds in specified positions with a specified geometrical configuration. Once again the methods are divided by the use of acetylene compounds or the Wittig reaction. An important question is what starting material to select and where to link the carbon chains. The synthesis of two representative sex pheromones is described below.

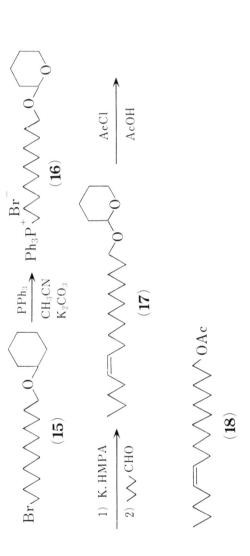

Scheme 6.3 (Z)-11-Hexadecen-1-yl acetate ((Z)-11-HDA). Tetrahydropyranyl ether of 11-hydroxyundeca-1-yl bromide (**15**) is treated with PPh₃ in CH₃CN in the presence of K₂CO₃ to prepare (11-tetrahydropyranyloxyundecyl)-triphenyl phosphonium bromide (**16**), which undergoes the Wittig reaction with pentenal in HMPA potassium base to synthesize *cis*-olefin (**17**). The compound (**17**) is acetylated with CH₃COCl to synthesize (Z)-11-HDA (**18**).

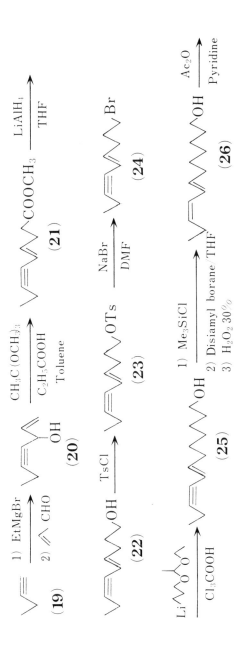

Scheme 6.4 (*E*,*Z*)-7,9-Dodecadien-1-yl acetate ((*E*,*Z*)-7,9-DDA). 1-Butyne (**19**) is reacted with EtMgBr and acrolein to synthesize 1-hepten-4-yn-3-ol (**20**), which is reacted with trimethyl orthoacetate to give methyl-(*E*)-4-nonen-6-ynoate (**21**). The compound (**21**) is reduced with LiAlH$_4$ to (*E*)-4-nonen-6-yn-1-ol (**22**), then 1-tosyloxy-(*E*)-4-nonen-6-yne (**23**) is quantitatively synthesized, using *p*-toluene-sulphonyl chloride. Compound **23** is brominated with NaBr in DMF to yield 1-bromo-(*E*)-4-nonen-6-yne (**24**), which is coupled with 3-[(1-ethoxy)ethoxy]propyl lithium, using Li$_2$CuCl$_4$ as a catalyst in THF. Compound **25** is synthesized after removal of the 1-ethoxyethyl protective group. Next, after trimethyl silylation of the hydroxyl group of **25**, (*E*,*Z*)-7,9-dodecadien-1-ol (**26**) is prepared by *cis* partial hydrogenation with disiamyl borane and acetylation gives (*E*,*Z*)-7,9-DDA (**27**).

Scheme 6.5 (Z,E)-9,11-Tetradecadien-1-yl acetate ((Z,E)-9,11-TDDA). 1,9-Nonandiol (**28**) is monobrominated and acetylated to produce 9-acetoxynon-1-yl bromide (**30**), which is heated with PPh₃ in CH₃CN to obtain the corresponding phosphonium salt (**31**) quantitatively. The salt is next subjected to the Wittig reaction with (E)-2-pentenal, using dimsyl anion prepared from DMSO and NaH as the base, to synthesize the Z,E isomer of 9,11-tetradecadien-1-yl acetate (**32**) with 85–90% stereoselectivity. The 10–15% E,E isomer can be removed as a Diels–Alder adduct.

Many ways are known to synthesize (Z,Z)- and (Z,E)-7,11-hexadecadien-1-yl acetate, the sex pheromone of the pink bollworm, *Pectinophora gossypiella* Saunders. Anderson and Henrick (1976) and Muehowski and Venuti (1980) synthesize it using the Wittig reaction. Mori and Tominaga (1976) and Ishihara and Yamamoto (1981) have reported a coupling reaction using acetylene compounds. The method of Friedman and Chanan (1976), using 1,5-hexadiyne (**33**) as the starting material, is described in Scheme 6.6.

Among the methods of synthesizing (Z,Z)-3,13-octadecadien-1-yl acetate, the sex pheromone of the peachtree borer, *Synanthedon exitiosa* Say, is a method developed by Uchida *et al.* (1980) using tosylate derivatives of (Z)-3-hexenol, and a method taking the (Z)-10-alkenol route (Mori, 1980). A third method developed by Ishihara *et al.* (1981, 1982), using Grignard coupling, is given in Scheme 6.7.

6.2.4 Alcohols

(E,E)-8,10-Dodecadien-1-ol, the sex pheromone of the codling moth, *Laspeyresia pomonella* L., is one of the pheromones whose synthesis has been most extensively studied. The key point in synthesizing it is how to introduce the E,E conjugated diene. Since natural sorbic acid (and its derivatives) have the E,E configuration, most known methods of synthesizing this pheromone use that structure unaltered. Henrick and Siddall (1974a, 1974b, 1974c, 1976) use sorbic acid and sorbyl aldehyde as the starting material; Samain *et al.* (1979, 1980) use sorbyl alcohol. There are also methods to make the E,E configuration from scratch: a Wittig reaction method using the triphenyl phosphonium salt of crotyl bromide has been developed by Roelofs *et al.* (1974a) and a method of synthesis from cyclopropenyl aldehyde has been developed by Descoins and Henrick (1974). The method developed by Henrick and co-workers, starting from sorbic acid, is described in Scheme 6.8.

6.2.5 Aldehydes and ketones

Many lepidopteran pests are known to have aldehydes and ketones as sex pheromones. In particular, (Z)-11-hexadecen-1-al, (Z)-9-hexadecen-1-al, (Z)-9-tetradecen-1-al and other similar alkenals are major components of the sex pheromones of *Heliothis* species, considered the world's most serious insect pest.

In the alkadienal family, (Z,Z)-11,13-hexadecadien-1-al (Bishop and Morrow, 1983; Carney and Henrick, 1980a, 1980b), the sex pheromone of the navel orange worm, *Amyelois transitella*, is well known, as is (Z,E)-9, 11-hexadecadien-1-al (Carney and Henrick, 1980a), the sex pheromone of the sugarcane borer, *Diatraea saccharalis* F., and in the ketone family,

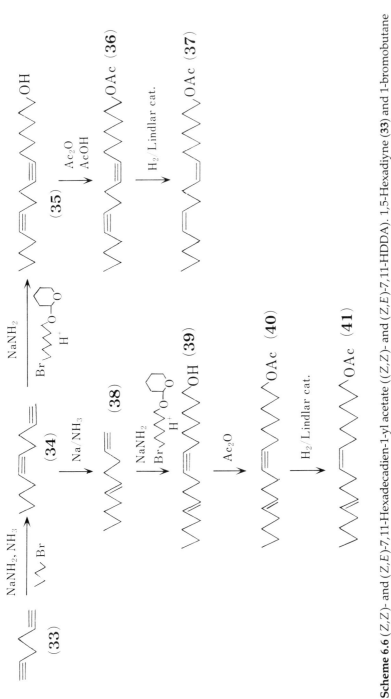

Scheme 6.6 (Z,Z)- and (Z,E)-7,11-Hexadecadien-1-yl acetate ((Z,Z)- and (Z,E)-7,11-HDDA). 1,5-Hexadiyne (**33**) and 1-bromobutane are coupled to synthesize 1,5-decadiyne (**34**) as a key intermediate. 7,11-Hexadecadiyn-1-ol (**35**) is obtained by coupling with THP ether of hexamethylene bromohydrin, on the route to the Z,Z isomer. Acetylation of **35** with anhydrous acetic acid followed by *cis* partial hydrogenation with the Lindlar catalyst yields (Z,Z)-7,11-HDDA (**37**). For the Z,E isomer, Birch reduction is applied to 1,5-decadiyne (**34**) to make (E)-5-decen-1-yne (**38**), and this is coupled with THP ether of hexamethylene–bromohydrin and acetylated to prepare (E)-11-hexadecen-7-yn-1-yl acetate (**40**). Finally, this is partially hydrogenated with the Lindlar catalyst to obtain (Z,E)-7,11-HDDA (**41**).

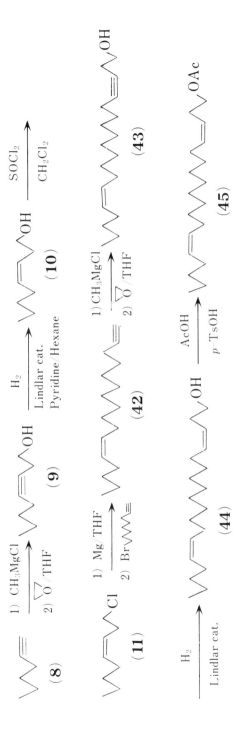

Scheme 6.7 (Z,Z)-3,13-Octadecadien-1-yl acetate ((Z,Z)-3,13-ODDA). 1-Hexyne (**8**) and Grignard reagent are reacted with ethylene oxide to produce 3-octynol (**9**). After *cis* partial hydrogenation (**10**), the compound is chlorinated to produce (Z)-3-octen-1-yl chloride (**11**). The Grignard reagent of **11** is coupled with 1-bromo-7-octyne to give (Z)-11-hexadecen-1-yne (**42**) in about 60% yield. A reaction among **42**, Grignard reagent and ethylene oxide produces (Z)-13-octadecen-3-yn-1-ol, from which (Z,Z)-3,13-ODDA (**45**) is obtained by *cis* partial hydrogenation with Lindlar catalyst and acetylation.

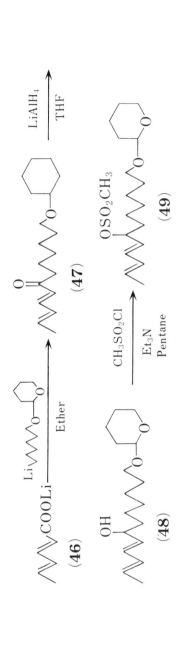

Scheme 6.8 (*E*,*E*)-8,10-Dodecadien-1-ol ((*E*,*E*)-8,10-DDDOL). The lithium salt of sorbic acid (**46**) is reacted with 2-[(6′-lithium hexyl)oxy]-tetrahydropyran in ether at −10 °C to synthesize 1-(tetrahydropyran-2′-yloxy)-(*E*,*E*)-8,10-dodecadien-7-one (**47**). This ketone (**47**) is next reduced to alcohol (**48**) with LiAlH₄; then CH₃SO₂Cl is used to produce 7-mesylate (**49**). The mesylate (**49**) is reduced with LiAlH₄ and the tetrahydropyranyl group of **49** is removed using *p*TsOH. (*E*,*E*)-8,10-DDDOL (**50**) is obtained.

(Z)-6-heneicosen-11-one (Ishihara *et al.*, 1980; Ito *et al.*, 1983; Uchida *et al.*, 1983), the sex pheromone of Douglas-fir tussock moth, *Orgyia pseudotsugata*. A number of patent applications have been made for the synthesis of these sex pheromones.

In the synthesis of aldehydes and ketones, the alkenyl group and alkadiene group can be produced in essentially the same ways as the acetates and other compounds described above. The problem is how to introduce the carbonyl group. The most orthodox method is to start by synthesizing the corresponding alcohol and to oxidize it in the final process. Various oxidizing agents have been employed, including chromium trioxide–pyridine (Carney and Henrick, 1980a; Roelofs, 1975; Research Corp., 1975), *N*-halosuccinimides (such as NBS and NCS) (Carney and Henrick, 1980b; Carney, 1982), and pyridinium chlorochromate (PCC) (Koshihara *et al.*, 1984). Methods of introducing the carbonyl group directly by the Grignard reaction are also known (Bishop and Morrow, 1983; Ishihara *et al.*, 1981). A method of aldehyde synthesis due to Mori *et al.* (1980) and Mori (1981), which uses an oxidative ring-opening reaction of a 1,2-epoxy derivative, is given in Scheme 6.9.

6.2.6 Epoxy compounds

The sex pheromone of the gypsy moth, *Lymantria dispar* (L.) is the well-known compound *cis*-7,8-epoxy-2-methyl octadecane (disparlure). It is synthesized by epoxidation of the corresponding olefin compound. A racemic compound can easily be obtained by oxidation with peroxide. To synthesize the actual sex pheromone, which is the optically active (+)-disparlure, however, it is necessary to use a more complex method of asymmetric synthesis. A method developed by Katsuki and Sharpless (1982) is described in Scheme 6.10.

6.3 DIPTERA AND OTHERS

6.3.1 Diptera

Most of the known pheromones of Diptera are aliphatic monoene compounds such as (Z)-9-tricosene, (Z)-9-heneicosene and (Z)-14-nonacosene. Of these, (Z)-9-tricosene, the sex pheromone of the housefly, *Musca domestica* L., has been synthesized commercially with the simplest process. The method of direct coupling of oleyl tosylate with lithium pentane developed by Hennart and Martin (1981) (Scheme 6.11), a method using the lithium salt of *n*-octyl acetylene developed by Eiter (1974) and a reaction method developed by Yoshida (1983) in which the Grignard reagent of oleyl chloride is cross-coupled with amyl halide, have been described in patents.

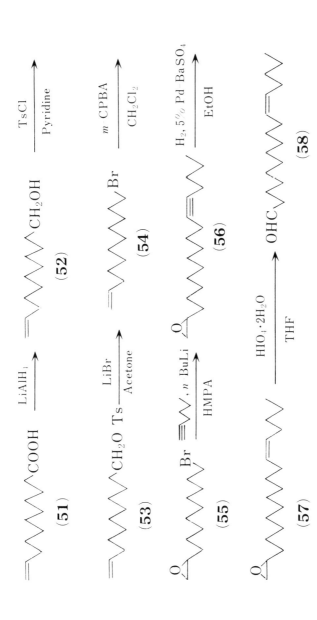

Scheme 6.9 (Z)-11-Hexadecen-1-al ((Z)-11-HDAL). Undecylenic acid (**51**) is reduced with LiAlH₄ to yield undecenyl alcohol (**52**). A reaction of **52** with *p*-toluenesulphonyl chloride then yields tosylate (**53**). Next, 10-undecenyl bromide (**54**) is obtained from **53**, using LiBr in acetone. Compound **54** is converted to epoxide with *m*-CPBA and coupled with 1-hexyne to synthesize 1,2-epoxy-12-heptadecyne (**56**). After *cis* partial hydrogenation of **56** using 5% Pd-BaSO₄ as a catalyst, the material (**57**) is treated with periodic acid at low temperature, giving (Z)-11-HDAL (**58**).

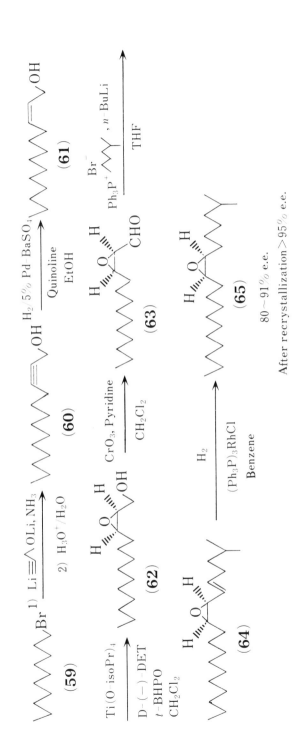

Scheme 6.10 cis-(+)-7,8-Epoxy-2-methyl octadecane (disparlure). 1-Bromodecane (**59**), the starting material, is coupled with propargyl alcohol in ammonia, then hydrogenated to obtain (Z)-2-tridecen-1-ol (**61**). Asymmetric epoxidation of **61** with t-butyl hydroperoxide (t-BHPO) is performed in the presence of D-(−)-diethyl tartrate (DET) and titanium tetra-isopropoxide to obtain the epoxy alcohol (**62**), which is oxidized by CrO₃ to make the epoxy aldehyde (**63**). The Wittig reaction of **63** with 4-methylpentyl triphenylphosphonium bromide, followed by partial hydrogenation catalysed by (Ph₃P)₃RhCl then results in (+)-disparlure (**65**), with 80–91% enantiomeric excess.

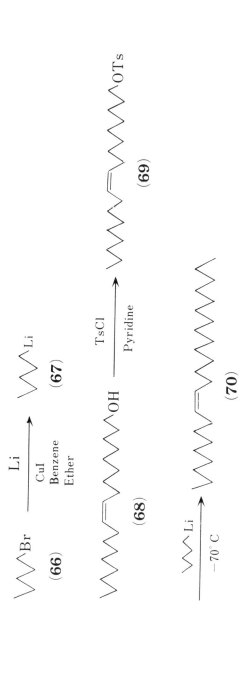

Scheme 6.11 (Z)-9-Tricosene. Lithium pentane (**67**) is prepared from 1-bromopentane (**66**) with lithium in the presence of CuI. Oleyl tosylate (**69**), prepared from oleyl alcohol and *p*-toluenesulphonyl chloride, is reacted with **67** at −70 °C, to give (Z)-9-tricosene (**70**) in high yield (94%).

Zoecon workers developed conditions for the production of 150 kg batches of Z-9-tricosene by addition of catalytic quantities of lithium chlorocyanocuprate to such Grignard coupling reactions.

6.3.2 Hemiptera

A fairly large number of methods for synthesis of the sex pheromones of scale insects, Hemiptera, have been patented. There is a consistency among the sex pheromone compounds of scale insects; many of them are C_{10}–C_{18} terpene derivatives. In many cases, accordingly, the synthesis begins from a terpene compound: for instance, Negishi *et al.* (1984) have synthesized dimethyl-1,5-heptadienyl acetate (the sex pheromone of *Pseudococcus* species) from 6-methyl-5-hepten-2-one, and Anderson (1979) has synthesized 3-methyl-6-isopropenyl-9-decen-1-yl acetate, the sex pheromone of California red scale, *Aonidiella aurantii* Maskell, from *dl*-citronellol and (*S*)-(+)-carvone.

Several methods for 3,7-dimethyl-(*Z*)-2,7-octadien-1-yl propionate, the sex pheromone of San Jose scale, *Quadraspidiotus perniciosus* Comst., have been presented (Anderson and Henrick, 1978; Szantay *et al.*, 1983). Szantay and co-workers use nerol (terpene alcohol) as the starting material. An example using the Wittig reaction is the synthesis of 3,9-dimethyl-6-isopropyl-(*E*)-5,8-decadien-1-yl acetate, the sex pheromone of yellow scale, *Aonidiella citrina*, by Anderson and Henrick (1978), which is shown in Scheme 6.12.

6.3.3 Coleoptera

The largest number of insect species belong to Coleoptera (beetles and weevils). A difference between Coleoptera and other orders is that, in addition to the sex pheromones of Coleoptera, their aggregating pheromones and simple attractant pheromones are known and have been put to use. Whereas the sex pheromones of Lepidoptera and Diptera have the chemical structure of derivatives of unsaturated aliphatic straight-chain hydrocarbons, and the sex pheromones of Hemiptera are consistently terpene derivatives, the pheromones of the Coleoptera are diverse, and many of them are quite individual. The sex pheromone of the black carpet beetle, *Attagenus megatoma*, for instance, is (*E,Z*)-3,5-tetradecadienoic acid, which can be synthesized by much the same methods as used for the sex pheromones of Lepidoptera (Burkholder, 1970). The well-known Coleoptera pheromones shown in Sections 6.3.3.1 to 6.3.3.3, however, are optically active compounds, and special techniques are required to introduce asymmetric carbons.

6.3.3.1 Boll weevil

The sex pheromone of the boll weevil, *Anthonomus grandis* Boheman, known as grandlure, is a mixture of four compounds. Synthesis methods for two of

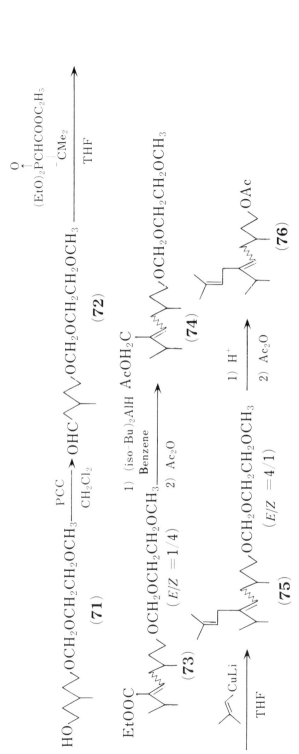

Scheme 6.12 3,9-Dimethyl-6-isopropyl-(*E*)-5,8-decadien-1-yl acetate. The monoether of 3-methyl-1,5-pentanediol (**71**) is oxidized with pyridinium chlorochromate (PCC) to obtain the aldehyde (**72**). The aldehyce (**72**) is treated with NaH, diethyl-1-ethoxycarbonyl-2-methylpropylphosphonate to obtain the ester (**73**) (*E*/*Z* = 1/4). This is reduced and acetylated to **74**; it then undergoes a coupling reaction with 2-methylpropenyl lithium cuprate to yield a diene compound (**75**). The *E*/*Z* ratio at this point is 4/1 (the apparent change is just a change in priority sequence of the substituents). Next, after the alcohol protective group is removed, this compound is acetylated to obtain 3,9-dimethyl-6-isopropyl-5,8-decadien-1-yl acetate (**76**), in which the *E* isomer is the main product, with an *E*/*Z* ratio of 4/1.

these, *cis*- and *trans*-3,3-dimethyl-$\Delta^{1,\alpha}$-cyclohexane ethanol, have been given in patents (Tumlinson, 1974; Traas, 1979). A method developed by Zurfluh and Siddall (1971) for the optically active (+)-*cis*-2-(1′-methyl-2′-isopropenyl-cyclobutyl) ethanol (grandisol) is described in Scheme 6.13.

6.3.3.2 *Japanese beetle*

The sex pheromone of the Japanese beetle, *Popillia japonica* Newman, is (*R*)-(*Z*)-5-(1-decenyl)dihydro-2-(3H)-furanone. Since the optically active *R* isomer has roughly 30 times the attractant effect of the racemic compound, this is one of the sex pheromones for which asymmetric synthesis is particularly important. An example of its preparation by Tumlinson *et al.* (1979) is given in Scheme 6.14.

6.3.3.3 *Spruce bark beetle*

Spruce bark beetle, *Ips typographus* L., is a serious threat to forests in northern Europe. Large-scale mass trapping is carried out there using a mixture of ipsdienol, *cis*-verbenol and methylbutenol. A method of synthesizing ipsdienol due to Ohloff and Giersch (1979) is given in Scheme 6.15. Of the enantiomers, (*S*)-(−)-ipsdienol is considered to have the strongest attractant effect.

6.4 MANUFACTURERS AND PRICES

At present, sex pheromones are mass-produced and marketed by Shin-Etsu, Agrisense and Bedoukian. A number of other firms produce pheromones in small quantities for monitoring purposes. The manufacturers are listed in Table 6.1.

The world demand for pheromones is less than 2 tons per year at present, so no manufacturer has a plant dedicated to producing them commercially. Even Shin-Etsu Chemical and the other major producers carry out sex pheromone synthesis using liquid ammonia reactors and/or acetylene reactors originally owned for other purposes. While the market price of the product is still comparatively high, the price of gossyplure has been reduced from $2 per gram formerly to less than 50¢ per gram (Table 6.2). Since this price includes the cost of delivery from Japan, taxes, dealers' commissions and a profit margin, it may reasonably be inferred that the current production cost is at least 30% below the market price, or less than 40¢ per gram. If demand reaches 100 tons per year, the market price can be expected to fall another 40% to under 30¢ per gram.

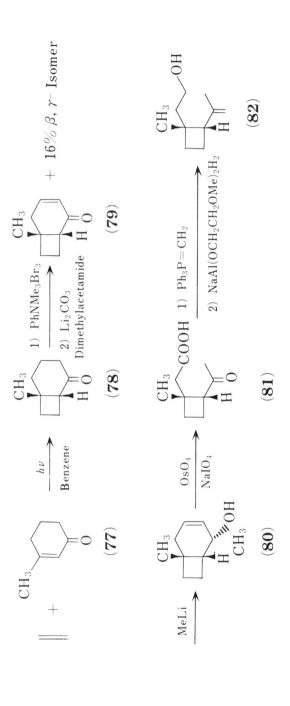

Scheme 6.13 (+)-*cis*-2-(1'-Methyl-2'-isopropenyl-cyclobutyl) ethanol (grandisol). A photochemical reaction of 3-methyl-2-cyclohexenone (**77**) and ethylene gives 6-methyl bicyclo[4,2,0]octan-2-one (**78**) with 55% yield. The compound (**78**) is brominated with PhNMe₃Br₃, and the ensuing dehydrobromination reaction. After quantitative methylation by alkylation with MeLi, the unsaturated compound (**79**) is given by the ensuing dehydrobromination reaction. After quantitative methylation by alkylation with MeLi; the unsaturated bonds are oxidatively cleaved with OsO₄–NaIO₄ to obtain the keto acid derivative (**81**). Next, the Wittig reaction of **81** with Ph₃P=CH₂ followed by reduction gives (+)-*cis*-2-(1'-methyl-2'-isopropenyl-cyclobutyl) ethanol (**82**) in 80% yield.

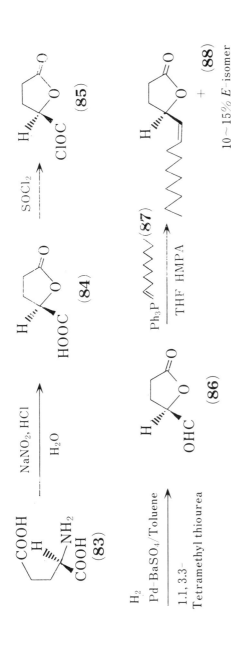

Scheme 6.14 (R)-(Z)-5-(1-Decenyl) dihydro-2(3H)-furanone. (R)-(−)-Glutamic acid (**83**) is dissolved in water, and HCl and NaNO₂ are added. This reaction gives a 70–80% yield of a lactone compound (**84**), which is reacted with SOCl₂ to produce the acid chloride derivative (**85**). This is reduced in toluene, using Pd-BaSO₄ as a catalyst, to produce the corresponding aldehyde (**86**). The Wittig reaction in THF–HMPA between this aldehyde (**86**) and the phosphorane compound (**87**) prepared from n-BuLi and the phosphonium salt of nonylbromide then yields (R)-(Z)-5-(1-decenyl) dihydro-2(3H)-furanone (**88**), having a Z/E ratio of 9/1 (optical rotation $[\alpha]_D^{26}$ of −69.6 °), with 42% yield.

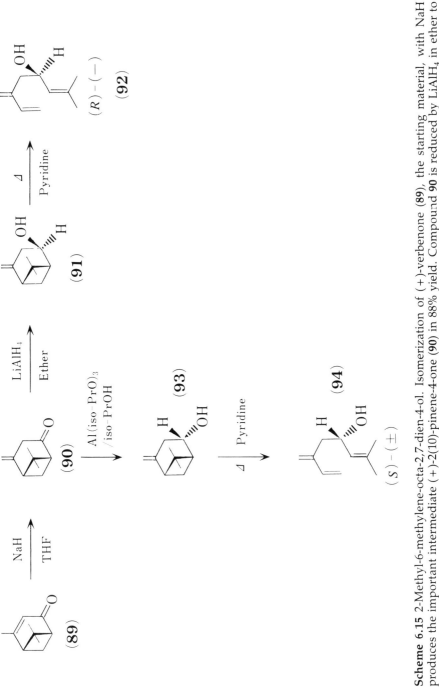

Scheme 6.15 2-Methyl-6-methylene-octa-2,7-dien-4-ol. Isomerization of (+)-verbenone (**89**), the starting material, with NaH produces the important intermediate (+)-2(10)-pinene-4-one (**90**) in 88% yield. Compound **90** is reduced by LiAlH₄ in ether to produce (+)-2(10)pinene-4-ol (**91**), which is pyrolized in the presence of pyridine to yield (R)-(−)-2-methyl-6-methylene-octa-2,7-dien-4-ol (**92**). Compound **90** is alternatively reduced by aluminum isopropoxide in iso-PrOH to produce trans-(−)-2(10)-pinene-4-ol (**93**), which is pyrolized to yield (S)-(+)-2-methyl-6-methylene-octa-2,7-dien-4-ol (**94**).

TABLE 6.1 Pheromone manufacturers.

Firm	Products
Shin-Etsu Chemical	Largest product line, mainly lepidopteran pheromones
Agrisense	(Z)-11-hexadecenal, (E/Z)-9-dodecenyl acetate, gossyplure
Orsynex	Bought Albany facilities; lepidopteran pheromones
Zoecon	Synthesizes for monitoring
Istituto Guido Donegani	Synthesizes for monitoring
BASF	Synthesizes for in-house disruption work
International Pheromones	Synthesizes *Ips* and other pheromones
Bedoukian Research	Boll weevil and lepidopteran pheromones

TABLE 6.2 Market prices of gossyplure in 1985

Quantity	Shin-Etsu market price to USA (¢/g)	Estimated production cost (¢/g)*
< 1 kg	100	70
1 kg and up	80	56
10 kg and up	66	46
100 kg and up	57	40
500 kg and up	51	36
1000 kg and up	49	34

* The production cost is estimated by multiplying the market price by 0.7.

In considering the economics of pheromones, the required degree of purity is, along with volume, another important factor, since the low levels of purity required when they are used for communication disruption should make it possible to supply pheromones at even lower prices.

ABBREVIATIONS USED IN SCHEMES

HMPA (Scheme 6.3) hexamethylphosphoramide
DMF (Scheme 6.4) N,N-dimethylformamide
THF (Scheme 6.4) tetrahydrofuran
DMSO (Scheme 6.5) dimethylsulfoxide
THP (Scheme 6.6) tetrahydro-4H-pyran-2-yl
ODDA (Scheme 6.7) octadecadienyl acetate
p-TsOH (Scheme 6.8) *para*-toluenesulfonic acid
DDDOL (Scheme 6.8) dodecadienol
m-CPBA (Scheme 6.9) *meta*-chloroperbenzoic acid

REFERENCES AND PATENTS

Key to Patents

U.S.   USA patent
G.B.   UK patent
D.E.   German patent
Ger. off. German Offenlengungsschrift
FR.   French patent
E.P.   European patent
J.P.   Japanese patent
J. Kokai Japanese patent publication
CA.   Canada patent

Anderson, R. J. and Henrick, C. A. (1976). 4,8-Tridecadien-1-ol,4,*cis*,8,*cis*,*trans*; U.S.3.953.532. Synthesis of pink bollworm sex pheromone; U.S.3.919.329. Synthesis of tridecadienoic esters; U.S.3.987.073. Synthesis of 8-oxo-4-*cis*-octenoic acid ester; U.S.3.989.729.
Anderson, R. J. (1979). Intermediates for insect pheromone; U.S.4.158.096.
Anderson, R. J. and Henrick, C. A. (1978). Unsaturated carboxylic acids and esters; U.S.4.264.518.
Anderson, R. J. and Henrick, C. A. (1981). Intermediates for a sex pheromone for yellow scale; U.S.4.276.363.
Bishop, C. E. and Morrow, G. W. (1983). Synthesis of Z,Z-11,13-hexadecadienal-sex pheromone of navel orange worm; U.S.4.400.550.
Burkholder, W. E., Silverstein, R. M., Rodin, J. O. and Gorman, J. E. (1970). Sex attractant of the black carpet beetle; U.S.3.501.566.
Carney, R. L. and Henrick, C. A. (1980a). (11Z,13Z)-Hexadecadien-1-ol and derivatives; U.S.4.198.533.
Carney, R. L. and Henrick, C. A. (1980b). (11Z,13Z)-11,13-Hexadecadiyn-1-ol and (11Z,13Z)-11,13-hexadecadien-1-ol and trimethylsilyl ethers thereof; U.S.4.228.093.
Carney, R. L. (1982) Insect pheromone; U.S.4.357.474. Descoins, C. and Samain, D. (1977). Nouveau procede de synthèse stereoselective de pheromones sexuelles d'insectes; FR.2.341.548.
Descoins, C. E. and Henrick, C. A. (1974). Synthesis of 1-bromo-*trans*-3,*trans*-5-heptadiene; U.S.3.825.607.
Descoins, C. and Samain, D. (1977). Nouveau procède de synthèse stéreoselective de phéromones sexuelles d'insectes; FR.2.341.548.
Eiter, K. (1974). Process for the preparation of 9-*cis*-tricosene; U.S.3.851.007.
Friedman, L. and Chanan, H. H. (1976). Intermediate for gossyplure, the sex pheromone of the pink bollworm; U.S.3.996.270.
Goto, G. (1975). 9Z,11E-Tetradecadien-1-ol esters or ethers, compounds are sex attractants for insect; *J. Kokai* 50-53312.
Hennart, C. and Martin, G. (1981). Process for the preparation of higher alkenes; U.S.4.249.029.
Henrick, C. A. and Siddall, J. B. (1974a). Synthesis of codling moth attractant; U.S.3.783.135.
Henrick, C. A. and Siddall, J. B. (1974b). Synthesis of codling moth attractant; U.S.3.856.866.

Henrick, C. A. and Siddall, J. B. (1974c). Synthesis of codling moth attractant; U.S.3.818.049.

Henrick, C. A. and Siddall, J. B. (1976). Synthesis of codling moth attractant; U.S.3.943.157.

Henrick, C. A. (1977). The synthesis of insect sex pheromones. *Tetrahedron*, **33**, 1845.

Hoechst, AG. (1975). Verfahren zur Herstellung von Heptacosen-(13) und Nonacosen-(14) bzw deren Gemischen; *Ger. off.* 2.355.534.

Holan, C. (1975). Process for the preparation of *cis*-8-dodecen-1-ol acetate; U.S.3.906.035.

Ishihara, T., Yamamoto, A., Takasaka, N. and Oshima, M. (1980). 1-Halo-4-decene compounds; U.S.4.324.931, E.P.29.575 (1981).

Ishihara, T., Yamamoto, A. and Taguchi, K. (1981). ω-Halo-1-bromo-alkyne compounds. Useful intermediates for sex pheromone; *J. Kokai* 56-166129.

Ishihara, T., Yamamoto, A. and Tagichi, K. (1982). Method for the preparation of *cis*-11-hexadecen-1-yne; E.P.44558, U.S.3.783.135 (1981).

Ito, S., Saito, I., Hatada, K. and Asano, T. (1983). 6- or 7-alkyne-11-one production from 3-oxo carboxylic ester- and halogenated alkyne, used as intermediate for sex pheromone for the douglas fir tussock moth; *J. Kokai* 58-134047.

Katsuki, T. and Sharpless, K. B. (1982). Asymmetrical synthesis; E.P.46.033.

Koshihara, T., Yamada, I., Takahashi, N. and Ando, N. (1984). Attracting imago of male cabbage web-worm using sex pheromone (E,E)-11,13-hexadecadienal; J.P. 59-40801.

Labovitz, J. N. and Henrick, C. A. (1976). Synthesis of non-4-en-6-ynoic acid ester; U.S.3.985.813. Sulfonate esters of non-4-en-6-yn-1-ol; U.S.3.994.896. Synthesis of 1-bromonon-4-en-6-yne; U.S.4.014.946.

Mori, K. and Uchida, M. (1981). Unsaturated cyclopropyl derivatives — useful as intermediate in pheromone production; J.P.56-37988.

Mori, K. and Tominaga, M. (1976). Sex-pheromone active 7,11-hexadecadienyl acetate preparation from E.G. 1-bromo-heptene and propargyl magnesium halide; *J. Kokai* 51-8204.

Mori, K. (1980). Z-10-Alkenal — useful as intermediate for synthesis of sex pheromone of *Synanthedon hector* Bulter; *J. Kokai* 55-136239.

Mori, K. (1981). The synthesis of insect pheromones. In *The Total Synthesis of Natural Products* Vol. 4, pp. 1–183. New York, John Wiley.

Muehowski, J. M. and Venuti, M. C. (1980). Preparation of unsaturated aliphatic insect pheromones using cyclic phosphonium ylide; U.S.4.296.042.

Negishi, T., Asano, M. and Uchida, M. (1984). 2,6-Dimethyl-1,5-heptadienyl acetate — having sex pheromone activity, and useful for attracting insects; J.P. 59-21853.

Ohloff, G. and Giersch, W. K. (1979). Process for the preparation of ipsdienol; U.S.4.157451, D.E.2.750.604 (1978).

Passardo, P., Cassani, G. and Piccardi, P. (1979). *cis*-*cis* and *cis*-*trans* Aliphatic conjugated diene preparation by reaction of alkynyl-alkene with alkyl magnesium halide and reduction with stereospecific reagent; *J. Kokai* 54-59205.

Research Corp. (1975). Preparation et utilisation d'agents attractifs pour la destruction d'insectes les plants de tabac; FR.2.267.298.

Roelofs, W., Kochansky, J. and Carde, R. (1974a). *trans*-7,*cis*-9-Dodecadien-1-yl acetate; U.S.3.845.108.

Roelofs, W., Comeau, A. and Hill, A. (1974b). *trans*-8,*trans*-10-Dodecadin-1-ol as an attractant; U.S.3.852.419, CA.961-862 (1975).

Roelofs, W. (1975). Sex-attraktans fur die mannchender Tabak-Motte; *Ger. off.* 2.515.371.

Rossi, R. (1981). Dienoic sex pheromones. *Tetrahedron*, **37**, 2617.

Samain, D., Descoins, C. and Kunesch, G. (1980). Process for the stereospecific preparation of sexual pheromone; U.S.4.189.614.

Samain, D., Kunesch, G. and Descoins, C. (1979). Conjugated diene alcohol sexual pheromone preparation by Grignard reaction giving high *trans-trans* stereoselectivity; *J. Kokai* 54-119406.

Szantay, C., Novak, L., Toth, M., Balla, J. and Stefko, B. (1979). Composite insect attractant for male cabbage moths and a process for preparating its active agents; U.S.4.243.660.

Szantay, C., Novak, L., Baan, B., Kistamas, A., Jurak, F. and Ujuray, I. (1982). A process for the preparation of 11-dodecen-1-ol and derivatives; D.E.3.142.114, G.B.2.085.881.

Szantay, C., Novak, L., Kistamas, A., Majoros, B., Ujvary, I., Jurak, F. and Poppe, L. (1983). Z-3,7-dimethyl-2,7-octadien-1-yl propanoate. Economic preparation from nerol for use as sex pheromone for San José scale; *J. Kokai* 58-110541.

Taguchi, K., Yamamoto, A., Ishihara, T., Takasaka, N. and Shimizu, H. (1981). Preparation of sex pheromone, *cis*-9-tetradecenyl acetate from *cis*-undec-6-enyl chloride in high yield and purity; E.P.32.396.

Traas, P. C. (1979). Process for the preparation of bollweevil sex pheromone components; U.S.4.152.355.

Tumlinson, J. H. (1974). Boll weevil sex attractant; U.S.3.813.443.

Tumlinson, J. H., Klein, M. G. and Doolittle, R. E. (1979). Sex pheromone produced by the female Japanese beetle: specificity of male response to enantiometers; U.S.4.179.446.

Uchida, M. (1980). *cis*-3-Hexene-diol derivatives with antibacterial and antagonist activity, useful as intermediates for pheromone; *J. Kokai* 55-33460.

Uchida, M., Matsui, M. and Mori, K. (1983). (1)-Hexadecynone-(6) used as an intermediate for sex hormone *cis*-(6)-heneicosenone-(11); J.P.58-31330.

Uchida, M. and Nakagawa, K. (1984). Preparation of 1,2-epoxy derivatives. An intermediate for Z-11-hexadecenal; *J. Kokai* 59-46513.

Uchida, M. and Nakagawa, K. (1980). Preparation of 1,2-epoxy derivatives. An intermediate for Z-11-hexadecenal; *J. Kokai* 55-157578.

Yamamoto, A., Ishihara, T. and Taguchi, K. (1981). Pheromone *cis*-alkenyl acetate preparation by reacting 3-alkenyl-magnesium chloride with dibromo-alkane and the resulting *cis*-alkenyl bromide with metal acetate; E.P.38.052.

Yoshida, S. (1983). Highly selective preparation of *cis*-tricosene and *cis*-9-heneicosene by coupling Grignard reagent obtained. From metallic magnesium and 1-halo-*cis*-9-octadecene with *n*-amyl or *n*-propyl halide; J.P.58-4007.

Yushima, K. (1976). Tetradeca-9,11-dienyl-Acetat und hieraus hergestelltes Insekten anzienehendes Praparat; *Ger. off.* 2.406.259.

Zurfluh, R. C. and Siddall, J. B. (1971). *cis*-1-methyl-2-isopropenyl-cyclobutyl acetic acid; U.S.3.701.800.

Insect Pheromones in Plant Protection
Edited by A. R. Jutsum and R. F. S. Gordon
©1989 John Wiley & Sons Ltd

7

Plastic Laminate Dispensers

Alberto R. Quisumbing and Agis F. Kydonieus

Hercon Laboratories Corporation, South Plainfield, New Jersey 07080, USA

7.1 INTRODUCTION

The effects of pheromones and other behaviour-modifying chemicals on insect behaviour have been described in Chapter 2, and the various uses of such molecules have been addressed in Chapters 3, 4 and 5. Each of these chapters has highlighted the fact that specialized formulation is necessary to exploit fully these volatile and often unstable chemicals.

The objective of this chapter is to review one of these formulations—the Hercon, controlled-release, multilayered dispensing system; to examine the various factors affecting the release rates of pheromones from these laminates; and to discuss the various pheromone formulations and application equipment currently used in pest management programmes.

7.2 THE HERCON CONTROLLED-RELEASE DISPENSING SYSTEM

The basic dispenser for pheromones consists of the active ingredient implanted and protectively sealed in a layer between outer plastic barrier layers (Fig. 7.1). The specially formulated inner layer serves as the reservoir of the insect pheromone, which then migrates continuously, owing to imbalance of chemical potential, through one or more initially inert, permeable barrier layers. Once at the exposed surface, the active ingredient becomes available for biological action and is eventually removed by volatilization, ultraviolet light degradation, oxidation, acid or alkaline hydrolysis, or mechanical removal by wind, rain, insects, humans or other agents.

The Hercon dispenser manufacturing process was described previously (Kydonieus and Quisumbing, 1980). To make the laminated pheromone

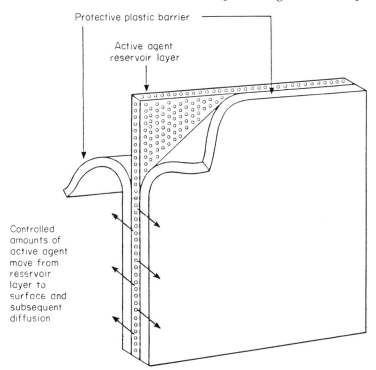

Fig. 7.1 Schematic illustration of Hercon multilayered dispenser, showing inner reservoir layer containing the active ingredient (e.g. insect behaviour-modifying chemical) between protective barrier layers

dispenser, a sheet of polymeric film is coated on one side with an attractant-impregnated polymeric mixture, and then overlayed with a second sheet of polymeric film. This three-layered structure is then allowed to set under suitable conditions of heat and pressure until an integral, firmly bonded product is obtained.

The laminated sheet is then processed into its final form, e.g. by slitting it into strips, ribbons or 'squares'; or die-cut into the smaller 'flakes' or confetti.

Some of the insect pests for which pheromones have been incorporated into plastic laminate formulations are described in Section 7.5, but in addition to pheromones other chemicals which have been dispensed by multilayered laminates include drugs, e.g. nitroglycerin from transdermal patches for angina patients (Kydonieus *et al.*, 1984); antibacterial agents (Hyman *et al.*, 1972; Kydonieus *et al.*, 1976c); insecticides (Kydonieus *et al.*, 1976a,b,e; Quisumbing *et al.*, 1978); and fragrances and perfumes (Kydonieus, 1978; Kydonieus *et al.*, 1976b).

The multilayered dispensers are available in various forms, such as rolled ribbons of varying widths and lengths, strips (e.g. the 2.54 cm × 11.16 cm Insectape insecticide dispensers) or 'squares' (e.g. the 2.54 cm × 2.54 cm or 1.27 cm × 1.27 cm pheromone lures for traps), or uniformly diced flakes, confetti or 'chips' which are 0.63 cm × 0.63 cm pheromone dispensers often broadcast from aircraft equipped with specially built applicators.

Quisumbing *et al.* (1976) successfully tested an insecticide (resmethrin) dispenser which had attached to it a separate controlled release dispenser of vanillin, a food attractant of the housefly, *Musca domestica* (L.). More recently, multilayered dispensers have been produced with the pheromone and insecticide within the same reservoir for control of the pink bollworm, *Pectinophora gossypiella* (Saunders), and *Periplaneta* cockroach species (Bell *et al.*, 1984).

Dispensers of insect sex or aggregation pheromones which are to be used as lures in traps or as manually applied pheromone point sources in disruption of mating programmes are called Luretape. Examples include products used as lures to trap the boll weevil, *Anthonomus grandis* Boheman, the gypsy moth, *Lymantria dispar* L., the Japanese beetle, *Popillia japonica* (Newman), the codling moth, *Laspeyresia pomonella* (L.), the tobacco budworm, *Heliothis virescens* (Fabricius), and the corn earworm, tomato fruitworm or bollworm, *Heliothis zea* (Boddie).

The aerially applied confetti or flakes are called Disrupt and formulations available include those for the gypsy moth, the pink bollworm, the artichoke plume moth, *Platyptilia carduidactyla* (Riley), the western pineshoot borer, *Eucosma sonomana* Kearfott, and the spruce budworm, *Choristoneura fumiferana* (Clemens).

Prior to aerial application, the confetti is mixed with an inert acrylic- or polybutene-based adhesive, which allows the tiny dispensers to adhere on contact with foliage. Formulations which contain both the pheromone and insecticide are called Lure N Kill products and utilize pyrethroids such as permethrin.

7.3 FACTORS AFFECTING PHEROMONE RELEASE RATES

The mathematics affecting transport and the various factors that influence the release rates of chemicals dispensed by multilayered laminates have been discussed previously (Kydonieus, 1977; Kydonieus and Quisumbing, 1980; Quisumbing and Kydonieus, 1982). In addition, other publications dealt with the emission rates of pheromone components from specific products, e.g. dispensers of pheromones for the gypsy moth, corn earworm, tobacco budworm and citrus scale insects (Leonhardt and Moreno, 1982; Plimmer and Leonhardt, 1983).

Basically, the concentration of the implanted or stored behaviour-modifying chemical (e.g. pheromone) and the dispenser's composition and/or construction of the plastic layer components control the release rate of the pheromone.

Solution–diffusion membranes, i.e. non-porous, homogeneous polymeric films, are used as barrier films in the multilayered controlled-release system. The pheromone is able to pass through the membrane material, which does not have any holes or pores, by absorption, solution and diffusion down a gradient of thermodynamic activity, and final desorption. The permeation process is governed primarily by Henry's law and Fick's first law (Crank and Park, 1968; Richards, 1973). Equations for diffusion of chemicals under both 'steady-state' and 'unsteady-state' conditions have been described in Kydonieus (1977) and Kydonieus and Quisumbing (1980).

Molecular and structural factors control the release rate of pheromones from the laminated dispensers. For a given combination of polymer structure and active agent, where energy to free rotations, free volume and intermolecular attractions are constant, the two most important parameters that regulate rate of transfer are (a) pheromone concentration in the reservoir and (b) the thickness of the barrier membrane.

Other factors that also affect the transport of pheromones from the inner reservoir layer through the barrier membranes are (c) polymer stiffness of the reservoir and/or barrier membranes, (d) codiffusants, (e) the molecular weight of the diffusant, i.e. the pheromone, (f) the chemical functionality of the pheromone, and (g) environmental factors, such as temperature and air flow (wind). The size of the dispenser itself and how this pheromone carrier is applied in the field will also affect pheromone release.

7.3.1 Pheromone concentration in the reservoir

In general, the mass of pheromone released within a given time from a dispenser with a higher initial pheromone concentration will be greater than the mass released from a similarly constructed dispenser with less pheromone. Zero-order release rates will be obtained if the concentration gradient across the barrier remains constant and there is a large reservoir of chemicals in the dispenser. However, since pheromones are eventually exhausted or the concentration in the reservoir will fall continuously with time, first-order delivery is often observed. In the latter case, the amount of pheromone released will also vary as a function of time.

Bierl-Leonhardt (1982) compared the dispensers of disparlure, the adult gypsy moth pheromone, and determined the half-life after ageing the 5 mm × 25 mm units outdoors. Dispensers that contained an initial load of 4.3 mg disparlure had a half-life (depleted half of the pheromone load), of 12 weeks. Dispensers with 1.2 and 0.22 mg disparlure per 5 mm × 25 mm unit had half-lives of 8 and 6.5 weeks, respectively. All dispensers were formulated using 2 mil (0.05 mm) polyvinyl chloride (vinyl) barrier films.

7.3.2 Thickness of barrier membranes

The mass of the pheromone released is inversely proportional to the thickness of the barrier films used. Film thickness is expressed in 'mil' where 1 mil = 0.001 in or 0.0254 mm. Fig. 7.2 illustrates how greater amounts of dodecenyl acetate and hexadecyl acetate are released through dispensers with 2 mil vinyl layers than through laminates with either 13 mil or 20 mil vinyl barrier films.

Similar results were obtained from 3 mm × 3 mm disparlure dispensers constructed using either 2 mil or 5 mil vinyl films to 'sandwich' the pheromone reservoir. When aged in the greenhouse, laminates with 2 mil vinyl exhausted half of its original disparlure load in 29 days, while those with the thicker 5 mil vinyl barriers had a half-life of 54 days (Bierl-Leonhardt, 1982).

7.3.3 Polymer stiffness of the reservoir and barrier membranes

The amount of pheromone that diffuses from the reservoir through the outer membranes will decrease as polymer stiffness increases. During diffusion,

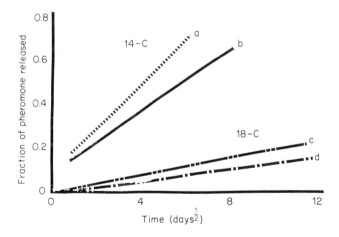

Fig. 7.2 Effect of barrier membrane thickness on release rates of 14-C dodecenyl acetate (a and b) and 18-C hexadecyl acetate (c and d). Release rates a and c were obtained from dispensers constructed using 2 mil polyvinyl chloride barrier films, while b and d were obtained from laminates with either 20 mil or 13 mil vinyl barrier layers, respectively

the pheromone molecule reorients several segments of the reservoir or barrier polymer chain in order to migrate. However, with stiff polymers (e.g. those that are glassy or highly crystalline), pheromone movement will be slower because the polymer segments will be more difficult to rearrange.

Kydonieus (1977) demonstrated that chemical migration is fastest through flexible polyvinyl chloride films, followed by rigid polyvinyl chloride, acrylic, polypropylene and nylon, and least through non-permeable polyester.

Plimmer and Leonhardt (1983) studied the emission rates of two aldehydes, (Z)-11-hexadecenal and n-tetradecanal, from four differently constructed dispensers which were aged in a greenhouse in Maryland (mean temperature approx. 32 °C) and outdoors in Texas (maximum daytime temperature 20–25 °C). The four constructions for aldehyde components which were evaluated were: (a) flexible vinyl on both sides, (b) rigid vinyl on both sides, (c) flexible vinyl on one and polyester (Mylar) on the other, and (d) acrylic on both sides.

Dispensers with acrylic barrier films provided the longest half-lives, while the flexible vinyl dispensers released the chemicals the fastest. Using Mylar on one side also reduced the emission rate of the dispenser, since Mylar, being non-permeable to pheromones, forced the diffusion to occur only through the other barrier (flexible vinyl) and from the dispenser's perimeter or edges.

Generally, the dispensers exposed outdoors (where temperatures were relatively higher) had shorter half-lives than samples aged in the greenhouse. The 16-carbon (Z)-11-hexadecenal dispensers also outlasted the n-tetradecanal, especially in the greenhouse where conditions (e.g. temperature and air flow) remained constant. Outdoors, the effects of environmental factors seemed to be more significant on the unsaturated 16-carbon aldehyde than on the saturated 14-carbon n-tetradecanal.

In addition to flexible and rigid vinyl barrier films, acrylic and polyester or Mylar, the following polymers may also be used as protective layers of the laminate: cellophane, nylon, polycarbonate, polyethylene, polypropylene and urethane.

7.3.4 Codiffusants

Although seldom used in controlled-release dispensers of pheromones, codiffusants are important additives that can affect diffusivity and permeation rate by altering the structure of barrier films—swelling, softening and/or dissolving the polymer matrix. The presence of phenol or ethylene glycol phenyl ether (which are codiffusants) in the dispenser either as an additive, or as a behaviour-modifying chemical, must be studied since their presence can increase emission rates, thus shortening the dispenser's usefulness.

7.3.5 Molecular weight of the diffusant

Molecular weight is inversely related to diffusivity. Graham's law states that diffusion is inversely proportional to the square root of the molecular weight of the diffusant, i.e. as the molecular weight of the behaviour-modifying chemical increases, its release rate decreases.

The study reported by Plimmer and Leonhardt (1983) which showed the *n*-tetradecanal dispensers having a shorter half-life than the (Z)-11-hexadecenal samples, especially under greenhouse conditions, demonstrated the effect of molecular weight on release rates.

In a multilayered dispenser, pheromone migration and diffusion occur when the pheromone molecule successfully reorientates several segments of the polymer chain to allow its passage from site to site. A pheromone with a low molecular weight will diffuse faster since it will have fewer segments to rearrange than one with a high molecular weight.

The effects of pheromone concentration, barrier membrane thickness and molecular weight of pheromones on release rates were observed in a 1979 study involving the large-scale application of gossyplure and virelure pheromones for control of pink bollworm, *Pectinophora gossypiella*, and tobacco budworm, *Heliothis virescens*, in cotton (Henneberry *et al.*, 1981a,b). Gossyplure, a 1 : 1 mixture of (Z,Z)- and (Z,E)-isomers of 7,11-hexadecadienyl acetate (Hummel *et al.*, 1973; Bierl *et al.*, 1974), was formulated using flexible vinyl films as protective barriers; pheromone load was 10 mg gossyplure a.i. per 2.54 cm^2 of laminated product. Virelure, a 16 : 1 mixture of (Z)-11-hexadecenal and (Z)-9-tetradecenal, two components of the tobacco budworm pheromone (Roelofs *et al.*, 1974; Tumlinson *et al.*, 1975), was dispensed from rigid vinyl plastic laminates; pheromone load was 15 mg virelure a.i. per 2.54 cm^2 of product.

After 28 days of exposure outdoors and unsheltered in Arizona, both gossyplure and virelure dispensers were analysed for residual content. The tests showed that 86% of the virelure load was released (compared to 78% of the gossyplure) despite the virelure dispenser being constructed with a rigid and thicker (16 mil rigid vinyl) polymer. The greater loss of virelure may be due to the chemicals' lower molecular weights and ready degradability.

7.3.6 Chemical functionality of the pheromone

The 'like dissolves like' rule is equally applicable to polymers. Dissolution of the polymer matrix by the diffusing molecules of the behaviour-modifying chemical is important in the transport process because the distribution coefficient is increased by dissolution. Fig. 7.3 shows the varying effects of chemical functionality on the release rates of three 16-carbon compounds: epoxyhexadecane, hexadecanone and hexadecenal. Because of its epoxy

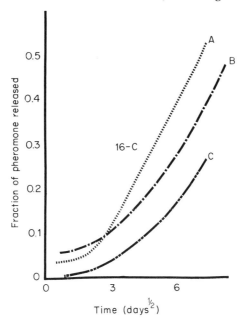

Fig. 7.3 Effect of chemical functionality of three 16-carbon chemicals (namely, A, epoxyhexadecane, B, hexadecanone, and C, hexadecenal) on their release rates from controlled-release laminated dispensers

group, more epoxyhexadecane was released from its dispenser compared to those of hexadecanone and hexadecenal.

7.3.7 Environmental factors

In addition to the preceding variables (sections 7.3.1 to 7.3.6) directly arising from the chemical nature of the diffusant and the specific dispenser's construction, environmental factors such as temperature, wind or air movement, and humidity will also influence pheromone emission rates.

Generally, pheromone release will increase as temperature and/or air flow increase. Bierl-Leonhardt (1982) reported that an increase of 16.5 °C in temperature, e.g. from 21.1 to 37.6 °C (70 to 100 °F), not only increases the vapour pressure of most pheromones by 200% but also affects the physical parameters of the dispensers' polymeric barrier films. The release rates of disparlure increased from 0.28 to 0.84 μg h^{-1} when oven temperatures were raised from 28 to 38 °C (Bierl *et al.*, 1976).

Bierl-Leonhardt *et al.* (1979) provided evidence showing that disparlure loss increased in proportion to air flow. As the gypsy moth pheromone was removed from the dispenser surface by wind, more disparlure moved from the reservoir to maintain the disparlure concentration on the surface, until it was likewise removed. Replenishment will continue until the disparlure concentration in the reservoir is significantly reduced.

The observations that environmental factors influence the release rates of the behaviour-modifying chemicals underline the need to study those formulations that performed well in the laboratory or greenhouse under actual field conditions where the effects of sun, rain, humidity and wind can be determined.

Since release rate is also influenced by the ratio of the dispenser's perimeter or edge to its total surface area, the smaller confetti or flake forms would lose the pheromone faster than larger dispensers made of the same construction. This 'edge effect' is the reason why hand-applied ribbon and strip dispensers (Luretape) containing the pheromones of the gypsy moth, peachtree borer, *Synanthedon exitiosa* (Say), and European grape moth, *Eupoecilia ambiguella* Hubner, tend to outlast the confetti (Disrupt) versions. In preliminary gypsy moth mating disruption work, when higher amounts of disparlure were desired to be released during the insect's peak flight period, smaller 5.7 mm^2 confetti was applied instead of the previously tested but larger 9 mm^2 units (Leonhardt and Moreno, 1982).

Confetti dispensers are applied with an adhesive coating so they will cling to foliage in the insect's mating zone, thus emitting the behaviour-modifying chemical where it is most effective. These stickers may also affect the product's duration of effectiveness.

During the early pink bollworm mating disruption research, prototype applicators required the gossyplure dispensers to be premixed with the sticker before loading. It was noted that if considerable time elapsed between pre-mixing and actual application, gossyplure would leach out or be absorbed by the sticker. Further, since the amount of sticker coating the dispenser often varied from flake to flake, release rates varied significantly owing to a thick adhesive layer reducing the pheromone release.

The development of Hercon's new spraying equipment significantly reduced problems of product effectiveness arising from non-uniform adhesive coating or extended periods between mixing and application. The leaching effect, however, must still be considered, particularly with other controlled-release formulations which require a sticker but do not provide adequate protection of the pheromone within the product.

Depending on the time of application and the dispenser construction, the sticker may also be utilized to improve product effectiveness. In the winter of 1982, the effect of the sticker Phero-tac was desirable to ensure that large amounts of (Z)-11-hexadecenal would be in the air when the artichoke plume moths were mating. Releasing 73% of the pheromone from small, Phero-tac-coated confetti is understandably more efficient and less wasteful than using large, uncoated dispensers which will retain 94% of the chemical even after 2 weeks.

The current commercially available artichoke plume moth disruptant product is formulated using 8 mil flexible vinyl protective layers. Each application is effective for approximately 14 days regardless of whether it is made to suppress the summer or winter/spring generations of the insect.

7.4 AERIAL SPRAYING EQUIPMENT

An aerial dispensing apparatus was specially designed for various sizes of multilayered flakes or confetti (Kodadek and Welch, 1984). Fig. 7.4 is a schematic illustration of the applicator showing the major sections of the equipment.

The apparatus is detachably mounted under the wing (A) of an aircraft. In practice, a set of two applicators—one on each wing—is used to apply confetti formulations for mating disruption of the pink bollworm, gypsy moth, artichoke plume moth, tobacco budworm, corn earworm and spruce budworm, *Choristoneura fumiferana*.

The apparatus itself consists of an aerodynamically styled adhesive chamber (B) secured to the forward end of the detachable support frame, and a hopper (E) which holds the confetti. The hopper is located at the rear portion of the frame and may be pulled towards the trailing edge of the wing for convenient loading of the dispensers. The hopper has a bottom opening which communicates with an electrically driven auger (G) for metering a supply of pheromone dispensers to the cylindrical mixing chamber (H). The

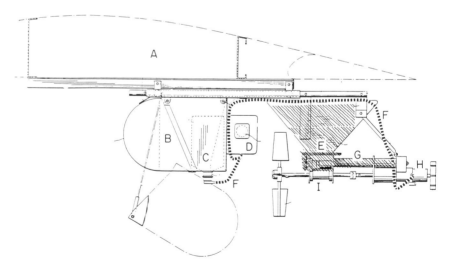

Fig. 7.4 Schematic illustration of patented Hercon aerial spraying equipment. See text for explanation

TABLE 7.1 List of insects whose behaviour-modifying chemicals have been studied in Hercon multilayered controlled-release dispensers (as of May 1988)

ORDER Family	Scientific name[1]	Common name	Reference[2]
I. COLEOPTERA Curculionidae	*Anthonomus grandis* Boheman	Boll weevil	Forey and Quisumbing (1987) Hardee *et al.* (1975) Johnson *et al.* (1976) Leonhardt *et al.* (1988) Lloyd *et al.* (1983) Marengo *et al.* (1987) Quisumbing *et al.* (1986) Rummell *et al.* (1987)
Scarabaeidae	*Popillia japonica* Newman	Japanese beetle	Klein (1981) Ladd and Klein (1982) Ladd and McGovern (1984)
Scolytidae	*Dendroctonus frontalis* Zimm. *Ips typographus* (L.)	Southern pine beetle Eight-toothed engraver beetle (spruce bark beetle)	Payne (1981) Bakke (1982) Bakke and Riege (1982) Lie and Bakke (1981)
	Scolytus multistriatus Marsham	Smaller European elm bark beetle	Cuthbert and Peacock (1979) Lanier (1979) Lanier and Jones (1985) Wollerman (1979)
II. DIPTERA Tephritidae	*Ceratitis capitata* (Wiedemann)	Mediterranean fruit fly	Leonhardt *et al.* (1984) Nakagawa *et al.* (1979) Rice *et al.* (1984)
III. HOMOPTERA Diaspididae	*Aonidiella aurantii* (Maskell) *Planococcus citri* (Risso) *Pseudococcus comstocki* (Kuwana)	California red scale Citrus mealybug Comstock mealybug	Leonhardt and Moreno (1982) Leonhardt and Moreno (1982) Leonhardt and Moreno (1982)
IV. LEPIDOPTERA Gelechiidae	*Pectinophora gossypiella* (Saunders)	Pink bollworm	Beasley and Henneberry (1984) Critchley *et al.* (1985) Henneberry *et al.* (1981a,b, 1982, 1984, 1986)

(Continued)

ORDER Family	Scientific name[1]	Common name	Reference[2]
	Sitotroga cerealella (Oliver)	Angoumois grain moth	Cogburn and Vick (1983)
			Vick et al. (1978, 1979)
Geometridae	Alsophila pometaria (Harris)	Fall cankerworm	Palaniswamy et al. (1986)
Lymantriidae	Lymantria dispar (L.)	Gypsy moth	Bailey (1983)
			Bailey and Kludy (1986)
			Beroza et al. (1974, 1975a,b)
			Bierl et al. (1976)
			Bierl-Leonhardt et al. (1979)
			Elkinton (1987)
			Leonhardt and Moreno (1982)
			Mastro et al. (1977)
			Plimmer and Leonhardt (1983)
			Plimmer et al. (1977, 1978, 1982)
			Schwalbe and Mastro (1988)
			Schwalbe et al. (1979, 1983)
			Webb et al. (1986, 1988)
Noctuidae	Orgyia pseudotsugata (McDunnough)	Douglas-fir tussock moth	Daterman and Sower (1977)
	Agrotis ipsilon (Hufnagel)	Black cutworm	Clement et al. (1981)
	Heliothis virescens (Fabricius)	Tobacco budworm	Hartstack et al. (1980)
			Hendricks (1982)
			Hendricks et al. (1977a,b, 1980, 1987)
			Henneberry et al. (1981a, 1982)
			Hollingsworth et al. (1978)
			Johnson (1983)
			Kydonieus et al. (1980)
			Leonhardt and Moreno (1982)
			Lopez and Witz (1988)
			Mitchell et al. (1976)
			Plimmer and Leonhardt (1983)
			Tingle and Mitchell (1982)
			Zvirgzdins and Henneberry (1983)
			Zvirgzdins et al. (1984)
	Heliothis zea (Boddie)	Bollworm, corn earworm, tomato fruitworm	Chowdhury et al. (1987)
			Hartstack et al. (1980)
			Shaver and Lopez (1982)
			Shaver et al. (1982)

Family	Scientific name[1]	Common name	References[2]
	Spodoptera frugiperda (J.E.S.)	Fall armyworm	Slosser *et al.* (1987), McLaughlin *et al.* (1981), Mitchell *et al.* (1976), Tingle and Mitchell (1978), Kehat *et al.* (1983)
	Spodoptera littoralis (Boisduval)	Egyptian cotton leafworm	
Olethreutidae	*Trichoplusia ni* (Hubner)	Cabbage looper	Hendricks *et al.* (1977a), Daterman (1982)
	Eucosma sonomana Kearfott	Western pineshoot borer	Sartwell *et al.* (1983), Sower and Daterman (1986), Sower *et al.* (1982)
Psychidae	*Grapholitha molesta* (Busck)	Oriental fruit moth	Gentry *et al.* (1976, 1980a, 1982)
	Thyridopteryx ephemeraeformis (Haworth)	Bagworm	Leonhardt *et al.* (1984)
Pyralidae	*Amyelois transitella* (Walker)	Navel orangeworm	Landolt *et al.* (1981, 1982), Gentry *et al.* (1980b)
Sesiidae	*Synanthedon exitiosa* (Say)	Peachtree borer	Snow and Gentry (1986), Yonce and Gentry (1982)
	Synanthedon pictipes (Grote and Robinson)	Lesser peachtree borer	Gentry *et al.* (1980b), Yonce and Gentry (1982)
Tortricidae	*Choristoneura fumiferana* (Clemens)	Spruce budworm	Alford and Silk (1983), Sanders and Silk (1982), Seabrook and Kipp (1986), Silk and Kuenen (1984), Wiesner and Silk (1982)
	Choristoneura occidentalis Freeman	Western spruce budworm	Daterman *et al.* (1985), Sower and Daterman (1986)
	Eupoecilia ambiguella Hubner	European grape moth	Arn *et al.* (1981), Neumann *et al.* (1986)
V. ORTHOPTERA			
Blattidae	*Periplaneta americana* (L.)	American cockroach	Bell *et al.* (1984)

[1] Scientific names are from *Common Names of Insects and Related Organisms* (Sutherland, 1978).
[2] References are publications on research which studied or utilized laminated dispensers of behaviour-modifying chemicals. Pheromones of over 30 more species have been formulated in controlled release laminates but studies have not been published as of May 1988.

sticker, inside plastic containers (C) stored in the adhesive chamber, is fed in metered amounts by an electrically driven pump (D) through hoses (F) to the mixing chamber (H).

A propeller-driven mixing blade is mounted in the mixing chamber to facilitate the coating of the individual dispensers with the adhesive, and for impelling the flakes through the open rear end of the mixing chamber. The spinner mounted adjacent to the open end of the chamber is designed to disperse the sticker-coated confetti behind the plane.

Electric power for the auger motor and adhesive pump motor is supplied through wiring and switches permanently installed in the aircraft. The pilot is in full control of the in-flight operation of the system through the use of cockpit-mounted control switches.

This application system has been used in the United States (Arizona and California for the pink bollworm mating disruption programmes; California for artichoke plume moth; Virginia and North Carolina for gypsy moth); in Canada for the spruce budworm; and in Egypt for pink bollworm (Von Ramm, 1986).

7.5 FIELD TRIALS WITH
LAMINATED PHEROMONE FORMULATIONS

Table 7.1 lists the insects whose pheromones were formulated in laminated dispensers which were field tested and evaluated in the field. Earlier research involving multilayered pheromone dispensers were reviewed previously (Kydonieus et al., 1976d; Kydonieus and Beroza, 1977, 1981; Kydonieus and Quisumbing, 1980; Quisumbing and Kydonieus, 1982).

In 1986, grandlure dispensers used in the south-eastern United States (North Carolina and South Carolina) boll weevil eradication programme were improved to withstand the high temperatures of the south-west (Arizona and California). This objective was obtained by using 16 mil thick barrier films, instead of the original 8 mil layers, and using a stiffer reservoir polymer (Quisumbing et al., 1986). Chemical analyses also showed that the four grandlure components in the laminate constantly remained at or close to the original attractive 30 : 40 : 15 : 15 ratio, while a cigarette filter alternative formulation was quickly depleted of the pheromone's aldehyde components. After 14 days at 32.2 °C, the grandlure component ratio was 29 : 37 : 18 : 16 in the multilayered dispenser and 41 : 46 : 6 : 7 in the cigarette filter (Quisumbing et al., 1986).

Mating disruption of a wild H. virescens population was obtained with the aerial application of laminated virelure dispensers at a dosage of 33 g a.i. ha^{-1} (Zvirgzdins et al., 1984). Approximately 13 000 virelure-emitting point sources, each 0.64 cm^2 in size and containing 2.5 mg active ingredient, were applied per hectare. Mating was reduced between 82 and 95%, and trap catch reduced 89 to 98%, for seven nights after treatment. Virelure confetti samples analysed 14 days after treatment showed that the laminates released 64% and 73% of the (Z)-11-hexadecenal and (Z)-9-tetradecenal (the two virelure components), respectively.

In 1984 and 1985, hand-applied dispensers of (Z,Z)-3,13-octadecadienyl acetate successfully controlled peachtree borer, *Synanthedon exitiosa*, infestations in peach orchards (Snow and Gentry, 1986). The pheromone was released from 2.54 cm^2 dispensers, each containing 43 mg a.i. A dosage of approximately 104 mg ha^{-1} day^{-1} was obtained by applying one pheromone dispenser per tree. The studies showed that treatment of wild hosts (plums and black cherries) adjacent to the peach orchards was not necessary to obtain control.

7.6 CONCLUSIONS

It is apparent that the laminated, multilayered dispensing system is an effective formulation for pheromones, particularly the unstable and ultraviolet-degradable aldehydes. The success of the laminated pheromone dispensers for boll weevil, spruce budworm and *Heliothis* species (all containing aldehydes) in field applications is encouraging. However, additional work is needed to develop products effective over a wide range of climates and for extended durations. The application of pheromones to control pest populations using mating disruption or the attracticide approach can become more popular and successful if, among other things, it is made price competitive with conventional insecticides. This can be achieved when pheromones are produced in much larger quantities (instead of the current minimal 'gram-quantity' requirements) and/or controlled release formulations will allow the products to be effective for periods longer than a conventional insecticide application.

7.7 REFERENCES

Alford, A. R. and Silk, P. J. (1983). Effect of pheromone-releaser distribution and release rate on the mating success of spruce budworm (Lepidoptera: Tortricidae). *J. Econ. Entomol.*, **76**, 774.

Arn, H., Rauscher, S., Schmid, A., Jaccard, C. and Bierl-Leonhardt, B. A. (1981). Field experiments to develop control of the grape moth, *Eupoecilia ambiguella*, by communication disruption. In *Management of Insect Pests with Semiochemicals — Concepts and Practice*, E. R. Mitchell (Ed.), Plenum Press, New York, p. 327.

Bailey, R. E. (1983). A Virginia eradication project using aerially applied disparlure. In *Proc. Nat. Gypsy Moth 1983 Annual Review*, M. Birmingham (Ed.), Albany, New York, p. 42.

Bailey, R. E. and Kludy, D. H. (1986). A Virginia gypsy moth eradication project using controlled release disparlure. In *Proc. 13th Intl. Symp. Controlled Release of Bioactive Materials*, I. A. Chaudry and C. Thies (Eds), Controlled Release Society, Inc., Lincolnshire, Illinois, p. 102.

Bakke, A. (1982). Mass trapping of the spruce bark beetle *Ips typographus* in Norway as a part of an integrated control program. In *Insect Suppression with Controlled Release Pheromone Systems* (Vol. II), A. F. Kydonieus and M. Beroza (Eds), CRC Press, Florida, p. 17.

Bakke, A. and Riege, L. (1982). The pheromone of the spruce bark beetle *Ips typographus* and its potential use in the suppression of beetle populations. In *Insect Suppression with Controlled Release Pheromone Systems* (Vol. II), A. F. Kydonieus and M. Beroza (Eds), CRC Press, Florida, p. 3.

Beasley, C. A. and Henneberry, T. J. (1984). Combining gossyplure and insecticides in pink bollworm control. *California Agriculture*, **38**, 22.

Bell, W. J., Fromm, J., Quisumbing, A. R. and Kydonieus, A. F. (1984). Attraction of American cockroaches (Orthoptera: Blattidae) to traps containing periplanone B and to insecticide-periplanone B mixtures. *Environ. Entomol.*, **13**, 448.

Beroza, M., Paszek, E. C., Mitchell, E. R., Bierl, B. A., McLaughlin, J. R. and Chambers, D. L. (1974). Tests of a 3-layer laminated plastic bait dispenser for controlled emission of attractants from insect traps. *Environ. Entomol.*, **3**, 926.

Beroza, M., Hood, C. S., Trefrey, D., Leonard, D. E., Knipling, E. F. and Klassen, W. (1975a). Field trials with disparlure in Massachusetts to suppress mating of the gypsy moth. *Environ. Entomol.*, **4**, 705.

Beroza, M., Paszek, E. C., DeVilbiss, D., Bierl, B. A. and Tardif, J. G. R. (1975b). A 3-layer laminated plastic dispenser of disparlure for use in traps for gypsy moths. *Environ. Entomol.*, **4**, 712.

Bierl, B. A., Beroza, M., Staten, R. T., Sonnet, P. E. and Adler, V. E. (1974). The pink bollworm sex attractant. *J. Econ. Entomol.*, **67**, 211.

Bierl, B. A., DeVilbiss, E. D. and Plimmer, J. R. (1976). The use of pheromones in insect control programs: Slow release formulations. In *Controlled Release Polymer Formulations*, ACS Symp. Series No. 3, D. R. Paul and F. W. Harris (Eds), American Chemical Society, Washington, DC, p. 265.

Bierl-Leonhardt, B. A. (1982). Release rates from formulations and quality control methods. In *Insect Suppression with Controlled Release Pheromone Systems* (Vol. I), A. F. Kydonieus and M. Beroza (Eds), CRC Press, Florida, p. 245.

Bierl-Leonhardt, B. A., DeVilbiss, E. D. and Plimmer, J. R. (1979). Rate of release of disparlure from laminated plastic dispensers. *J. Econ. Entomol*, **72**, 319.

Chowdhury, M. A., Chalfant, R. B. and Young, J. R. (1987). Comparison of sugarline sampling and pheromone trapping for monitoring adult populations of corn earworm and fall armyworm (Lepidoptera: Noctuidae) in sweet corn. *Environ. Entomol.* **16**, 1241.

Clement, S. L., Hill, A. S., Levine, E. and Roelofs, W. L. (1981). Trap catches of male *Agrotis ipsilon* with synthetic sex pheromone emitted from different dispensers. *Environ. Entomol.*, **10**, 521.

Cogburn, R. R. and Vick, K. W. (1983). *Agrotis ipsilon* males captured in traps baited with the synthetic sex pheromone of *Sitotraga cerealella*. *Southwestern Entomol.*, **8**, 98.

Crank, J. and Park, G. S. (1968). Methods of measurement. In *Diffusion of Polymers*, J. Crank and G. S. Park (Eds), Academic Press, New York.

Critchley, B. R., McVeigh, L. J., Campion, D. G., McVeigh, E. M., Cavanagh, G. G., Hosny, M. M., Nasr, El Sayed A., Khidr, A. A., and Naguib, A. A. (1985). Control of the pink bollworm, *Pectinophora gossypiella* (Saunders) (Lepidoptera: Gelechiidae) in Egypt by mating disruption using hollow-fibre, laminate flake and micro-encapsulated formulations of synthetic pheromone. *Bull. Ent. Res.* **54**, 329–345.

Cuthbert, R. A. and Peacock, J. W. (1979). The Forest Service program for mass trapping *Scolytus multistriatus. Bull. Entomol. Soc. Amer.*, **25**, 105.

Daterman, G. E. (1982). Control of western pineshoot borer damage by mating disruption—a reality. In *Insect Suppression with Controlled Release Pheromone Systems* (Vol. II), A. F. Kydonieus and M. Beroza (Eds), CRC Press, Florida, p. 155.

Daterman, G. E. and Sower, L. L. (1977). Douglas-fir tussock moth pheromone research using controlled release system. In *Proc. 1977 Intl. Controlled Release Pesticides Symp.*, R. Goulding (Ed.), Oregon State University, Corvallis, Oregon, p. 68.

Daterman, G. E., Sower, L. L. and Sartwell, C. (1985). Courtship disruption of western spruce budworm by aerial application of synthetic pheromone. In *Recent Advances in Spruce Budworm Research, Proc. CANUSA Spruce Budworm Res. Symp.*, C. J. Sanders, R. W. Stark, E. J. Mullins and J. Murphy (Eds), Can. For. Serv., Ottawa, Ontario, p. 386.

Elkinton, J. S. (1987). Changes in efficiency of the pheromone-baited milk-carton trap as it fills with male gypsy moths (Lepidoptera: Lymantriidae). *J. Econ. Entomol.* **80**, 754.

Forey, D. E. and Quisumbing, A. R. (1987). Newly designed boll weevil scout trap. *Proc. Beltwide Cotton Prod. Res. Conf.*, National Cotton Council, Memphis, Tennessee, p. 139.

Gentry, C. R., Bierl, B. A. and Blythe, J. L. (1976). Air permeation field trials with the oriental fruit moth pheromone. In *Proc. 1976 Intl. Controlled Release Pesticides Symp.*, N. F. Cardarelli (Ed.), University of Akron, Ohio, p. 3.22.

Gentry, C. R., Bierl-Leonhardt, B. A., Blythe, J. L. and Plimmer, J. R. (1980a). Air permeation tests with orfralure for reduction in trap catch of oriental fruit moths. *J. Chem. Ecol.*, **6**, 185.

Gentry, C. R., Bierl-Leonhardt, B. A., McLaughlin, J. R. and Plimmer, J. R. (1980b). Air permeation tests with (Z,Z)-3,13-octadecadien-1-ol acetate for reduction in trap catch of peachtree and lesser peachtree borer moths. *J. Chem. Ecol.*, **7**, 575.

Gentry, C. R., Yonce, C. E. and Bierl-Leonhardt, B. A. (1982). Oriental fruit moth: Mating disruption trials with pheromone. In *Insect Suppression with Controlled Release Pheromone Systems* (Vol. II), A. F. Kydonieus and M. Beroza (Eds), CRC Press, Florida, p. 107.

Hardee, D. D., McKibben, G. H. and Huddleston, P. M. (1975). Grandlure for boll weevils: Controlled release with a laminated plastic dispenser. *J. Econ. Entomol.*, **68**, 477.

Hartstack, A. W., Jr, Lopez, J. D., Klun, J. A., Witz, J. A., Shaver, T. N. and Plimmer, J. R. (1980). New trap designs and pheromone bait formulations for *Heliothis. Proc. Beltwide Cotton Prod. Res. Conf.*, National Cotton Council, Memphis, Tennessee, p. 132.

Hendricks, D. E. (1982). Polyvinyl chloride capsules: A new substrate for dispensing tobacco budworm (Lepidoptera: Noctuidae) sex pheromone bait formulations. *Environ. Entomol.*, **11**, 1005.

Hendricks, D. E., Hartstack, A. W. and Raulston, J. R. (1977a). Compatibility of virelure and looplure dispensed from traps for cabbage looper and tobacco budworm survey. *Environ. Entomol.*, **6**, 566.

Hendricks, D. E., Hartstack, A. W. and Shaver, T. N. (1977b). Effect of formulations and dispensers on attractiveness of virelure to the tobacco budworm. *J. Chem. Ecol.*, **3**, 497.

Hendricks, D. E., Perez, C. T. and Guerra, R. J. (1980). Effects of nocturnal wind on performance of two sex pheromone traps for noctuid moths. *Environ. Entomol.*, **9**, 483.

Hendricks, D. E., Shaver, T. N. and Goodenough, J. L. (1987). Development and bioassay of molded polyvinyl chloride substrates for dispensing tobacco budworm (Lepidoptera: Noctuidae) sex pheromone bait formulation. *Environ. Entomol.* **16**, 605.

Henneberry, T. J., Bariola, L. A., Flint, H. M., Lingren, P. D., Gillespie, J. M. and Kydonieus, A. F. (1981a). Pink bollworm and tobacco budworm mating disruption studies on cotton. In *Management of Insect Pests with Semiochemicals—Concepts and Practice*, E. R. Mitchell (Ed.), Plenum Press, New York, p. 267.

Henneberry, T. J., Gillespie, J. M., Bariola, L. A., Flint, H. M., Lingren, P. D. and Kydonieus, A. F. (1981b). Gossyplure in laminate plastic formulations for mating disruption and pink bollworm control. *J. Econ. Entomol.*, **74**, 376.

Henneberry, T. J., Gillespie, J. M., Bariola, L. A., Flint, H. M., Butler, G. D. Jr, Lingren, P. D. and Kydonieus, A. F. (1982). Mating disruption as a means of suppressing pink bollworm (Lepidoptera: Gelechiidae) and tobacco budworm (Lepidoptera: Noctuidae) on cotton populations. In *Insect Suppression with Controlled Release Pheromone Systems* (Vol. II), A. F. Kydonieus and M. Beroza (Eds), CRC Press, Florida, p. 75.

Henneberry, T. J., Beasley, C. A. and Butler, G. D., Jr (1984). Report of studies with gossyplure for pink bollworm control. *Proc. Beltwide Cotton Prod. Res. Conf.*, National Cotton Council, Memphis, Tennessee, p. 185.

Henneberry, T. J., Beasley, C. A. and Lingren, P. D. (1986). Potential of gossyplure for pink bollworm control in cotton. In *Proc. 13th Intl. Symp. Controlled Release of Bioactive Materials*, I. A. Chaudry and C. Thies (Eds), Controlled Release Society, Inc., Lincolnshire, Illinois, p. 157.

Hollingsworth, J. P., Hartstack, A. W., Buck, D. R. and Hendricks, D. E. (1978). Electric and non-electric moth traps baited with the synthetic sex pheromone of the tobacco budworm. US Dept. of Agriculture Publ. ARS-S-173.

Hummel, H. E., Gaston, L. K., Shorey, H. H., Kaae, R. S., Byrne, K. J. and Silverstein, R. M. (1973). Clarification of the chemical status of the pink bollworm sex pheromone. *Science*, **181**, 873.

Hyman, S., Bernstein, B. S. and Kapoor, R. (1972). US Patent 3.705.938. *Activated polymer materials and process for making same.*

Johnson, D. R. (1983). Relationship between tobacco budworm (Lepidoptera: Noctuidae) catches when using pheromone traps and egg counts in cotton. *J. Econ. Entomol.*, **76**, 182.

Johnson, W. L., McKibben, G. H., Rodriguez, J. and Davich, T. B. (1976). Boll weevil: Increased longevity of grandlure using different formulations and dispensers. *J. Econ. Entomol.*, **69**, 263.

Kehat, M., Dunkelblum, E. and Gothilf, S. (1983). Mating disruption of the cotton leafworm, *Spodoptera littoralis* (Lepidoptera: Noctuidae), by release of sex pheromone from widely separated Hercon-laminated dispensers. *Environ. Entomol.*, **12**, 1265.

Klein, M. G. (1981). Mass trapping for suppression of Japanese beetles. In *Management of Insect Pests with Semiochemicals—Concepts and Practice*, E. R. Mitchell (Ed.), Plenum Press, New York, p. 183.

Kodadek, R. and Welch, J. H. (1984). US Patent 4.453.675. *Aerial spraying apparatus.*

Kydonieus, A. F. (1977). The effect of some variables on the controlled release of chemicals from polymeric membranes. In *Controlled Release Pesticides*, ACS Symposium Series No. 53, H. B. Scher (Ed.), American Chemical Society, Washington, DC, p. 152.

Kydonieus, A. F. (1978). Hercon controlled release system: A novel way to dispense fragrances and room deodorizers. *Soap Cosmet. Chem. Spec.*, May.

Kydonieus, A. F. and Beroza, M. (1977). Insect control with multi-layered Luretape dispenser. In *Proc. Intl. 1977 Controlled Release Pesticide Symp.*, R. Goulding (Ed.), Oregon State University, Corvallis, Oregon, p. 78.

Kydonieus, A. F. and Quisumbing, A. R. (1980). Multilayered laminated structures. In *Controlled Release Technologies: Methods, Theory and Applications* (Vol. I), A. F. Kydonieus (Ed.), CRC Press, Florida, p. 183.

Kydonieus, A. F. and Beroza, M. (1981). The Hercon dispenser: Formulation and recent test results. In *Management of Insect Pests with Semiochemicals — Concepts and Practice*, E. R. Mitchell (Ed.), Plenum Press, New York, p. 445.

Kydonieus, A. F., Baldwin, S. and Hyman, S. (1976a). Hercon granules and powders for agricultural applications. In *Proc. 1976 Intl. Controlled Release Pesticides Symp.*, N. F. Cardarelli (Ed.), University of Akron, Ohio, p. 4.23.

Kydonieus, A. F., Quisumbing, A. R. and Hyman, S. (1976b). Application of a new controlled release concept in household products. In *Controlled Release Polymeric Formulations*, ACS Symposium Series No. 33, D. R. Paul and F. W. Harris (Eds), American Chemical Society, Washington, DC, p. 295.

Kydonieus, A. F., Rofheart, A. and Hyman, S. (1976c). Marketing and economic considerations for Hercon consumer and industrial controlled release products. In *Preprints of Symposium on Economics and Market Opportunities for Controlled Release Products*, ACS Chemical Marketing and Economics Division, American Chemical Society, Washington, DC, p. 140.

Kydonieus, A. F., Smith, I. K. and Beroza, M. (1976d). Controlled release of pheromones through multilayered polymeric dispensers. In *Controlled Release Polymeric Formulations*, ACS Symposium Series No. 33, D. R. Paul and F. W. Harris (Eds), American Chemical Society, Washington, DC, p. 283.

Kydonieus, A. F., Smith, I. K. and Hyman, S. (1976e). A polymeric delivery system for the controlled release of pesticides: Hercon Roach-Tape. In *Proc. 1976 Intl. Controlled Release Pesticides Symp.*, N. F. Cardarelli (Ed.), University of Akron, Ohio, p. 3.40.

Kydonieus, A. F., Bierl-Leonhardt, B. A., Plimmer, J. R., Barry, M. W. and Quisumbing, A. R. (1980). Cotton insects: dispenser development and disruption of mating trials. In *Proc. 6th Intl. Symp. Controlled Release Bioactive Materials*, R. Baker (Ed.), Academic Press, New York, IV-13.

Kydonieus, A. F., Kauffman, D., Lambert, C., Berner, B., Lambert, H., Geising, D., Frazier, W., Morris, R. and Niebergall, P. (1984). Transdermal delivery of nitroglycerin from a laminated reservoir system. In *Proc. 11th Intl. Symp. Controlled Release of Bioactive Materials*, W. E. Meyers and R. L. Dunn (Eds), Controlled Release Society, Inc., Lincolnshire, Illinois, p. 32.

Ladd, T. L., Jr and Klein, M. G. (1982). Trapping Japanese beetles with synthetic female sex pheromone and food-type lures. In *Insect Suppression with Controlled Release Pheromone Systems* (Vol. II), A. F. Kydonieus and M. Beroza (Eds), CRC Press, Florida, p. 65.

Ladd, T. L., Jr and McGovern, T. P. (1984). 2-Methoxy-4-propyl-phenol, a potent new enhancer of lures for the Japanese beetle (Coleoptera: Scarabaeidae). *J. Econ. Entomol.*, **77**, 957.

Landolt, P. J., Curtis, C. E., Coffelt, J. A., Vick, K. W., Sonnet, P. E. and Doolittle, R. E. (1981). Disruption of mating in the navel orangeworm with (Z,Z)-11,13-hexadecadienal. *Environ. Entomol.*, **10**, 745.

Landolt, P. J., Curtis, C. E., Coffelt, J. A., Vick, K. W. and Doolittle, R. E. (1982). Field trials of potential navel orangeworm mating disruptants. *J. Econ. Entomol.*, **75**, 547.

Lanier, G. N. (1979). Protection of elm groves by surrounding them with multilure-baited sticky traps. *Bull. Entomol. Soc. Amer.*, **25**, 109.

Lanier, G. N. and Jones, A. H. (1985). Trap trees for elm bark beetles augmentation with pheromone baits and chlorpyrifos. *J. Chem. Ecol.*, **11**, 11.

Leonhardt, B. A. and Moreno, D. S. (1982). Evaluation of controlled release laminate dispensers for pheromones of several insect species. In *Insect Pheromone Technology: Chemistry and Applications*, ACS Symp. Series No. 190, B. A. Leonhardt and M. Beroza (Eds), American Chemical Society, Washington, DC, p. 159.

Leonhardt, B. A., Rice, R. E., Harte, E. M. and Cunningham, R. T. (1984). Evaluation of dispensers containing trimedlure, the attractant for the Mediterranean fruit fly (Diptera: Tephritidae). *J. Econ. Entomol.*, **77**, 744.

Leonhardt, B. A., Harte, E. M., DeVilbiss, E. D. and Inscoe, M. N. (1984). Effective controlled-release dispensers for the pheromone of the bagworm. Release rate vs. trap catch. *J. Controlled Release*, **1**, 137.

Leonhardt, B. A., Dickerson, W. A., Ridgway, R. L. and DeVilbiss, E. D. (1988). Laboratory and field evaluations of controlled release dispensers containing grandlure, the pheromone of the boll weevil (Coleoptera: Curculionidae). *J. Econ. Entomol.*, **81**, 937.

Lie, R. and Bakke, R. (1981). Practical results from the mass trapping of *Ips typographus* in Scandinavia. In *Management of Insect Pests with Semiochemicals — Concepts and Practice*, E. R. Mitchell (Ed.), Plenum Press, New York, p. 175.

Lloyd, E. P., McKibben, G. H., Leggett, J. E. and Hartstack, A. W. (1983). Pheromones for survey, detection, and control. In *Cotton Insect Management with Special Reference to the Boll Weevil*, R. L. Ridgway, E. P. Lloyd and W. H. Cross (Eds), Agr. Hdbk. 589, US Dept. of Agriculture, p. 179.

Lopez, J. D., Jr (1980). Comparison of two types of boll weevil pheromone traps to monitor seasonal response. *J. Econ. Entomol.*, **73**, 324.

Lopez, J. D. Jr. and Witz, J. A. (1988). Influence of *Heliothis virescens* sex pheromone dispensers on captures of *H. zea* males in pheromone traps relative to distance and wind direction. *J. Chem. Ecol.* **14**, 265.

Marengo, L. R. M., Alvarez, L. A. and Whitcomb. W. H. (1987). *El picudo Mejicano del algodonero, Anthonomus grandis Boh.* Ministerio de Agricultura y Ganaderia Pub. Mis. 18, Asuncion, Paraguay, 94 pp.

Mastro, V. C., Richerson, J. V. and Cameron, E. A. (1977). An evaluation of gypsy moth pheromone-baited traps using behavioral observations as a measure of trap efficiency. *Environ. Entomol.*, **6**, 128.

McLaughlin, J. R., Mitchell, E. R. and Cross, J. H. (1981). Field and laboratory evaluation of mating disruptants of *Heliothis zea* and *Spodoptera frugiperda* in Florida. In *Management of Insect Pests with Semiochemicals — Concepts and Practice*, E. R. Mitchell (Ed.), Plenum Press, New York, p. 243.

Mitchell, E. R., Baumhover, A. H. and Jacobson, M. (1976). Reduction of mating potential of male *Heliothis* spp. and *Spodoptera frugiperda* in field plots treated with disruptants. *Environ. Entomol.*, **5**, 484.

Nakagawa, S., Harris, E. J. and Urago, T. (1979). Controlled release of trimedlure from a three-layer laminated plastic dispenser. *J. Econ. Entomol.*, **72**, 625.

Neumann, U., Vogt, H., Schropp, A., Englert, W. D. and Schruft, G. (1986). Control of grape berry moth (Cochylis) using the technique of mating disruption. *BASF Agricultural News*, 1/86, 5.

Palaniswamy, P., Underhill, E. W., Gillott, C. and Wong, J. W. (1986). Synthetic sex pheromone components disrupt orientation, but not mating, in the fall cankerworm, *Alsophila pometaria* (Lepidoptera: Geometridae). *Environ. Entomol.* **15**, 943.

Payne, T. L. (1981). Disruption of southern pine beetle infestations with attractants and inhibitors. In *Management of Insect Pests with Semiochemicals—Concepts and Practice*, E. R. Mitchell (Ed.), Plenum Press, New York, p. 365.

Plimmer, J. R. and Leonhardt, B. A. (1983). Formulation and insect pheromones: Application rates and persistence. In *IUPAC Pesticide Chemistry—Human Welfare and Environment*, J. Miyamoto *et al.* (Eds), Pergamon Press, New York, p. 233.

Plimmer, J. R., Bierl, B. A. and Schwalbe, C. P. (1977). Controlled release of pheromone in the gypsy moth program. In *Controlled Release Pesticides*, ACS Symposium Series No. 53, H. B. Scher (Ed.), American Chemical Society, Washington, DC, p. 168.

Plimmer, J. R., Schwalbe, C. P., Paszek, E. C., Bierl, B. A., Webb, R. E., Marumo, S. and Iwaki, S. (1978). Contrasting effectiveness of (+) and (±) enantiomers of disparlure for trapping native populations of gypsy moth in Massachusetts. *Environ. Entomol.*, **7**, 815.

Plimmer, J. R., Leonhardt, B. A., Webb, R. E. and Schwalbe, C. P. (1982). Management of the gypsy moth with its sex attractant pheromone. In *Insect Pheromone Technology: Chemistry and Applications*, ACS Symp. Series No. 190, B. A. Leonhardt and M. Beroza (Eds), American Chemical Society, Washington, DC, p. 231.

Quisumbing, A. R. and Kydonieus, A. F. (1982). Laminated structure dispensers. In *Insect Suppression with Controlled Release Pheromone Systems* (Vol. I), A. F. Kydonieus and M. Beroza (Eds), CRC Press, Florida, p. 213.

Quisumbing, A. R., Kydonieus, A. F., Calsetta, D. R. and Haus, J. B. (1976). Hercon 'Lure N Kill' Flytape: A non-fumigant insecticidal strip containing attractants. In *Proc. 1976 Intl. Controlled Release Pesticides Symp.*, N. F. Cardarelli (Ed.), University of Akron, Ohio, p. 3.40.

Quisumbing, A. R., Kydonieus, A. F. and Gauthier, N. L. (1978). Use of laminated controlled release contact-action insecticidal strips for cluster fly control. In *Proc. 5th Intl. Symp. on Controlled Release of Bioactive Materials*, F. E. Brinckman and J. A. Montemarano (Eds), Gaithersburg, Maryland, p. 5.62.

Quisumbing, A. R., Shevchuk, I., Laurencot, J., Gillespie, J. M., Matz, S., Zeoli, L. T. and Kydonieus, A. F. (1986). Luretape with grandlure: Effects of multilayered laminate formulation on pheromone release rate. In *Proc. 13th Intl. Symp. Controlled Release of Bioactive Materials*, I. A. Chaudry and C. Thies (Eds), Controlled Release Society, Inc., Lincolnshire, Illinois, p. 162.

Rice, R. E., Cunningham, R. T. and Leonhardt, B. A. (1984). Weathering and efficacy of trimedlure dispensers for attraction of Mediterranean fruit flies (Diptera: Tephritidae). *J. Econ. Entomol.*, **77**, 750.

Richards, R. W. (1973). *The Permeability of Polymers to Gases, Vapours and Liquids*, NTIS AD-767 627, US Dept. of Commerce, Springfield, Virginia.

Roelofs, W. L., Hill, A. S., Carde, R. T. and Baker, T. C. (1974). Two sex pheromone components of the tobacco budworm moth, *Heliothis virescens*. *Life Sci*, **14**, 1555.

Rummell, D. R., Carroll, S. C. and Shaver, T. N. (1987). Influence of boll weevil trap design on internal trap temperature and grandlure volatilization. *Southwest. Entomol.* **12**, 127.

Sanders, C. J. and Silk, P. J. (1982). Disruption of spruce budworm mating by means of Hercon plastic laminated flakes, Ontario 1981. Information Dept. O-X-335, Canadian Forest Service, Sault Ste. Marie, Ontario, 22 pp.

Sartwell, C., Daterman, G. E., Overhulser, D. L. and Sower, L. L. (1983). Mating disruption of western pineshoot borer (Lepidoptera: Tortricidae) with widely spaced releasers of synthetic pheromone. *J. Econ. Entomol.*, **76**, 1148.

Schwalbe, C. P. and Mastro, V. A. (1988). Gypsy moth mating disruption. *J. Chem. Ecol.* **14**, 581.

Schwalbe, C. P., Paszek, E. C., Webb, R. E., Bierl-Leonhardt, B. A., Plimmer, J. R., McComb, C. W. and Dull, C. W. (1979). Field evaluation of controlled release formulations of disparlure for gypsy moth mating disruption. *J. Econ. Entomol.*, **72**, 322.

Schwalbe, C. P., Paszek, E. C., Bierl-Leonhardt, B. A. and Plimmer, J. R. (1983). Disruption of gypsy moth (*Lepidoptera: Lymantriidae*) mating with disparlure. *J. Econ. Entomol.*, **76**, 841.

Seabrook, W. D. and Kipp, L. R. (1986). The use of a two-component blend of the spruce budworm sex pheromone for mating suppression. In *Proc. 13th Intl. Symp. Controlled Release of Bioactive Materials*, I. A. Chaudry and C. Thies (Eds), Controlled Release Society, Inc., Lincolnshire, Illinois, p. 128.

Shaver, T. N. and Lopez, J. D., Jr (1982). Effect of pheromone components in mating and trap catches of *Heliothis* spp. in small plots. *Southwestern Entomol.*, **7**, 181.

Shaver, T. N., Lopez, J. D., Jr and Hartstack, A. W., Jr (1982). Effects of pheromone components and their degradation products on the response of *Heliothis* spp. to traps. *J. Chem. Ecol.*, **8**, 775.

Silk, P. J. and Kuenen, L. P. S. (1984). Sex pheromones and their potential as control agents for forest Lepidoptera in Eastern Canada. In *Chemical and Biological Controls in Forestry*, ACS Symp. Series No. 238, W. Y. Garner and J. Harvey Jr. (Eds), American Chemical Society, Washington, DC, p. 35.

Slosser, J. E., Witz, J. A., Puterka, G. J., Price, J. R. and Hartstack, A. W. (1987). Seasonal changes in bollworm (Lepidoptera: Noctuidae) moth catches in pheromone traps in a large area. *Environ. Entomol.* **16**, 1296.

Snow, J. W. and Gentry, C. R. (1986). The controlled release of (Z,Z)-3,13-octadecadienyl acetate and its effect on the peachtree borer. In *Proc. 13th Intl. Symp. Controlled Release of Bioactive Materials*, I. A. Chaudry and C. Thies (Eds), Controlled Release Society, Inc., Lincolnshire, Illinois, p. 124.

Sower, L. L. and Daterman, G. E. (1986). Use of behavioral chemicals in western U.S. forests. In *Proc. 13th Intl. Symp. Controlled Release of Bioactive Materials*, I. A. Chaudry and C. Thies (Eds), Controlled Release Society, Inc., Lincolnshire, Illinois, p. 122.

Sower, L. L., Overhulser, D. L., Daterman, G. E., Sartwell, C., Laws, D. E. and Koerber, T. W. (1982). Control of *Eucosma sonomana* by mating disruption with synthetic sex attractant. *J. Econ. Entomol.*, **75**, 315.

Sutherland, D. W. S. (1978). *Common Names of Insects and Related Organisms (1978 Revision)*. Special Publ. 78-1, Entomological Society of America, Maryland, 132 pp.

Tingle, F. C. and Mitchell, E. R. (1978). Controlled release plastic strips containing (Z)-9-dodecen-1-ol acetate for attracting *Spodoptera frugiperda*. *J. Chem. Ecol.*, **4**, 41.

Tingle, F. C. and Mitchell, E. R. (1982). Effect of synthetic pheromone on parasitization of *Heliothis virescens* (F.) (Lepidoptera: Noctuidae) in tobacco. *Environ. Entomol.*, **11**, 913.

Tumlinson, J. H., Hendricks, D. E., Mitchell, E. R., Doolittle, R. E. and Brennan, M. M. (1975). Isolation, identification and synthesis of the sex pheromone of the tobacco budworm. *J. Chem. Ecol.*, **1**, 203.

Vick, K. W., Coffelt, J. A. and Sullivan, M. A. (1978). Disruption of pheromone communication in the Angoumois grain moth with synthetic female sex pheromone. *Environ. Entomol.* **7**, 528.

Vick, K. W., Kuenberg, J., Coffelt, J. A. and Steward, C. (1979). Investigation of sex pheromone traps for simultaneous detection of Indianmeal moths and Angoumois grain moth. *J. Econ. Entomol.*, **72**, 245.

Von Ramm, C. (1986). The use of pheromones to control pink bollworm on cotton in Egypt. *BASF Agricultural News* 2/86, 19.

Webb, R. E., Leonhardt, B. A., Kolodny-Hirsch, D., Cohen, D., Waghray, R. N. and Morris, C. L. (1986). Controlled release of racemic disparlure to suppress the gypsy moth—Current status and future potential. In *Proc. 13th Intl. Symp. Controlled Release of Bioactive Materials*, I. A. Chaudry and C. Thies (Eds), Controlled Release Society, Inc., Lincolnshire, Illinois, p. 104.

Webb, R. E., Tatman, K. M., Leonhardt, B. A., Plimmer, J. R., Boyd, V. K., Bystrak, P. G., Schwalbe, C. P. and Douglass, L. W. (1988). Effect of aerial application of racemic disparlure on male trap catch and female mating success of gypsy moth (Lepidoptera: Lymantriidae). *J. Econ. Entomol.* **81**, 268.

Wiesner, C. J. and Silk, P. J. (1982). Monitoring the performance of eastern spruce budworm formulations. In *Insect Pheromone Technology: Chemistry and Applications*, ACS Symp. Series No. 190, B. A. Leonhardt and M. Beroza (Eds), American Chemical Society, Washington, DC, p. 209.

Wollerman, E. H. (1979). Attraction of European elm bark beetles, *Scolytus multistriatus*, to pheromone-baited traps. *J. Chem. Ecol.*, **5**, 781.

Yonce, C. E. and Gentry, C. R. (1982). Disruption of mating of peachtree borer. In *Insect Suppression with Controlled Release Pheromone Systems* (Vol. II), A. F. Kydonieus and M. Beroza (Eds), CRC Press, Florida, p. 99.

Zvirgzdins, A. and Henneberry, T. J. (1983). *Heliothis* spp.: Sex pheromone trap studies. *Proc. Beltwide Cotton Prod. Res. Conf.*, National Cotton Council, Memphis, Tennessee, p. 176.

Zvirgzdins, A., Lingren, P. D., Henneberry, T. J., Nowell, C. E. and Gillespie, J. M. (1984). Mating disruption of a wild population of tobacco budworm (Lepidoptera: Noctuidae) with virelure. *J. Econ. Entomol.*, **77**, 1464.

Insect Pheromones in Plant Protection
Edited by A. R. Jutsum and R. F. S. Gordon
©1989 John Wiley & Sons Ltd

8

Hollow-fibre Controlled-release Systems

D. W. Swenson[1]

Morton Thiokol, Inc., Alfa Products, Danvers, Massachusetts 01923, USA

I. Weatherston

Département de Biologie, Université Laval, Québec G1K 7P4, Canada

8.1 INTRODUCTION

Albany International in the early 1970s had developed the ability to extrude polymeric hollow fibres as part of the ongoing research effort in the area of reverse osmosis technology. In mid-1974, as an outgrowth of this work, the hollow-fibre release device was conceived. Early studies indicated that hollow fibres released volatile materials such as insect pheromones at a controlled rate with virtual zero-order kinetics. This provided the basis for a patent application in 1975 and subsequent commercial development of hollow-fibre controlled-release products, with first governmental registration in the USA in 1978.

8.2 THEORETICAL ASPECTS OF THE RELEASE OF VOLATILE MATERIALS FROM CAPILLARIES

The use of pheromones and other behaviour-modifying chemicals in insect control strategies requires that they be disseminated from controlled release formulations. Ideally a controlled release device discharges all of its active material in a given time period, releasing it at a constant rate (zero-order kinetics). In the commercial application of behaviour-modifying chemicals it is unlikely that the ideal controlled release device will be developed.

---

[1]Formerly of Albany International, Controlled Release Division, Needham Hts, Massachusetts 02193, USA.

However, in relation to constancy of release rate, capillaries, after an initial burst, release at a relatively constant rate thereafter (pseudo zero-order kinetics) as reported by Brooks (1980). Capillaries, then, would appear to be superior to impregnated plastic and rubber matrices, plastic laminates, microcapsules, etc., all of which release with first-order release kinetics.

The mechanism of vapour release from capillaries, if trans-wall permeation is excluded, has been reported by Ashare et al. (1975, 1976), Brooks et al. (1977) and Brooks (1980) to be a simple three-stage process: (a) evaporation at the liquid–vapour interface, (b) diffusion through the vapour–air column to the end of the capillary, and (c) convection away from the end of the capillary. The diffusion stage is generally considered to be the rate-controlling factor, and the rate of diffusion can be predicted by using transport equations. The mathematical treatment of transport phenomena leading to an equation relating rate of vapour release from a capillary to the square root of time has been summarized by Brooks (1980).

$$\frac{dl}{dt} = \left[-\frac{McD}{2\rho} \ln\left(1 - \frac{P_{vap}}{P} \right) \right]^{\frac{1}{2}} t^{-\frac{1}{2}} \qquad \text{Eq. 1}$$

where M = molecular weight of liquid charge, c = molar density of vapour–air column, D = diffusion coefficient, ρ = density of the liquid, P_{vap} = vapour pressure of the liquid and P = atmospheric pressure.

The rate of meniscus regression (dl/dt), hence the release rate of the liquid charge, is predicted to be inversely proportional to the half power of time. The validity of this relationship has been reported for carbon tetrachloride (Ashare et al., 1976; Brooks et al., 1977; Brooks, 1980) and o-dichlorobenzene (Ashare et al., 1975) for capillaries made from polyester terephthalate. Whether the meniscus within the capillary is concave or convex on the vapour side determines whether the value of P_{vap} in Eq. 1 is less or greater than true equilibrium vapour pressure. The fibre material/active ingredient pairing influences the surface tension of which the meniscus curvature is a function. With the pairing normally encountered between fibre material and active ingredient the meniscus is concave on the vapour side, hence the vapour pressure value is reduced. The extent of this reduction is a function of the radius of curvature, which in turn is dependent on the capillary diameter.

Weatherston et al. (1984a, 1985a) re-examined the release of volatile materials from glass capillaries with regard to the kinetics of the release and the effect of the initial vapour–air column above the liquid, and developed a predictive model to facilitate formulation design of hollow-fibre semio-chemical-based insect control systems.

Data obtained by measuring the release rates of volatile materials loaded to various levels within the same diameter glass capillaries of different lengths indicated that the release rate is independent of the amount of material in a capillary, but that the initial length of the vapour–air column is a critical factor in controlling the release rate.

The results reported by Weatherston *et al.* (1984a, 1985a) are analogous to those of Brooks (1980 and references therein), with Weatherston also showing that the amount of material evaporating from the capillaries varies directly with the square root of time.

$$x = kt^{1/2} \qquad \text{Eq. 2}$$

Considering the capillary as a cylinder, the amount evaporated can also be related to the length of the vapour–air column (L) above the liquid surface within the capillary.

$$x = L \pi r^2 \rho \qquad \text{Eq. 3}$$

where r = inner radius of the capillary and ρ = the density of the liquid with which the capillary is charged.

$$\frac{dx}{dt} = \frac{1}{2} k t^{-1/2} \qquad \text{Eq. 4}$$

and

$$\frac{dx}{dl} = \pi r^2 \rho \qquad \text{Eq. 5}$$

Combining Eqs. 4 and 5:

$$\frac{dl}{dt} = \frac{k}{2 \pi r^2 \rho} t^{-1/2} \qquad \text{Eq. 6}$$

Equating Eq. 6 with that of Brooks *et al.* (Eq. 1):

$$k = \pi r^2 \left[-2 McD\rho \ln \left(1 - \frac{P_{\text{vap}}}{P} \right) \right]^{1/2}$$

Weatherston *et al.* (1984a, 1985a) were able to illustrate good agreement between calculated and experimentally derived values for k for five volatile materials emitted from glass capillaries, allowing them to calculate release rates instead of measuring them experimentally. By substituting Eq. 3 into Eq. 2 an expression for the half power of time can be realized.

$$t^{1/2} = \frac{L \pi r^2 \rho}{k} \qquad \text{Eq. 7}$$

Substituting Eq. 7 into Eq. 4:

$$\frac{dx}{dt} = \frac{k^2}{2\pi r^2}\frac{1}{L}$$

which by substituting the value of k leads to:

$$\frac{dx}{dt} = -McD\pi r^2 \ln\left(1 - \frac{P_{vap}}{P}\right)\frac{1}{L} \qquad\qquad \text{Eq. 8}$$

An algorithm was developed based on Eq. 8 which allowed the release rate of volatile materials from glass capillaries to be predicted with respect to length of capillary (L) and time, given for the volatile materials the various constants and parameters indicated in Eq. 8. In this manner the effects of capillary length, radius, temperature, different active materials and atmospheric pressure on the release rate can be rapidly tested.

Validity of the model has been shown for various volatile materials releasing from different diameter glass capillaries of different lengths and at several temperatures (Weatherston et al., 1985a). This is exemplified in Fig. 8.1 with the release of hexyl acetate from glass capillaries of 0.095 cm diameter at 15 and 23 °C. The utility of this model in the design of behaviour-modifying chemical formulations depends on its validity when tested with capillaries made of polymeric materials.

When the predictive model was tested with Celcon hollow fibres of the type used commercially in aerial broadcast control products it appeared to be invalid. Measurements indicated that the actual release rate from Celcon fibres was not only in poor agreement with the predicted rate, but that the inter-fibre rate variation was very large, most fibres releasing at a much faster rate than predicted (Weatherston et al., 1985b). Among the factors which could explain the differences between the predicted and observed release rates and also the inter-fibre variation are: (a) absorption of the active ingredient into the wall of the fibre; (b) irregular internal diameter of the lumen along the active length of the fibre; and (c) the topography of the lumen walls.

As stated above the fibre material/active ingredient pairings normally used result in a meniscus concave on the vapour side; this is true for Celcon/organic compound pairings, and the effect of the meniscus shape on the vapour pressure would be to reduce the vapour pressure below the equilibrium value, hence this cannot account for faster than predicted release rates. One of the reasons Celcon was selected as the polymer from which the fibres are made was its high degree of crystallinity, which is indicative that little or no absorption of the active ingredient into the walls would take place.

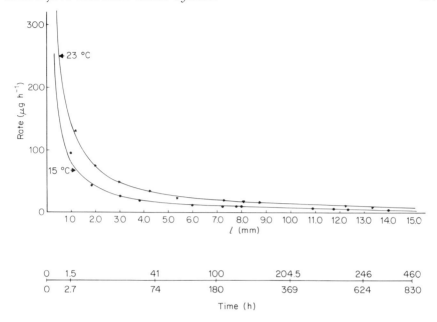

Fig. 8.1 Release of hexyl acetate from glass capillaries (1.5 cm long & 0.0475 cm radius) at 15 and 23 °C relative to l and time: predicted rates (——), observed rates (● and ▼). Time scale upper numbers for 23 °C, lower numbers for 15 °C

By measuring the internal diameter of the lumen of 0.020 in (0.0508 cm) fibres at several points along the active length it was found that the diameter was never greater than 0.0508 cm and never varied by more than 2.36% at the open end of the fibre nor by 0.98% along the active length (Weatherston *et al.*, 1985b). Again this would not account for the fact that the predicted release rate was slower than that observed, nor does it explain the inter-fibre variations.

The major contributing factor to the erratic release of volatile materials from Celcon fibres, causing the invalidity of the model, was elucidated when the internal surface of the fibres was compared by scanning electron microscopy to the smooth internal surface of the glass capillaries. It was found that the less smooth the internal wall surface of a fibre the faster that fibre released its active ingredient, since the receding liquid meniscus can be trapped in the interstices producing added release reservoirs, some of which have menisci, convex on the vapour side. This was shown to be true for 0.008-inch diameter (0.0203 cm) Celcon fibres charged with hexyl alcohol, and 0.020-inch diameter (0.0508 cm) fibres charged with hexyl acetate (Weatherston *et al.*, 1985b).

To test this hypothesis the same workers made fibres from microbore teflon tubing which was shown to have smooth lumen walls. Some teflon fibres were scored with an insect pin to roughen the internal surface of the walls. The release rates of butyl acetate from both the unscratched and scratched fibres were determined and compared. The release from the unscratched fibres followed the predictive model and exhibited very small inter-fibre variations, whereas the release of the butyl acetate from the scratched fibres did not fit the model, was at a faster rate than predicted, and showed large inter-fibre variations.

Provided that (a) the lumen wall is smooth, (b) the fibre diameter is uniform throughout its length, and (c) a fibre material is chosen such that there is little or no absorption of the active material into the fibre wall, the predictive model developed for the release of volatile materials from glass capillaries is valid for the release of volatile materials from capillaries (fibres) made of other substances and can be used to facilitate the design of semiochemical formulations used in insect control. The studies outlined above indicate that although hollow-fibre technology is amongst the best for the controlled release of insect pheromones, the current commercial formulations are inefficient in the utilization of the pheromones. By improving the Celcon fibres with regard to the smoothness of the lumen wall, or by using new fibre materials and the application of the predictive model in formulation design, pheromone utilization could be optimized with the concomitant reduction in the cost of formulations.

To illustrate briefly how the model may be used in formulation design, let us consider a hypothetical case where there is a requirement for a formulation to release (Z)-9-tetradecenyl acetate at 50 ng h^{-1} over 40 days at a temperature of 30 °C. With fibres of internal radius 0.0102 cm and an active length of 1.5 cm filled with this pheromone to a length of 1.45 cm the model predicts a longevity of 220 days, with a mean release rate of 57.3 ($\pm$13.6) ng h^{-1} after the first 40 days. However, the model also indicates that among the choices for the given conditions is one which utilizes 0.0102 cm internal radius fibres of active length 1.35 cm with an initial vapour–air column length of 1.1 cm (i.e. an initial pheromone column length of 0.25 cm). This formulation will release at 50.0 ($\pm$3.0) ng h^{-1} from the first day and have a longevity of 45 days.

Studies have been completed (Weatherston, unpublished) on the correlation between release rate and temperature in relation to the predictive model, and the further development of the model for predicting the release of binary mixtures.

The advantages and disadvantages of hollow fibres as controlled-release devices for semiochemicals are listed in Table 8.1.

Flexibility in design allows both the release rate and longevity to be varied over wide parameters, and constant release rates can be achieved.

TABLE 8.1 Advantages and disadvantages of hollow-fibre controlled-release devices

Advantages	Disadvantages
1. May be formulated to give constant release	1. High weight of delivery vehicle compared to active ingredient
2. Flexibility in design	2. Specialized application equipment required
3. Good point-source strength	
4. High efficiency of utilization of active material	3. Storage and transport necessary under controlled temperature
5. Low release capability in terms of mass released/fibre	

Since release kinetics are not first order there is a high efficiency of utilization of the active ingredient. This is especially important considering the high cost of most pheromones. The low release capability in terms of mass released per fibre is only advantageous because the high physiological activity of pheromone permits the use of low release rates. The mechanisms by which sex pheromones act in insect communication are not well elucidated. However, strong point sources are advantageous to either camouflage pheromone emanating from females or to lay down false trials of greater strength than those from female insects.

The specialized application equipment required for the hollow-fibre dispensers has both advantages and disadvantages. For example, with area-wide control systems, such as NoMate PBW, an application rate of 36 g ha^{-1} allows application of 160 ha with one payload of the aircraft. Depending on the types of aircraft used it could have to refuel before it requires another payload. As fuel costs rise this aspect increases in importance. At application sites no attendant nurse truck is necessary because there is no need for water. This specialized application equipment has the major disadvantages of high initial cost (approx. $10 000 US), restricted use, and of being unavailable in many potential use areas. The high vehicle/active ingredient weight ratio is disadvantageous if the active ingredient costs are not substantially greater than the vehicle raw material and processing costs.

8.3 HOLLOW-FIBRE CONTROLLED-RELEASE DEVICES

Insect control methods based on pheromones require that the active materials be disseminated from controlled-release formulations. In common with other release devices hollow fibres are but one component of the control system. It is well known that behaviour-modifying chemicals, particularly sex pheromones, have three distinct uses in crop protection. These include population surveying and monitoring, mass trapping, and area-wide dissemination to control pest species by either disruption of sexual communication, attract and kill strategies, or a bioirritant technique. These uses are also described in Chapters 3–5.

The hollow-fibre formulations used may be divided into two categories. The first encompasses monitoring and mass trapping and requires: (a) the complete pheromone blend; (b) the controlled-release device designed so as to release the blend at the correct release rate and component ratio, and have the desired field longevity; (c) an insect trap of optimized design; and (d) additives such as solvent for diluting the active ingredients to regulate longevity, antioxidants and ultraviolet stabilizers to protect the active materials from degradation and in some cases insecticides used as killing agents in the trap. The second category, the area-wide control system, requires: (a) the complete pheromone or a major component; (b) a controlled-release device; (c) an application device; (d) additives such as those described above, with the addition of an adhesive to stick the devices to the plants; and (e) a monitoring system incorporating the first category.

There are basically two types of pheromonal hollow-fibre products: (a) tape dispensers (Scentry™ products), and (b) chopped-fibre dispensers (NoMate™ products). The former type consists of a parallel array of fibres (2–100) affixed to an adhesive tape. These dispensers are primarily used as trap lures in monitoring and quarantine programmes, and for the timing of insecticide and/or pheromone applications. They may also be deployed by hand placement in mating disruption strategies with labour-intensive high-value crops which are grown in relatively small areas (e.g. control of peachtree borers on peaches; see Weatherston *et al.*, 1984b and references cited therein).

The chopped-fibre dispensers are used in area-wide broadcast control where they can be utilized in classical mating disruption strategies by interfering with the female to male sexual communication, or in the attract and kill technique in which small amounts of insecticide associated with the fibres increases the control by killing male insects attracted to the point sources. This assumes that the insect remains at the point source for sufficient time to absorb a lethal dose. Additionally there is some evidence (Kaissling, 1982; Floyd and Crowder, 1981) that sublethal doses of insecticide interfere with the perception of the sex pheromone by the male insects. However, data are not available which would relate this to the male's ability to respond to its conspecific pheromone thus preventing mating.

A third strategy using chopped-fibre formulations is known as the bioirritant technique, in which simultaneous applications of insecticide (at the label rate) and pheromone are made, giving better overall control than insecticide alone. Although it is claimed that this technique is successful, its mode of action is unknown. It is presumed that the pheromone causes more activation and movement of both sexes of the target insect, thus exposing them to a greater likelihood of contacting the insecticide. Another use for the chopped-fibre dispenser, analogous to trap cropping and applicable to the control of coleopteran species, particularly boll weevil, is to treat a small percentage of the crop area (5–10%) with the pheromone, forming barrier zones into which the target insects will be attracted. These zones should then be oversprayed with insecticide beginning the day after the pheromone application and continuing at intervals of several days during the field life of the pheromone formulation to ensure optimum kill of the attracted insects.

Brooks (1980) noted that although it may be desirable to vary the fibre material depending on the application there are several general considerations which must be taken into account:

(a) There must be compatibility between the fibre material and the various components of the liquid load; these entities must be mutually insoluble. In addition the fibre material must be impermeable to the active ingredient, i.e. exclude emission through the wall to ensure release according to the evaporation–diffusion–convection mechanism discussed above. It has been stated (Brooks, 1980) that the vapour flux (and hence release rate) is not necessarily directly proportional to the square of the internal radius of the capillary for all fibre material/active ingredient pairings. Data obtained by Weatherston *et al.* (1985b) tend to disprove this and to indicate a lesser effect on vapour pressure of the meniscus shape, which in turn is dependent on the nature of the liquid load and the fibre material.

(b) The fibre material should possess processing characteristics such that it can be extruded, sealed and cut to give fibres with smooth internal wall surfaces, consistent internal diameter throughout their active length and clean, unobstructed end openings.

(c) For application on food crops the fibre material must be acceptable for food contact use, preferably already possessing US Food and Drug Administration approval.

(d) Ideally the fibres should be bio-, photo-, or environmentally degradable where the devices are not recoverable in normal cultivation practices.

Based partly on compliance with the criteria set out above most of the development work on the hollow-fibre systems was undertaken with the poly-oxymethylene-co-oxyethylene copolymer Celcon™ (Celanese Corp., Chatham, NJ 07928, USA) or polyethylene terephthalate, although high-density polyolefins have also been investigated.

Prior to the development of the predictive model described above laboratory release rate studies were undertaken to define the release characteristics of the formulations. Although studies of the methodology for the determination of emission rates of pheromones from controlled-release devices have been adequately reviewed (Beroza *et al.*, 1975; Bierl *et al.*, 1976; Bierl-Leonhardt *et al.*, 1979; Cross, 1980; Cross *et al.*, 1980; Weatherston *et al.*, 1981a,b) it is cogent briefly to discuss release rate determination with regard to hollow-fibre devices. Initially a cathometer was used to follow the meniscus regression of the pheromone down the lumen (Ashare *et al.*, 1975, 1976). This method has an accuracy which translated into $\pm 1 \mu g$. Improvements in accuracy of greater than ten times occurred with the use of a Wilder Varibeam Optical Comparitor fitted with IKL digital positioners connected to an IKL Microcode Digital Readout system.

A sample of 25 or 50 fibres affixed to microscope slides was aged in environmentally controlled chambers and examined at regular time intervals. Release rates were calculated from the differences in the length of the liquid column, the internal diameter of the capillary and the density of the active ingredient. The disadvantges of this method are that it can only be applied to transparent and translucent fibres. It also does not yield any qualitative information concerning the active ingredient (Weatherston *et al.*, 1981a,b). The most accurate method for the measurement of release rates is to quantify the effluent vapours. In addition, by carrying out a residue analysis of the material remaining in the fibre it can be determined whether mass balance has been achieved.

After experimenting with various collection devices, adsorbents, flow rates and analytical methods and considering the efficiency of adsorption and desorption, the mini-flow device was designed (Weatherston *et al.*, 1981a,b; Golub *et al.*, 1983b). Briefly this method, which became the standard for release rate determinations in hollow-fibre formulations, was to prepare formulations containing radiolabelled pheromones, subject them to an airflow of $1 \, l \, min^{-1}$ in the mini-flow apparatus and trap the effluent vapour on glass beads. The beads were desorbed daily for 21 days, and the daily amount of pheromone collected determined by scintillation counting. After 21 days the amount of pheromone remaining in the fibre was also assayed by scintillation counting. For a detailed description of this method the reader is referred to the report of Golub *et al.* (1983b). It is possible by using two components individually labelled with ^{14}C and ^{3}H and an unlabelled component to determine the overall release rate and the individual release rates for three-component blends. The determination of release rates of multicomponent (greater than three components) mixtures, although possible, is time-consuming.

Protection of pheromone components from degradation during formulation, storage, and field use, is of vital importance since degradation products can affect the longevity, release rate and biological effectiveness of the formulation. It is obvious that if the pheromone is degraded then the longevity will be shortened. The effect on the release rate can take place in several ways: (a) if any of the degradation products are insoluble in the liquid load then there is the possibility of them precipitating out, thus altering the internal diameter of the fibre, causing interstices or completely occluding the lumen; (b) more volatile degradation products will increase the overall release rate of the active material while less volatile decomposition products will tend to decrease the overall release rate (Richardson, 1959; Weatherston *et al.*, 1985c); and (c) if one component is preferentially degraded not only will the overall release rate be altered but the blend ratio will be changed, with a resultant decrease or loss of biological effectiveness.

Most of the pheromones used commercially in insect control strategies are mono- or diunsaturated aliphatic compounds of molecular weight less than 300, and possessing alcohol, acetate or labile aldehyde functional groups. The labile nature of aldehydes and their susceptibility to oxidative decomposition has contributed to most work on pheromone degradation being directed at this class of compounds (Shaver and Ivie, 1982; Shaver, 1983; Golub *et al.*, 1983a). In 1980 problems with aldehyde degradation in Celcon fibres being field tested against *Heliothis* species caused formulations designed to have a longevity of 21 days to become ineffective after 3 days. This promoted an investigation of how best to protect fibre formulations of aldehydes (Weatherston, unpublished). It was shown in the laboratory by bubbling oxygen through neat aldehyde that the use of antioxidants containing either butylated hydroxytoluene or butylated hydroquinone offered protection. However, aldehydes plus antioxidant formulated in Celcon fibres again gave evidence of degradation after only three days of field testing.

The effect of ultraviolet (UV) light on aldehyde degradation was studied using (Z)-11-hexadecenal in stoppered quartz flasks exposed to Arizona sunshine. The data obtained showed that with aldehyde alone 41% decomposed within the first two days, and that 80% had decomposed by the twentieth day. The addition of 1% Escalol 507 (2-ethylhexyl-1-dimethylaminobenzoate), a sunscreen, retarded the onset of decomposition for at least six days. The use of 1% 2-hydroxy-4-iso-octoxybenzophenone, however, inhibited degradation over the 41-day duration of the experiment. Having established that UV-catalysed oxidative decomposition was the primary cause of aldehyde degradation in hollow-fibre formulations under field use conditions, there were two ways by which protection could be offered to such formulations. The first was the addition of UV stabilizer such as 2-hydroxy-4-iso-octoxybenzophenone to the pheromone solution prior to filling the fibres. The second was to modify the fibre material to preclude the UV radiation reaching the active ingredient contained in the lumen. Modification of the Celcon could have been accomplished by adding either a UV stabilizer or carbon (lampblack) to the polymer melt before extrusion. Tests (partially reported by Golub *et al.*, 1983a) indicated that protection was best afforded using black fibres, therefore all aldehyde-containing products were subsequently formulated with 1% antioxidant in black Celcon fibres.

There is also cause for concern both in storage of aldehydes as raw materials and finished products as well as in field use owing to their ability to form cyclic trimers (Starr and Vogl, 1978; Kubisa *et al.*, 1980; Dunkelblum *et al.*, 1984). Trimer impurities cannot be detected by gas chromatography since trioxanes dissociate in the chromatograph, hence characterization must be by infrared and nuclear magnetic resonance spectroscopy.

The olefinic nature of the majority of pheromones makes them susceptible to reaction with singlet oxygen. Using (Z,E)-9,11-tetradecadienyl acetate Shani and Klug (1980) have shown that this compound is first isomerized to the E,E isomer, which in turn is oxidized to an endoperoxide that is photodecomposed and dehydrated to form a 2,5-disubstituted furan. This oxidative decomposition can, however, be inhibited by antioxidants. The pheromone (E,Z)-3,13-octadecadienyl acetate in hollow fibres was shown (Sharp and Cross, 1980) to be 50% degraded within 2 months. By comparing formulations of (Z,Z)-3,13-octadecadienyl acetate with and without antioxidant Weatherston *et al.* (1984b) reported that, whereas degradation products appeared in field-aged samples without antioxidant between 14 and 29 days, the antioxidant Banox 20BA (Swift & Co., Chicago, Illinois) gave protection for at least 70 days.

It is appropriate at this juncture to comment generally on the use of antioxidants in hollow-fibre formulations. Prior to 1979 it was normal practice to dissolve in the distillation solution 1% (by weight based on the amount of pheromone) solid butylated hydroxytoluene. This was discontinued to avoid potential problems such as precipitation of the antioxidant on concentration as the pheromone was released. Liquid antioxidants, usually hindered phenols in vegetable or corn oil, offer a better solution. However, the choice must be made judiciously since some commercial antioxidants contain organic acids which have the potential to catalyse pheromone decomposition.

The NoMate hollow-fibre system for area-wide control has a requirement for an adhesive to stick the fibres of the formulation to the host plant of the target insect. This adhesive, sold under the trade name Bio-Tac™, is polybutene, and was originally available in three viscosity grades. The field-use temperature determines which grade should be used. This adhesive is mixed in the ratio of 100 g of the fibre formulation to 1 l of adhesive prior to loading into the application equipment.

The efficacy of pheromonal hollow-fibre release devices is outlined in Section 8.5. However, the reader is also referred to the reports of Doane and Brooks (1981) and Doane *et al.* (1982) and the references contained therein.

8.4 MANUFACTURE OF HOLLOW-FIBRE CONTROLLED-RELEASE DEVICES

The hollow-fibre controlled-release devices for insect monitoring (Scentry system) and area-wide control (NoMate system) differ significantly in their methods of manufacture. For this reason the manufacturing processes are presented in two sections.

8.4.1 NoMate systems for area-wide control

The fibre primarily used was the highly crystalline Celanese polyacetal copolymer Celcon. This extruded hollow fibre was supplied in a configuration such as to be compatible with the equipment designed for the sealing and cutting process. Outlined in Fig. 8.2 is a flow diagram of the overall manufacturing process.

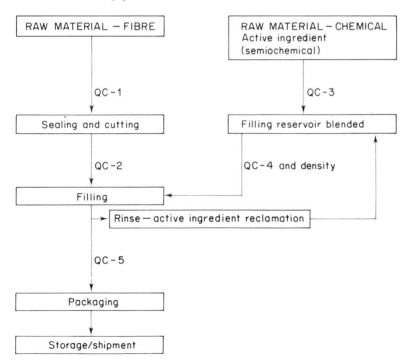

Fig. 8.2 Manufacturing flow chart—NoMate system. (QC=quality control)

Multiple continuous lengths of hollow fibres are mechanically fed across the anvil of an ultrasonic welding mechanism. At specified increments a signal is given to the ultrasonic device to crimp via pressure, using a metal ultrasonic transmitter called a 'horn', so that the internal walls of the hollow fibre meet. Once this has occurred a second signal is sent to the ultrasonic equipment to activate the vibrational motion of the horn. This vibrational energy transfers to the crimp wall to wall interface, which then fuses the walls together at that contact point. After cooling the predetermined welding programme is completed and the horn raises to allow the next incremental advance of the continuous hollow-fibre length. The fusing process may also be achieved using resistance heating elements as the sealing device.

Once the seal points are established the parallel array of continuous hollow fibres progresses to a cutting device. This device consists of a top cutting roll into which blades are set at predetermined arc lengths. The cutting roll maintains continuous contact with a urethane-coated bottom roll forming a nip point into which the incrementally sealed continuous fibre lengths pass. The fibres, once cut, are collected into large containers by a vacuum system and are then ready to be filled.

The fibre-filling procedure normally used consists of a batch process whereby a weighed quantity of empty chopped fibres is placed into a pheromone and solvent solution containing 1% (weight based on active ingredient) of an antioxidant, and vacuum-filled.

To ensure the efficient utilization of the expensive active ingredients it is critical to maintain an allowable range of pheromone/solvent ratios in the filling solution (reservoir). The active ingredient percentage is directly related to fibre weight, fibre length and, most importantly, liquid weight within a fibre. Therefore, a facile method of monitoring pheromone concentrations and density of the filling reservoir was developed. Knowledge of fibre parameters and the reservoir density allows the active ingredient concentration to be predicted and maintained within label specifications. A table of density versus concentration can be developed and utilized as one tool in monitoring the production process to achieve the designated active ingredient level in the formulation.

The filled fibres are transferred from the filling vessel and rinsed with solvent to remove excess active ingredients adhering to the outer surfaces. These fibres are then air-dried and packaged in variously sized hermetically sealed containers. As the filled fibre is rinsed, the excess active ingredient is removed from the outer surfaces of the fibre and mixes with the solvent. Since this is a batch process the active ingredient concentration of these washes gradually increases. To ensure adequate rinsing, this solution must be changed at specified intervals.

Batch rinses from the filling of the same pheromone are vacuum-distilled to recover the active material which is again utilized in the filling process.

Once packaged the appropriate label is placed on the can of filled hollow fibres. The product is then ready for subsequent storage and for shipment.

It is of paramount importance that regular safety measures be incorporated during the filling process owing to the large volume of flammable solvents used.

As in all manufacturing processes continuing final product quality can only be attained by ensuring the quality at intermediate stages. The manufacturing process described earlier in Fig. 8.2 for NoMate products is subjected to five quality control steps (QC-1 to QC-5), as noted in Fig. 8.2.

The manufactured fibre is examined microscopically for uniformity of outside and inside diameter. After the fibres are sealed and chopped, seal quality, seal placement and effective active length are checked (QC-1 and QC-2).

Using gas–liquid chromatography, infrared spectroscopy and other appropriate techniques, the overall purity and isomer content of the starting behaviour-modifying chemicals are verified (QC-3). Following the filling operation the reservoir contents are analysed to re-establish the semiochemical characteristics as compared to the raw material (QC-4). This check also confirms that the correct active ingredient/solvent ratio is maintained.

The final quality control verification (QC-5) is performed on the finished product to ensure compliance with label specification. For this analysis samples of the filled fibres are subjected to centrifugation in an appropriate solvent to extract the liquid completely, which is then analysed by gas chromatography. In addition, actual weight loss measurements are completed.

8.4.2 Scentry systems for insect monitoring

The manufacturing processes for these systems are illustrated in Fig. 8.3. In contrast to the NoMate system where only one fibre type was utilized, the Scentry monitoring systems employ two fibre types: Celcon, as described above, and polyester terephthalate. Both of these fibre types are manufactured and supplied in continuous lengths. A parallel array of the chosen fibre type containing the desired number of individual fibres is fixed to an adhesive substrate. Normally a 33 m length of this tape is selected for filling. The fibres of one end of this parallel array are placed into a length of stainless-steel tubing and epoxy-sealed, making certain that sufficient length of fibre is available to be subsequently placed into the semiochemical/solvent solution. The entire length of the fibre tape is filled using a pressure vessel. This is verified by the observation of the liquid being forced out of the opposite end of the fibre. The filling apparatus is then disassembled and both ends of the filled fibres are sealed for transport to the sealing equipment.

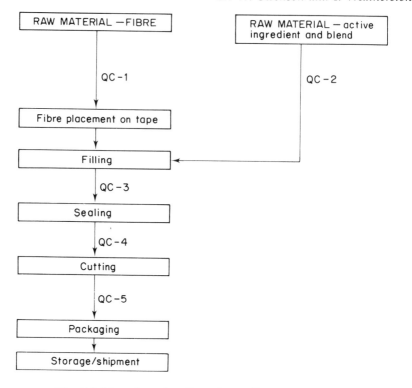

Fig. 8.3 Manufacturing flow chart — Scentry system

The sealing and cutting procedures are as described above, but the sealing process in the monitoring system requires different energy levels than that for individually sealed fibres because: (a) the tape substrate absorbs energy from the ultrasonic system; and (b) the liquid within the fibre acts as a 'heat sink' when welding the wall to wall interface at the crimp point. This is different from the NoMate area-wide control product as the sequence of filling and sealing in the monitoring product is reversed from the area-wide control product. For these reasons sealing problems become much more of a concern.

Finished lure products are packaged in hermetically sealed laminate packaging materials. This laminate is composed of paper, polyethylene, aluminium and polyethylene layers. Individual or multiple discrete tape units are placed into a laminate pouch and sealed.

As in the case of the NoMate area-wide control systems, quality control testing occurs at intermediate steps in the manufacturing process (QC-1 to QC-5 in Fig. 8.3). Identical testing (QC-1 and QC-2) is performed on the raw material — fibre and chemical respectively — as described above.

Following the mixture of the active ingredients, solvent and antioxidant (where applicable), quality control examination confirms the respective ratios (QC-3). In addition visual examination on the filled fibre array verifies completion of the filling step. Seal quality and the absence of polymer occlusion at the open ends are microscopically examined prior to packing (QC-4 and QC-5).

8.5 PRODUCTS DEVELOPED/MARKETING STRATEGIES

Given the methods and equipment to manufacture the hollow-fibre system in sufficient quantities for large area consumption one must then decide the products, markets and market strategies which should be utilized to establish a profitable controlled-release company. During the embryonic company development period the theoretical release concepts continued to be experimentally proven. Having accomplished this, the release device required specific applications which could develop into significant products. Concurrent with the release device development, work by USDA and other entomological researchers on the identification and application of insect communication chemicals increased. A key limiting factor to further development of these communication chemicals was the need for a system whereby they could be effectively disseminated.

Marketing strategy has been defined as the mixture of policy decisions that serve as a framework for optimum product launching (Kydonieus and Beroza, 1982). At the initial stage of the marketing of hollow-fibre release devices, product determination was obviously not a function of wide experimentally based decisions. Early product development criteria, while important, could not be classified as complete marketing strategies. These early criteria were: (a) the communication chemical had been elucidated; (b) existing insecticides or pest control costs per acre were high; (c) compounds used in formulations were thought to be stable; (d) initially single semiochemical component systems were viewed as optimum; (e) a progressive product distributor would be available in the targeted agricultural area (Brooks and Swenson, 1976).

The pheromone for pink bollworm (gossyplure) initially met these early criteria. The pink bollworm (PBW) infests about one-third of the world's cotton acreage. It is estimated that approximately 12 million hectares (out of a total of 32 million planted hectares) are infested with pink bollworm as a major pest. Of these 12 million hectares, approximately 4 million hectares would offer potential for a pink bollworm product. More specifically PBW was a major pest in the desert south-west of the United States cotton-growing area where a progressive distributor was interested in a product for this key pest. In addition the growing awareness of the environmental impact of insecticides was expected to provide an opportunity for easier registration of a pheromone product for PBW with the US Environmental Protection Agency (EPA).

This assumption was initially erroneous, as the first mating disruption product (Gossyplure H.F.—now called NoMate PBW) was treated exactly like a toxic chemical when presented for registration to the US EPA. The irony here, of course, is that a product has been presented to the EPA for registration which completely fulfils one of their objectives of protecting the public and the environment by reducing the use of toxic chemicals. With much debate and subsequent product registrations some of the early obstacles to registration have been reduced and presently a pheromone product is often considered in a more favourable light than toxic insecticides (see Chapter 12).

As one reviews the case of NoMate PBW, the key use strategies were developed usually in response to a perceived market opportunity. Table 8.2 summarizes these use strategies, with their respective advantages and disadvantages.

From the early decision criteria mentioned above other marketing considerations were developed as follows:

(1) Pest complex—products would be developed for the pest complex in a given crop.

(2) Crop target—a specific crop should be chosen and worldwide marketing of a product for an insect in a crop should be promoted.

(3) Geographical considerations—products should be developed by regional agricultural crops so as to utilize efficiently all distributors and sales people throughout the year.

(4) Chemical similarity—new products should be developed based on a targeted insect's pheromone being the same as, or similar to, a product already developed.

(5) Identified chemical—no new product should be anticipated where basic research is required to identify a semiochemical or blend of semiochemicals, i.e. only use materials already identified.

(6) Crop size—only large-acreage crops should be chosen for new product development. Cotton, corn, wheat, etc. would be target crops but artichoke, apples and tomatoes would not.

A broad range of monitoring and disruption products were developed during the company's short existence. The key area-wide control products with labelled composition and referenced efficacy information is presented in Table 8.3.

The development of insect pheromone products utilizing hollow fibres must compete with contemporary insecticide technologies. These hollow-fibre products must prove their merit not only efficaciously and ecologically but also economically. Key cost factors in a hollow-fibre product include:

TABLE 8.2 NoMate system use strategies

Strategies	Advantages	Disadvantages
Mating disruption	Use pheromone only Technically sound Economics, in some cases, competitive with conventional practice Reduces secondary insect pest potential Environmentally sound No field re-entry problems Reduces conventional chemical treatments	Low level of market acceptance to date Most effective against low levels of insect population Large areas required for product efficacy Unconventional application method and equipment
Attract and kill	Synergistic use of technologies — disruption and insecticides Increases market acceptance Improves cost/performance ratio Applicable to higher insect populations Potentially patentable Maintains environmental advantage	May have some field re-entry delays Not compatible with all insecticides
Bioirritant	Minimizes field maintenance and field scouting Provides greater options to grower	Concept has not been scientifically proven Does not reduce insecticide usage per application Field re-entry delays

(a) polymer and production cost of hollow fibre; (b) active ingredients costs; (c) level of dilution of the active material; (d) fibre internal volume; (e) manufacturing labour costs to produce a product; (f) overheads such as administration, marketing, sales and technical service. Since formulation costs are a key basis on which prices are established, it is most important that a formulation should use the shortest fibre length and the least active ingredient quantity, to achieve desired longevity and vapour emission. Recent work by Weatherston *et al.* (1985a,b) has shown that these parameters were not optimally developed in the hollow-fibre product developed from 1975 to 1984. Typical product costs for existing hollow fibre products range from $0.10 g^{-1} to $0.35 g^{-1}, with the active ingredient being 60–90% of total formulated cost. Pricing, however, was usually done in such a way as to project a grower's total anticipated use of a given product for a season and price that anticipated quantity of hollow-fibre product per acre equal to or just slightly in excess of what that grower would normally pay for

TABLE 8.3 Composition of hollow-fibre products (NoMate systems)

Product name	Target insect	Hollow-fibre design*	Chemical formulation	Chemical %	Efficacy references
NoMate PBW	*Pectinophora gossypiella*	1,7,9	(Z,Z)-7,11-Hexadecadienyl acetate	35.5	Brooks *et al.* (1979) Brooks *et al.* (1980) Doane *et al.* (1981)
			(Z,E)-7,11-Hexadecadienyl acetate	35.5	Doane *et al.* (1982)
			Hexane	29.0	Critchley *et al.* (1985)
NoMate TPW (Gusano)	*Keiferia lycopersicella*	1,7,9	(E)-4-Tridecenyl acetate	65.3	Van Steenwyck and Datman (1983)
			(Z)-4-Tridecenyl acetate	2.7	
			Hexane	32.0	
NoMate Chokegard	*Platyptilia carduidactyla*	2,7,9	(Z)-11-Hexadecenal	85.0	Haworth *et al.* (1982)
			Hexane	15.0	
NoMate Vantage	*Heliothis virescens* *Heliothis zea*	2,7,9	(Z)-11-Hexadecenal	37.6	Bartlett (1984)
			(Z)-9-Tetradecenal	2.4	
			Hexane	60.0	

NoMate Blockaide (1983 version) *Anthonomus grandis*

Grandlure portion	3,7,10	(Z)-2-Isopropenyl-1-methyl cyclobutane ethanol	25.4	Bartlett (1984)
		(Z)-3,3-Dimethyl-$\Delta^{1,\beta}$-cyclohexane ethanol	34.0
		(Z)-3,3-Dimethyl-$\Delta^{1,\alpha}$-cyclohexane ethanal	12.8
		(E)-3,3-Dimethyl-$\Delta^{1,\alpha}$-cyclohexane ethanal	12.8
Plant volatile portion	5,8,9	Hexane	15.0
		Myrcene	25.0
		1-Limonene	25.0
		β-Caryophyllene	25.0
		α-Pinene	25.0

NoMate Blockaide (1984 version) *Anthonomus grandis*

Grandlure portion	2,6,10	See 1983 version	See 1983 version
Plant volatile portion	4,7,9	See 1983 version	See 1983 version

*1. White Celcon 0.008 inch (0.0203 cm) inside diameter. 2. Black Celcon 0.008 inch (0.0203 cm) inside diameter. 3. Black Celcon 0.020 inch (0.0508 cm) inside diameter. 4. Yellow Celcon 0.008 inch (0.0203 cm) inside diameter. 5. Yellow Celcon 0.020 inch (0.0508 cm) inside diameter. 6. 0.96 cm long. 7. 1.41 cm long. 8. 2.00 cm long. 9. End sealed. 10. Centre sealed.

conventional insecticides. Product pricing to a distributor would range from $0.20 g^{-1} to $0.95 g^{-1} depending on product. Again, owing to variation in the marketing function in the USA, this pricing philosophy would change from that as stated to the concept that a new technology and product must be price advantageous to the eventual end user to obtain market share. With this price philosophy vacillation it was difficult to position any of the fibre products effectively.

8.6 CONCLUSIONS

From a technical standpoint the hollow-fibre release devices are highly effective for area-wide dissemination of insect pheromones. Continuing research could improve not only the efficacy of the hollow-fibre release systems but also the economics of their manufacture and commercial use.

Many of the difficulties of commercialization of pheromones as viable insect control agents are mirrored in the commercialization efforts of hollow-fibre technology in the USA. Since the late 1960s insect pheromones have been lauded as showing great potential for insect control, but for various reasons (e.g. lack of patent protection of active ingredient, perceived insecticide incompatibility, etc.) few major agrochemical companies have made an investment in insect semiochemical commercialization, and much of the 'great potential' has yet to be realized.

The companies which have attempted to commercialize pheromones have, without exception, done so based on their particular controlled release technology. This situation left them vulnerable in the areas of marketing and sales because the controlled-release products required different use education and therefore were perceived as more difficult to use. This, in time, became an obstacle to market penetration in the highly competitive agrochemical industry.

Hollow-fibre technology has not been successfully commercialized because of investors' non-flexibility in viewing return on initial investment. Conventional technologies often allow pay-backs from research and development to profit centre designation and complete return on investment in about seven years. However, because of the evolving technology and the significant difference in insect treatment strategies required with semiochemical use and the times required to optimize product design and develop user education strategies, return on investment would probably require 15–20 years.

Although it may already be too late for behaviour-modifying chemicals to be developed to their full potential, ever-growing environmental concerns, particularly following the pesticide tragedy in Bhopal, India, may provide the incentive to major agrochemical companies to commit to a long-term investment in alternative strategies such as semiochemical development.

8.7 REFERENCES

Ashare, E., Brooks, T. W. and Swenson, D. W. (1975). Controlled release from hollow fibres. In *Proc. 1975 Controlled Release Pesticide Symp.*, F. W. Harris (Ed.), pp. 42–49.

Ashare, E., Brooks, T. W. and Swenson, D. W. (1976). Controlled release from hollow fibres. In *Controlled Release Polymeric Formulations*, D. R. Paul and F. W. Harris (Eds), American Chemical Society, Washington, pp. 273–282.

Bartlett, R. H. (1984). NoMate™ Blockaide and NoMate Vantage, new behaviour modifying chemicals for cotton insect control, *Proceeding Beltwide Cotton Production–Mechanization Conference*, Atlanta, pp. 75–77.

Beroza, M., Bierl, B. A., James, P. and DeVilbiss, E. D. (1975). Measuring emission rates of pheromones from their formulations. *J. Econ. Entomol.*, **68**, 369–372.

Bierl, B. A., DeVilbiss, E. D. and Plimmer, J. R. (1976). Use of pheromones in insect control programs: Slow release formulations. In *Controlled Release Polymeric Formulations*, D. R. Paul and F. W. Harris (Eds), American Chemical Society, Washington, pp. 265–272.

Bierl-Leonhardt, B. A., DeVilbiss, E. D. and Plimmer, J. R. (1979). Rate of release of disparlure from laminated plastic dispensers, *J. Econ. Entomol.*, **72**, 319–321.

Brooks, T. W. (1980). Controlled vapor release from hollow fibres: Theory and applications with insect pheromones. In *Controlled Release Technologies: Methods and Applications* (Vol. II), A. F. Kydonieus (Ed.), CRC Press, Boca Raton, Florida, pp. 165–193.

Brooks, T. W. and Swenson, D. W. (1976). Market and economic factors in development of controlled release pest control products based on hollow fibres. *Chemical Marketing and Economic Reprints of the American Chemical Society*, pp. 166–180.

Brooks, T. W., Ashare, E. and Swenson, D. W. (1977). Hollow fibres as controlled vapor release devices. *Textile and Paper Chemistry and Technology*, J. C. Arthur (Ed.), American Chemical Society, Washington, pp. 111–126.

Brooks, T. W., Doane, C. C. and Staten, R. T. (1979). Experience with the first commercial pheromone communication disruptive for suppression of an agricultural pest. In *Chemical Ecology: Odor Communication in Animals*, F. W. Ritter (Ed.), Elsevier/North Holland Biomedical Press, Amsterdam, pp. 375–388.

Brooks, T. W., Doane, C. C. and Harworth, J. K. (1980). Suppression of *Pectinophora gossypiella* with a sex pheromone. *Proc. 1979 British Crop Protection Conference*, Vol. 3, pp. 853–860.

Critchley, B. R., Campion, D. G., McVeigh, L. J., McVeigh, E. M., Cavanagh, G. G., Hosny, M. M., Nasr, El-Sayed A., Khidr, A. A. and Naguib, A. A. (1985). Control of the pink bollworm, *Pectinophora gossypiella* (Saunders) (Lepidoptera: Gelechiidae), in Egypt by mating disruption using hollow-fibre, laminate flake and micro-encapsulated formulations of synthetic pheromone. *Bull. Ent. Res.*, **54**, 329–345.

Cross, J. H. (1980). A vapor collection and thermal desorption method to measure semiochemical release rates from controlled release formulations. *J. Chem. Ecol.*, **6**, 781–787.

Cross, J. R., Tumlinson, J. H., Heath, R. R. and Burnett, D. E. (1980). Controlled release formulations of semiochemicals: Apparatus and procedure for measuring release rate. *J. Chem. Ecol.*, **6**, 759–770.

Doane, C. C. and Brooks, T. (1981). Research and development of pheromones in insect control with emphasis on pink bollworm. In *Management of Insect Pests with Semiochemicals; Concepts and Practice*, E. R. Mitchell (Ed.), Plenum Press, New York, pp. 285–303.

Doane, C. C., Brooks, T. W. and Weatherston, J. (1982). Insect control through mating disruption with pheromones. In *Insect Pheromones and Their Application*, R. A. Galbreath (Ed.), DSIR Entomology Division Report 2, pp. 89–97.

Dunkelblum, E., Kehat, M., Klug, J. T. and Shani, A. (1984). Trimerization of *Earias insulana* sex pheromone (*E,E*)-10,12-hexadecadienal, a phenomenon affecting trapping efficiency. *J. Chem. Ecol.*, **10**, 421–428.

Floyd, J. P. and Crowder, L. A. (1981). Sublethal effects of permethrin on pheromone response and mating of male pink bollworms. *J. Econ. Entomol.*, **74**, 634–638.

Golub, M., Haworth, J., Alves, N., Phelan, S., Puck, P. R. and Weatherston, I. (1983a). NoMate Chokegard: the formulation of Z-11-hexadecenal in hollow fibres for use in the mating suppression of the artichoke plume moth. *Proc. 10th International Symp. Contr. Release of Bioactive Materials*, 234–237.

Golub, M., Weatherston, I. and Benn, M. H. (1983b). Measurement of release rates of gossyplure from controlled release formulations by the mini-airflow method. *J. Chem. Ecol.*, **9**, 323–333.

Harworth, J. K., Puck, R. P., Weatherston, J., Doane, C. C. and Ajaska, S. (1982). Les Mediateurs Chemignes Agissant Sur Le Comportement Des Insects. *Les Collogues de l'INRA*, No. 7, pp. 342–356.

Kaissling, K. E. (1982). Action of chemicals, including (+) *trans*-permethrins and DDT as insect olfactory receptors, *Insect Neurobiol. Pestic. Action, Proc. Soc Chem. Ind. Symp. 1979*, Soc. Chem. Ind., London, pp. 351–358.

Kubisa, P., Neeld, K., Starr, J. and Vogl, O. (1980). Polymerization of higher aldehydes. *Polymer*, **21**, 1433–1447.

Kydonieus, A. and Beroza, M. (1982). Marketing and economics in use of pheromones for suppression of insect populations. In *Insect Suppression with Controlled Release Pheromone Systems* (Vol. II), CRC Press, Boca Raton, Florida, pp. 187–199.

Richardson, J. F. (1959). The evaporation of two-component liquid mixtures. *Chem. Eng. Sci.*, **10**, 234–242.

Shani, A. and Klug, J. T. (1980). Photoxidation of (*Z,E*)-9,11-tetradecadienyl acetate, the main component of the sex pheromone of the female Egyptian cotton leafworm (*Spodoptera littoralis*). *Tetrahedron Lett.*, **21**, 1563–1564.

Sharp, J. L. and Cross, J. H. (1980). Longevity and attractivity to male lesser peachtree borer of *E,Z*-ODDA evaporated from various dispensers. *Environ. Entomol.*, **9**, 818–820.

Shaver, T. N. (1983). Environmental fate of (Z)-11-hexadecenal and (Z)-9-tetradecenal, components of a sex pheromone of the tobacco budworm (Lepidoptera: Noctuidae). *Environ. Entomol.*, **12**, 1802–1804.

Shaver, T. N. and Ivie, G. W. (1982). Degradation products of Z-11-hexadecenal and Z-9-tetradecenal, components of a sex pheromone of the tobacco budworm. *J. Agric. Food Chem.*, **30**, 367–371.

Starr, J. and Vogl, O. (1978). Higher aliphatic aldehyde polymers. V. Cyclic trimers of C_9 to C_{12} normal aliphatic aldehydes. *J. Macromol. Sci. Chem.*, **A12**, 1017–1039.

Van Steenwyck, R. A. and Datman, E. R. (1983). Motion disruption of tomato pinworm [Lepidoptera—Gelechiidae] as measured by pheromone traps, foliage and fruit damage. *J. Econ. Entomol.*, **76**, 80–84.

Weatherston, I., Golub, M. A., Brooks, T. W., Huang, Y. Y. and Benn, M. H. (1981a). Methodology for determining the release rates of pheromones from hollow fibres. In *Management of Insect Pests with Semiochemicals*, E. R. Mitchell (Ed.), Plenum Press, New York, pp. 425–443.

Weatherston, I., Golub, M. A. and Benn, M. H. (1981b). Release rates of pheromones from hollow fibres. In *Insect Pheromone Technology: Chemistry and Applications*, B. A. Leonhardt and M. Beroza (Eds), American Chemical Society, Washington, pp. 145–157.

Weatherston, I., Miller, D. and Dohse, L. (1984a). Release of volatile materials from glass capillaries — a reinvestigation. *Proc. 11th International Symp. Contr. Release of Bioactive Materials.*, pp. 114–115.

Weatherston, I., Neal, J. J. and Golub, M. A. (1984b). Efficiency of pheromone utilization by hollow fibre controlled release devices containing (Z,Z)-3,13-octadecadienyl acetate. *Acta Ent. Bohemoslov.*, **81**, 331–336.

Weatherston, I., Miller, D. and Dohse, L. (1985a). Capillaries as controlled release devices for insect pheromones and other volatile substances — a re-evaluation. 1. Kinetics and the development of a predictive model for glass capillaries. *J. Chem. Ecol.*, **11**, [8], 953–965.

Weatherston, I., Miller, D. and Lavoie-Dornik, J. (1985b). Capillaries as controlled release devices for insect pheromones and other volatile substances — a re-evaluation. 2. Predicting release rates from Celcon™ and teflon capillaries. *J. Chem. Ecol.*, **11**, [8], 967–978.

Weatherston, I., Miller, D. and Lavoie-Dornik, J. (1985c). Commercial hollow fibre pheromone formulations: the degrading effect of sunlight on Celcon fibres causing increased release rates of the active ingredient. *J. Chem. Ecol.*, **11**, [12], 1631–1644.

Insect Pheromones in Plant Protection
Edited by A. R. Jutsum and R. F. S. Gordon
Published 1989 by John Wiley & Sons Ltd

9

Microcapsules

D. R. Hall
*Overseas Development Natural Resources Institute, Central Avenue,
Chatham Maritime, Chatham, Kent ME4 4TB, UK*
G. J. Marrs
*ICI Agrochemicals, Jealott's Hill Research Station,
Bracknell, Berks, RG12 6EY, UK*

9.1 INTRODUCTION

In microencapsulation, core material in the form of particles or droplets is encased in a capsule wall to give microcapsules typically 1–1000 μm in diameter. The process can be said to have originated in 1954 with the production of carbonless copy papers, but it has since been utilized in many other areas and the basic techniques and applications have been extensively reviewed (e.g. Herbig, 1968; Gutcho, 1972, 1976, 1979; Vandegaer, 1974; Sliwka, 1975; Nixon, 1976; Patwardhan and Das, 1983; Finch, 1985). In the broadest sense, microencapsulation provides for the packaging of materials in a microgranular form suitable for the specific application. The encapsulated material is isolated to a greater or lesser extent from the surrounding environment, but it can be made available at a later stage by rupture of the capsule wall or passage through the wall.

When the technique of using synthetic pheromones to disrupt intraspecific communication of insect pests became a feasible proposition in the early 1970s, the technical problems were seen to be those of applying expensive, unstable, water-immiscible, volatile synthetic pheromones over large areas so that these materials were protected from environmental factors for periods of at least several weeks, during which time they were released slowly into the atmosphere. Microencapsulation seemed to be an exceedingly appropriate process to enable this technique to be put into practice. Thus, microencapsulation could be carried out on a large scale using established processes. A suspension of the microcapsules in a suitable aqueous or non-aqueous medium could be applied over large areas with the types of equipment used for spraying conventional insecticides. The microcapsule walls would provide some protection from environmental factors for the contents, and stabilizing agents could be mixed in with the contents or incorporated in the microcapsule walls. The physical and chemical properties of the microcapsules and their walls could be modified to provide appropriate release rates of the contents.

At the time, it was thought that for maximum effect in communication disruption, permeation of the atmosphere with pheromone should be as uniform as possible, and distributing the pheromone as an even spray of minute point sources should achieve just this. It has subsequently been shown that, at least in some cases, a more discontinuous spread of larger point sources of pheromone can be equally or more effective than a uniform 'fog' (e.g. Sanders, 1982; see also Bartell, 1982) but by using agglomerations of microcapsules the size of the resulting point sources can be varied over a wide range.

Microencapsulation has also been used to formulate conventional insecticides for similar reasons to those above, with the added advantage of reducing the toxicity of these materials to non-target organisms. The use of microencapsulation for formulation of agricultural chemicals in general has also been reviewed (e.g. DeSavigny and Ivy, 1974; Bakan, 1975, 1976, 1980; Scher, 1977a, 1983; Koestler, 1980; Somerville and Goodwin, 1980; Meghir, 1984).

This chapter aims to summarize first the methods that have been or could be used to microencapsulate pheromones and other behaviour-modifying chemicals; second, studies of the physicochemical properties of such formulations that have been carried out in order to optimize their performance in the field; and third, some of the operational and design features of microencapsulated formulations determining their use in the field. The results of field studies are described where relevant; many of these will be described in detail elsewhere, but a tabular summary of published results of the use of microencapsulated pheromone formulations in mating disruption is included.

9.2 METHODS OF MICROENCAPSULATION

Microencapsulation processes can be categorized operationally into two general types: physicochemical and mechanical (Herbig, 1968). Physico-chemical microencapsulation processes are typically carried out in a liquid medium and involve a chemical change and/or phase separation. They are characterized by batch operation with conventional equipment of high capacity and provide for a wide range of capsule types and sizes. Mechanical microencapsulation processes are generally carried out in the gas phase and the capsule wall is created by some physical process such as loss of solvent or change in temperature. Such processes are typically carried out on a continuous basis with specialized equipment of relatively low capacity, and the range of capsule types is more restricted.

The most significant advances in the use of microencapsulated pheromones for control of insect pests have been achieved with microcapsules made by physicochemical processes, although recently other processes for micro-encapsulation of pheromones have been reported. The processes described below have been used to encapsulate pheromones or could in principle be used for pheromones and other behaviour–modifying chemicals. References to specific patented processes are given, although classified summaries of the patent literature on microencapsulation until 1979 have been published by Gutcho (1972, 1976, 1979).

9.2.1 Physicochemical methods of microencapsulation

9.2.1.1 Wall formation in the presence of polymeric wall material

In these processes, the material to be encapsulated is dispersed in a continuous phase containing the wall-forming polymer and the latter is then deposited at the interface by coacervation, change in pH, chemical reaction, change in solvent or change in temperature.

9.2.1.1.1 Coacervation processes
The conversion of a colloid or polymer, uniformly distributed in a solvent, into a solid precipitate can proceed through an intermediate stage known as a coacervate. In appropriate systems, coacervate droplets will coalesce to form polymer-rich films around particles or droplets, with a consequent lowering of surface free energy. Many of the patents covering formation of microcapsules by such processes are assigned to the National Cash Register Co. (NCR).

Simple coacervation of a sol can be induced by addition of salts or other precipitants, by dilution or by change in pH, for example. Wall materials include gelatin, gum arabic, poly(vinyl alcohol), cellulose esters and ethers, styrene–maleic anhydride copolymers, poly(ethylene oxides), polychlorotri-fluoroethylene, vinylidene chloride–vinyl chloride copolymers and other polymers possessing some solvent solubility (Green, 1955, 1960; Green and Schleicher, 1956).

The majority of gelatin-based, microencapsulated formulations of phero-mones utilized in field applications have been prepared by complex coacervation involving mutual precipitation of two oppositely charged sols or polymers. Typical polymers used are gelatin, which is amphoteric with the isoelectric point at pH 8, and the negatively-charged gum arabic. Mixing the two in dilute solution at low pH causes the gelatin to become positively charged and coalesce with the gum arabic. Coacervation is carried out above the gelation temperature of the gelatin (37 °C) and subsequent cooling of the coacervate causes it to gel.

In a typical process, the material to be encapsulated is emulsified in a dilute aqueous solution of gelatin and gum arabic at pH 7 and 50 °C. The pH is lowered to pH 4.5 with acetic acid to induce coacervation around the droplets of core material and then the temperature is lowered to 5–10 °C to cause the coacervate to gel. The capsule wall can be hardened by addition of formaldehyde, glutyraldehyde, glyoxal or acrolein and adjusting the pH to 9–10 (Green and Schleicher, 1957). In a modification of this process, a second deposition of a similar film former such as an ethylene–maleic anhydride copolymer is effected by adding a solution of this after the initial coacervation, followed by reacidification (Brynko and Scarpelli, 1965).

9.2.1.1.2 Change in pH
The Shin-Etsu Chemical Co. Ltd has patented a process for micro-encapsulating pheromones with cellulose derivatives containing carboxyl groups. In one example, the pheromone is emulsified in an alkaline aqueous solution of hydroxypropylmethylcellulose acetate succinate; on acidification the latter is precipitated as a wall around the droplets (Shin-Etsu Chemical Co. Ltd, 1983).

9.2.1.1.3 Chemical reaction
In one example, the core material is dispersed in an aqueous solution of a hydrolysable cellulose derivative. The mixture is made alkaline and the dissolved cellulose derivative is saponified, leading to the precipitation around the droplets of a cellulose film insoluble in water or the core material (Harbort, 1966). This process could, in principle, be used for pheromones, although probably not with those which react with aqueous alkali such as esters or aldehydes.

9.2.1.1.4 Change in solvent
The core material is emulsified in water. A solution of a polymer such as an acrylic acid–methacrylic acid copolymer in chloroform and isopropanol is added and, if appropriate mixtures are used, this distributes itself around the droplets of core material. The solvents are redistributed between the core, water and wall phases, and the polymer precipitates to form capsule walls. The solvents can be removed by distillation and the wall material cross-linked by addition of formaldehyde, dialdehydes or polyamines as appropriate (Baum et al., 1971, 1972). This process could also be used to

microencapsulate pheromones, providing a suitable solvent system could be found.

9.2.1.1.5 Change in temperature

A process potentially useful for encapsulating lipophilic, behaviour-modifying chemicals involves emulsifying the core material in water and adding a hot solution of a polymer such as low-density polyethylene or poly(vinyl chloride) having a crystalline melting point greater than 90 °C in an organic solvent. On cooling, the polymer comes out of solution as a film around the core droplets (Pelah and Marcus, 1982).

9.2.1.2 Wall formation from monomeric or oligomeric starting materials

9.2.1.2.1 Interfacial polycondensation

Processes of this type were pioneered by the Pennwalt Corporation and have been extensively used for microencapsulation of both conventional insect-icides and pheromones (Koestler, 1980). The basic principle involves emulsification of the core material and one monomer in an aqueous solution of the other monomer; condensation then occurs at the core/water interface (Ruus, 1969; Vandegaer and Meier, 1969; Vandegaer, 1971; DeSavigny, 1976). In a typical process, the core material mixed with adipoyl chloride is emulsified in water containing poly(vinyl alcohol) as emulsifying agent. An aqueous solution of hexamethylene diamine and sodium carbonate is then added so that a capsule wall of polyamide is formed at the interface, the sodium carbonate neutralizing the acid liberated (Vandegaer, 1971).

Interfacial polycondensation is easily carried out on a large scale and a wide variety of monomers is readily available to give capsule walls of different compositions. Diamines (e.g. hexamethylene diamine, ethylene diamine) react with diacid chlorides (e.g. adipoyl, sebacoyl, terephthaloyl chlorides) to give polyamides or with diisocyanates (e.g. toluene diiso-cyanate, hexamethylene diisocyanate) to give polyureas. Bisphenols (e.g. bisphenol A) react with diacid chlorides or diisocyanates to give polyesters and polyurethanes, respectively. Cross-linking is achieved by using poly-functional isocyanates (e.g. poly(methylene(polyphenyl)) isocyanate), poly-functional acid chlorides (e.g. mesitoyl trichloride) and/or polyfunctional amines (e.g. diethylene triamine, tetraethylene pentamine).

9.2.1.2.2 In situ interfacial polycondensation

This is a variation of the above process patented by the Stauffer Chemical Co. The core material is mixed with a polyisocyanate and emulsified in water. Addition of alkali causes hydrolysis of the isocyanate to the carbamic acid, which decomposes with liberation of carbon dioxide to form the corresponding amine. The latter then reacts with unreacted isocyanate to form a polyurea (Scher, 1975, 1981). In the original process, the alkaline emulsion had to be heated to over 50 °C, but with addition of a quarternary ammonium phase transfer catalyst, the hydrolysis, decarboxylation and polymerization can be carried out rapidly at room temperature (Scher, 1979). This process gives capsules with an inner structure of polyurea, making them more pressure resistant. It also avoids the use of an excess of aliphatic polyamines required in conventional interfacial processes, and this is important for agricultural applications as these polyamines can be phytotoxic. These capsules can be further protected by a graft-copolymer formed by combination of hydrolysed maleic anhydride–methyl vinyl ether copolymer and poly (vinyl alcohol) (Rodson and Scher, 1984).

The presence of unreacted isocyanate groups in the final product could lead to toxicity problems and might react with certain core materials during storage. These can be minimized by treating the capsules with ammonia (Scher, 1977b) or by using for wall formation polyurethane–polyisocyanate prepolymers formed by reacting an excess of a polyisocyanate with a polyhydroxy compound such as 1,2,6-hexanetriol (Heinrich *et al.*, 1979, 1981).

9.2.1.2.3 *Interfacial polymerization*

The National Cash Register Co. has also patented microencapsulation processes in which the capsule wall is formed by addition polymerization at liquid–liquid interfaces. The core material is mixed with a monomer such as styrene, ethyl acrylate, methyl methacrylate or vinyl acetate, and the mixture is emulsified in a polar medium such as water. Polymerization at the interface is induced by thermal decomposition of an initiator such as azobisisobutyrylnitrile, di-*tert*-butyl peroxide or acetyl peroxide introduced into either of the phases (Brynko, 1961; Brynko and Scarpelli, 1961). This procedure could be used to microencapsulate pheromones which are not affected by the radical-driven polymerization process.

In another process of this type, the core material is dispersed in an aqueous solution of a urea–formaldehyde precondensate. Acidification causes curing and the resulting insoluble polymer precipitates around the droplets of core material (Matson, 1962).

9.2.2 Mechanical methods of microencapsulation

9.2.2.1 *Spray drying/congealing*

Processes of these types have been used for many years in the food industry to produce 'locked-in' flavours. In spray drying, the core material is dissolved in a solution of a film-forming polymer. The dispersion is atomized into a stream of hot, inert gas which removes the solvent so that the polymer is deposited on the core droplets (Macaulay, 1962). In spray congealing, the core material is dispersed in a polymer melt which is sprayed as droplets into cooler air or liquid so that the polymer solidifies.

The Nitto Electrical Industry Co. has patented a spray-drying process for microencapsulation of pheromones. In one example, a solution of a pheromone and the polysulphone P-1700 in dichloromethane is sprayed into warm air as 30–40 μm droplets so that the solvent evaporates leaving microcapsules 2–20 μm in diameter with 0.2–2 μm thick walls and containing 40% pheromone (Nitto Electrical Industry Co. Ltd, 1984).

The Zoecon Co. has developed a process in which a pheromone is dissolved in an oil and emulsified in a water-soluble acrylic resin. The dispersion is sprayed to form droplets 3–5 mm in diameter which are allowed to dry, forming capsules containing up to several thousand pheromone/oil droplets encased in a resin matrix. Different resins can be used, and these can be further modified by addition of polyamines which cross-link the resin during the drying process (Evans, 1984).

9.2.2.2 Multiple-orifice techniques

In these techniques, the core material is passed through a film of the wall-forming material either as droplets or as a continuous stream which subsequently breaks up into droplets surrounded by the wall material. These pass into a region where the wall material is consolidated by cooling, drying or hardening as appropriate. In one well-established technique of centrifugal microencapsulation, the core material is fed into a rotating disc which breaks the stream into small droplets. These are propelled through orifices in the periphery of the centrifugal bowl which are continuously supplied with a film of wall-forming material. In breaking through this film the droplets are encased in wall-forming material (Somerville, 1962, 1967, 1968; Somerville and Goodwin, 1980). Capsule size can be varied from a few microns to several millimetres by varying the speed of rotation of the disc and the size of the orifices; and wall thickness can be varied by changing the supply to the orifices. Wall materials can be selected from a range of soluble or liquefiable natural or synthetic polymers, including gelatin, waxes, cellulose, styrene/butadiene copolymers and acrylic polymers, and these processes would seem to be potentially applicable to microencapsulation of pheromones.

9.2.2.3 Electrostatic processes

Liquids as well as small particles can be encapsulated by atomizing core materials and wall material separately and giving them opposite electric charges with high-voltage, corona discharge devices before they pass into a mixing chamber where they combine. The coating substance can be molten, e.g. a phthalate ester or microcrystalline wax, or a solution of a synthetic resin, e.g. polyamide, polyvinyl esters, ethers, polystyrene or acrylates. Alternatively, the core and coating materials can contain appropriate monomers so that interfacial polycondensation occurs when they combine (Langer and Yamate, 1964, 1966; Berger *et al.*, 1965).

These processes are, in principle, applicable to pheromones, but the apparatus required is complex and the surface tensions and conductivities of the core and wall materials must meet certain criteria.

9.2.3 Other methods of microencapsulation

Other microencapsulation processes are probably not applicable to lipophilic, liquid behaviour-modifying chemicals as such. These processes include organic phase separation and meltable dispersion techniques for encapsulating hydrophilic materials, and pan coating and fluidized-bed spray coating for encapsulating solids.

9.3 PROPERTIES OF MICROCAPSULES

9.3.1 General considerations

Microencapsulation provides a convenient form of packaging pheromones, converting a fairly volatile, lipophilic liquid into a stable aqueous dispersion suitable for application with conventional spray equipment. Equally importantly, it is also a means of protecting the pheromone from environmental factors and of providing for controlled release of the pheromone.

Microencapsulation provides protection of the pheromone by virtue of the capsule wall restricting access of light and oxygen, but it also enables stabilizing additives to be incorporated into both the capsule wall and the contents during the manufacturing process.

Studies of the controlled release of pheromones and other agrochemicals from microcapsules have generally assumed that the predominant release mechanism is by diffusion of the contents through the capsule wall, and they have neglected the possibility of release by rupture of the capsule wall.

Release by diffusion is governed by Fick's law (Baker and Lonsdale, 1974; Thies, 1975; Scher, 1977a; Roth *et al.*, 1981).

$$\text{rate of release} = \frac{KDA\Delta C}{l} \qquad \text{Eq. 1}$$

where K = distribution coefficient for core material between wall and surrounding material; D = diffusion coefficient for core material in wall; A = area for diffusion; ΔC = concentration difference across wall; l = thickness of wall.

For a microcapsule of external radius r_o and internal radius r_i (both assumed to be small):

$$\text{rate of release} = \frac{4KD\Delta Cr_or_i}{(r_o - r_i)} \qquad \text{Eq. 2}$$

The rate of release per unit mass of core material (assuming density = 1)

$$= \frac{3KD\Delta C}{r_o\,(r_o - r_i)} \qquad \text{Eq. 3}$$

Thus, if the concentration difference across the capsule wall (ΔC) is kept constant, a steady, zero-order release should take place (Roth *et al.*, 1981). This might be achieved by encapsulating only the active material, but, in practice, the active material is usually mixed with solvents and/or additives so that the internal concentration of active material falls during the release process and a decreasing, first-order release is observed.

The effects of microcapsule design features on the rate of release by diffusion can be predicted with the help of the above equations, but, as shown in Table 9.1, the effects of changes in capsule parameters and composition of the capsule wall are qualitatively the same for release by diffusion or by rupture. Release by capsule rupture could also follow a first-order profile, the rate being proportional to the number of capsules remaining.

9.3.2 Properties of gelatin-based microcapsules

Gelatin-based microcapsules, manufactured by the National Cash Register Co. complex coacervation process, have been used with the pheromones of several lepidopteran pests, but all the developmental work was carried out at the USDA Beltsville laboratories with microencapsulated formulations of disparlure, the pheromone of the gypsy moth, *Lymantria dispar*. Disparlure, *cis*-7,8-epoxy-2-methyloctadecane, is a very unusual lepidopterous sex pheromone: it has no unsaturation and the only functionality is the epoxide grouping, whereas the majority of lepidopteran sex pheromones are unsaturated compounds, typically esters, alcohols or aldehydes.

9.3.2.1 Laboratory studies

In laboratory studies, a measured aliquot of the formulation, with additives if appropriate, was spread evenly on a microscope slide or metal planchet. The formulation was then maintained in a laboratory device under constant temperature, humidity and windspeed, or in the field exposed to selected

TABLE 9.1 Effects of changing microcapsule characteristics on release by diffusion and release by capsule rupture.

	Effect on rate of release	
	Diffusion	Rupture
CAPSULE PARAMETERS		
Capsule size increase	DECREASE Decreased surface/ volume ratio (Eq. 3)	DECREASE Decreased surface/ volume ratio, decreased rate of rupture per unit mass
Wall thickness increase	DECREASE (Eq. 3)	DECREASE Stronger wall
Regularity of walls increase	DECREASE Smooth walls minimize surface area and likelihood of thin patches	DECREASE minimize potential weak points
WALL COMPOSITION		
Density increase	DECREASE Decrease D,K	DECREASE Increased strength
Crystallinity increase	DECREASE Decrease D	DECREASE Increased strength
Cross-linking increase	DECREASE Decrease D	DECREASE Increased strength
Plasticizer increase	INCREASE Increase D	INCREASE (?)
Use good solvents in wall preparation	DECREASE Smaller micropores	DECREASE Increased strength

environmental factors such as wind, sun and rain. Release rates were measured by gas chromatographic (GC) analysis of the residual pheromone at intervals in replicate samples, and by measuring actual pheromone released by samples of different ages. The latter measurements were made by passing air over the sample at 100 ml min^{-1} (linear velocity approx. 0.003 m s^{-1}) and then through hexane to give a solution of the entrained pheromone that could be assayed by GC analysis (Beroza *et al.*, 1975a).

All the formulations tested had a core : wall ratio of 10 : 1. Increasing the amount of plastic coating formed by addition of hardener during manufacture caused the expected decrease in release rate (Plimmer *et al.*, 1976). Coating percentages of 0%, 25% and 50% were compared, and 25% was chosen for subsequent formulations. It was reported that varying capsule size had little effect on release rate (Plimmer *et al.*, 1976), but this was not extensively tested. Formulations used in the field had capsule diameters in the range 50–400 μm or 50–250 μm.

The formulations tested contained the pheromone in solution in xylene or, later, 3 : 1 xylene/amyl acetate. Lower concentrations of pheromone gave an initial rapid release which then slowed as the pheromone content dropped, while higher concentrations gave a similar initial release rate, i.e. lower percentage of contents released, which was more sustained (Plimmer *et al.*, 1976). Typical data for formulations aged in a shelter outdoors are shown in Table 9.2 (Plimmer *et al.*, 1977).

The relatively large capsule size meant that a sticker had to be used with these formulations. Originally Rhoplex-B-15 (Rohm and Haas Co.) was used, but this was later replaced by RA-1645 Latex (Monsanto Corp.) with 10% Triton X-202 surfactant. Emission rates were reduced to less than half by addition of 2% sticker, presumably due to coating and agglomeration of the capsules under the experimental conditions used (Plimmer *et al.*, 1976).

Increase in temperature caused a marked increase in release rate. At temperatures of 27 °C, 32 °C and 38 °C release rates from one formulation were reported to be 0.14, 0.38 and 1.0 μg h^{-1} mg^{-1} of formulated pheromone (Bierl *et al.*, 1976).

It was thought that humidity might have a significant effect on the rate of release of pheromone from microcapsules based on the hydrophilic gelatin, and microscopic observation suggested that more collapsed capsules were observed under conditions of high humidity. Actual release rates measured in the entrainment apparatus with moist air were initially increased by 13–26% relative to the rates in dry air. After 20 days the rate in moist air was 30–40% less than that in dry air, presumably owing to the relatively lower pheromone content in the samples aged in moist air (Bierl *et al.*, 1976).

TABLE 9.2 Release rate data on microencapsulated formulations of disparlure[1] aged in a shelter outdoors (from Plimmer *et al.*, 1977).

Designation	a.i. (%)	Size (μm)	Pheromone content (mg) Initial	9 weeks	% lost	Emission rate (μg h^{-1} mg^{-1}) 4 days	4 weeks	9 weeks
1975-NCR-2	2	50–400	3.7	1.0	73	0.4	0.1	0.005
1976-NCR-2	2	50–250	2.5	0.6	75	0.6	0.2	0.2
1976-NCR-10	10	50–250	9.6	6.8	29	0.2	0.08	0.05
3 part 1976-NCR-2 + 1 part 1976-NCR-10	4	50–250	4.9	1.9	60	0.3	0.2	0.02

[1] Microcapsules contained the pheromone in solution in 3 : 1 xylene/amyl acetate and the ratio of core : wall material was 10 : 1

The difficulty of relating laboratory experiments to true field conditions was demonstrated by comparing the half-lives, based on residual pheromone content, of samples of a microencapsulated formulation spread on microscope slides and aged under exposure to various environmental factors. Typical results are shown in Table 9.3 (Bierl *et al.*, 1976). The more rapid loss of pheromone from samples exposed outdoors relative to those exposed in the laboratory may be due to differences in temperature and/or windspeed, but exposure to sun and rain also greatly increases the rate of loss of pheromone, presumably owing to effects on the capsule wall.

9.3.2.2 Field studies of aerial concentrations

Three extensive experiments have been reported in which microencapsulated disparlure was sprayed onto grass (Caro *et al.*, 1977) or forest (Plimmer *et al.*, 1978; Caro *et al.*, 1981) and subsequent atmospheric concentrations of pheromone measured by air sampling with polymeric adsorbents. Application rates of $250 \, g \, ha^{-1}$ or $500 \, g \, ha^{-1}$, more than ten times higher than those used in mating disruption experiments, were used in order to make possible accurate measurements of the low concentrations of pheromone in the air.

In all the experiments, the atmospheric concentrations fell rapidly during the initial few days, and representative data from Caro *et al.* (1977) are summarized in Table 9.4. Although release of pheromone had apparently almost ceased by the end of the experiment, calculations of total pheromone fluxes suggested that only 2.3% and 12.2% of the pheromone applied had been released from the two formulations.

Caro *et al.* (1981) compared atmospheric concentrations obtained with microencapsulated, plastic laminate and hollow-fibre formulations of disparlure applied to forest. With all the formulations, atmospheric concentrations of pheromone fell off rapidly after the first few days (Table 9.5). Analysis of residual pheromone in the plastic laminate flakes and hollow fibres after 30 days showed that they had lost only 15% and 24% of their starting pheromone content, respectively, although corresponding measurements were not carried out on the microencapsulated formulation.

TABLE 9.3 Half-life of microencapsulated disparlure under various conditions (Bierl *et al.*, 1976).

Position	Exposed to			Half-life (days)
	Rain	Sun	Wind	
Indoors	–	–	–	123
Outdoors	–	–	+	34
Outdoors	–	+	+	15
Outdoors	+	+	+	10

TABLE 9.4 Midday concentrations of disparlure 1 m above grass sprayed with microencapsulated disparlure at 500 g ha^{-1} (from Caro *et al.*, 1977).

Day	Mean temp. (°C)	Pheromone concentration (ng m^{-3}) 2% Formulation	10% Formulation
1	18.0	170	37
2	15.5	26	17
3	13.5	4.8	6.2
7	8.0	2.1	3.0
15	30.3	3.1	33
25	21.0	5.8	33
44	14.5	0.4	1.1
66	31.2	0	0.4

TABLE 9.5 Mean concentrations of disparlure 2 m above 4 ha plots of woodland treated with three different formulations of disparlure (after Caro *et al.*, 1981)

Formulation	Application rate (g ha^{-1})	Concentration (ng m^{-3}) Day 1	Day 35
Microcapsules: NCR gelatin-based; 20–60 μm diameter; 2.2% disparlure in 3 : 1 xylene/amyl acetate; 1% RA 1645 sticker	500	27	2.2
Plastic laminate: 7–35 mm^2; 80 μm outer vinyl layer; 9.1% disparlure; 1 : 1 : 1 mixture of laminate : RA 1645 sticker hydroxyethylcellulose	500	6.5	1.2
Hollow fibre: 8–10 mm long, 0.15 mm diameter; 11.5% disparlure; 1 : 2 fibres : Biotac sticker	330 (500)*	28 (42)*	0.4 (0.6)*

Pro rata conversion to facilitate comparisons

Concentrations of pheromone in the atmosphere were greatest in the afternoon and early evening, but it is impossible to interpret these results directly in terms of the behaviour of the formulations since they are the result of numerous interacting factors on the formulations and the pheromone after release. For example, temperatures were probably highest at midday, causing rapid release of pheromone from the formulations, but winds were also stronger at this time, causing more rapid dissipation of the pheromone.

9.3.2.3 Field applications

Published results of field trials with pheromones formulated in gelatin-based microcapsules are summarized in Section 9.5.1.

Many trials have been reported on mating disruption of gypsy moth, *L. dispar*, with the pheromone formulated in gelatin-based microcapsules. Typically, over 90% reduction in mating of female moths was achieved throughout the 6-week season with an application rate of 20 g ha^{-1}. However, in comparative trials carried out in 1979, the microencapsulated formulation did not perform as well as previously, and better reductions in trap catches and in mating were obtained with hollow-fibre and plastic laminate formulations (Plimmer *et al.*, 1982).

In trials with gelatin-based, microencapsulated formulations of unsaturated acetates, significant levels of trap catch reduction have been reported for several species. With 8-dodecenyl acetate ($Z : E$ 94 : 6) against oriental fruit moth, *Grapholitha molesta* Busck., (Gentry *et al.*, 1980) and (Z,Z)-3,13-octadecadienyl acetate against peachtree borers, *Synanthedon* spp. (Gentry *et al.*, 1981; Yonce and Gentry, 1982) lower application rates were required with hollow-fibre and plastic laminate formulations compared with the microencapsulated formulations. A microencapsulated formulation of the aldehyde, (E)-11-tetradecenal, has been used against spruce budworm, *Choristoneura fumiferana* Clemens (Sanders, 1976), and a similar formulation of the alcohol, (E,E)-8,12-dodecadien-1-ol, has been used against codling moth, *Laspeyresia pomonella* L. (Rothschild, 1982).

9.3.3 Properties of microcapsules formed by interfacial processes

Microcapsules with polyamide and/or polyurea walls have been used to formulate the pheromones of several lepidopterous pests. Examples of such formulations were included in the evaluation programme for formulations of the gypsy moth pheromone, disparlure, but most developmental work has been reported with formulations of unsaturated acetates more typical of lepidopterous sex pheromones.

9.3.3.1 Initial release rate studies

In the gypsy moth programme, formulations tested included ones from the Pennwalt Co. formed by interfacial polycondensation and ones from the Stauffer Co. formed by *in situ* interfacial polycondensation. These were evaluated by the methods used for gelatin-based microcapsules, measuring residual pheromone and actual emission rates for samples spread on microscope slides or planchets.

The formulations contained 2.2% or 11% solutions of disparlure in xylene, and capsule diameter was typically 5–40 μm. In contrast to theory, capsules 1–5 μm in diameter released the pheromone more slowly (Plimmer *et al.*, 1976). Addition of stickers caused some decrease in release rate, but the effect was negligible after 2 weeks (Plimmer *et al.*, 1976). It was shown that the loss of pheromone was much more rapid when the formulation was exposed outdoors to sun, temperature and wind. For one formulation 94% of the initial pheromone content was lost after exposure for 7 weeks outdoors and 60% after 7 weeks indoors (Plimmer *et al.*, 1977).

Cardé *et al.* (1975) evaluated a formulation of 11-tetradecenyl acetate ($Z:E$ 89:11) in polyamide capsules, 30–50 μm in diameter. The undiluted formulation was spread on a glass plate, allowed to dry, and release rates were measured by drawing air over the sample at $0.0017 \, \text{m s}^{-1}$ and trapping the entrained pheromone on Porapak Q resin. Emission rates from a sample containing approximately 30 mg pheromone were reported to be 9.6 μg h^{-1} on the first day, falling to 0.2 μg h^{-1} on day 16, after which it was calculated that only 3.7% of the pheromone content had been released. One reason for incomplete release of pheromone in these studies may have been that the formulation was applied in an agglomerated mass with many layers of microcapsules.

In collaborative work by ICI Agrochemicals and the Overseas Development Natural Resources Institute (ODNRI), formulations were evaluated after spraying onto silicone-treated filter papers, simulating as closely as possible the situation in the field when the formulation is sprayed onto vegetation. The formulations contained an unsaturated acetate in polyurea microcapsules of diameter 1–5 μm prepared by interfacial polycondensation between toluene diisocyanate and a mixture of ethylene diamine and diethylene triamine. Polyamide cross-linking in the capsule walls was produced by adding mesitoyl trichloride at levels of approx. 0.5% or 5% of the toluene diisocyanate. The standard formulation contained equal weights of core and wall materials, giving a wall thickness approximately 20% of the total capsule radius. Other capsules were prepared with core : wall ratios of 1 : 0.6 and 1 : 1.5. With these very small microcapsules, no additional sticker was required other than the poly(vinyl alcohol) (2%) used as emulsifying agent in the encapsulation process and hydroxyethylcellulose (0.5%) used to reduce settling out of the formulation (Fig. 9.1).

The formulations were sprayed onto filter papers and aged in a wind tunnel at constant temperatures and realistic windspeeds up to $2.2 \, \text{m s}^{-1}$. Residual pheromone in replicate samples was quantified at intervals by GC analysis, and these studies showed the expected lower release rates from formulations with thicker capsule walls and those with cross-linked walls (Campion *et al.*, 1978). Formulations that had performed well in the wind tunnel were then evaluated in the field by both biological and chemical means. For the latter, the formulation was sprayed onto filter papers pinned to the crop, and loss of pheromone under these conditions was very much more rapid than under apparently similar conditions of temperature and windspeed in the laboratory (Campion *et al.*, 1978; Marks, 1978).

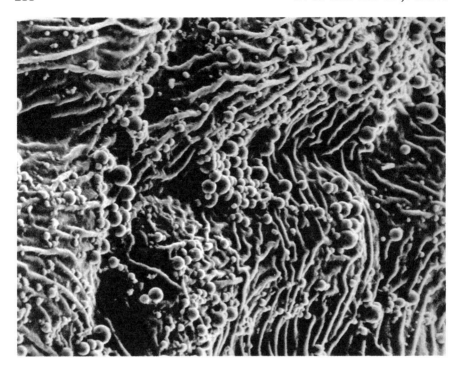

Fig. 9.1 Electronmicrograph of ICI microcapsules on the surface of a cotton leaf (×1800).

Degradation of the pheromone and/or microcapsule wall in sunlight was shown to be the cause of the difference between laboratory and field results by means of several experiments. Thus, more frequent sampling and analysis of sprayed formulations showed that loss of pheromone was more rapid during daytime than night-time (Campion *et al.*, 1981). Analysis of residual pheromone showed that loss of pheromone was very much more rapid from samples fully exposed to sunlight than from samples shielded from sunlight but otherwise exposed to the same environmental conditions. Typical half-lives of (Z)-9-tetradecenyl acetate were 1 day in exposed samples and 12 days in shielded samples (Campion *et al.*, 1981). These analytical results agreed well with those from a biological assay in which traps baited with the synthetic attractant for *Spodoptera littoralis*, (Z,E)-9,11-tetradecadienyl acetate, were surrounded by filter papers sprayed with microencapsulated formulations of the 'inhibitor', (Z)-9-tetradecenyl acetate. Non-cross-linked formulations caused significant reductions in catches of male moths for over 7 days when shielded from sunlight but for less than 2 days when exposed to direct sunlight (Campion *et al.*, 1981).

In order to determine the factors involved in these degradative processes, a microencapsulated formulation containing a mixture of compounds of similar volatilities but different functionalities was sprayed onto filter papers and exposed to sunlight. Half-lives were approximately 1 day for the conjugated diene, (Z,E)-9,11-tetradecadienyl acetate, 4 days for the mono-unsaturated (Z)-9-tetradecenyl acetate and (Z)-4-octadecene and 8 days for the saturated tetradecyl acetate and octadecane, indicating that the rate of degradation was dependent upon the degree of unsaturation (Hall *et al.*, 1982).

9.3.3.2 Stabilization of pheromone and microcapsule

The above results were confirmed by exposing the same compounds to sunlight as thin films in open petri dishes. Comparison of results from samples exposed to direct sunlight and those from samples shielded from sunlight but otherwise exposed to the same environmental conditions showed that saturated acetates were not significantly degraded in sunlight. Thus a saturated acetate could be used as a light-stable 'internal standard' against which to measure loss by degradation of corresponding unsaturated compounds (Hall *et al.*, 1982).

Candidate stabilizing materials were tested initially for their effect on photochemical degradation of (Z)-9-tetradecenyl acetate relative to tetradecyl acetate in simple thin films exposed to sunlight. Materials screened included antioxidants such as hindered phenols, gallate esters and phenylenediamines, ultraviolet screeners such as hydroxybenzophenones and aminobenzoates, dyes and carbon black.

Of the antioxidants tested, by far the best were those of the phenylene-diamine type, such as the commercially available antiozonant *N*-phenyl-*N'*-(2-octyl)-*p*-phenylenediamine (UOP 688). However, because of the unsubstituted secondary nitrogens which react rapidly with acyl chlorides and isocyanates, these compounds could not be encapsulated by interfacial polycondensation methods. However, dimethylation of UOP 688 produced a compound, *N,N'*-dimethyl-*N*-phenyl-*N'*-(2-octyl)-*p*-phenylenediamine, which could be encapsulated and which still retained useful stabilizing activity (Hall *et al.*, 1982).

Most of the ultraviolet screeners tested showed little protective action, although aminobenzoates have been reported to protect unsaturated pheromones against photochemical degradation (Bruce *et al.*, 1979; Bruce and Lum, 1981). However, combinations of dyes and carbon black, as in the Waxoline dyes (ICI), showed a significant stabilizing effect (Hall *et al.*, 1982).

Subsequent evaluation of the stabilizers in microencapsulated formulations was carried out with formulations of unsaturated pheromones containing 10–100% of the corresponding saturated compound. The formulation was sprayed evenly onto silicone-treated filter papers pinned to a rotating board. The board was then divided into halves and fixed outdoors with the filter papers on one half exposed to direct sunlight and those on the other half shielded from direct sunlight but otherwise exposed to the same weather conditions. The residual pheromone and saturated derivative in samples taken at intervals were quantified by GC analysis. The difference between the rate of disappearance of the unsaturated compound and that of the light-stable, saturated compound provided a measure of the degradation of the unsaturated compound. The difference between the rate of disappearance of the saturated compound in samples exposed to sunlight and those shielded from sunlight provided a measure of the degradation of the formulation itself.

In unstabilized, microencapsulated pheromone formulations, both photochemical degradation of the unsaturated pheromone and of the formulation were much in evidence (Table 9.6). Degradation of both pheromone and formulation were reduced in formulations containing N,N'-dimethyl-N-phenyl-N'-(2-octyl)-p-phenylenediamine or Waxoline Black, and the two stabilizers together showed an additive effect, as indicated in Table 9.6 for formulations containing a 1 : 1 mixture of (Z)-9-tetradecenyl acetate and tetradecyl acetate (Hall *et al.*, 1982).

Subsequent work by the ICI/ODNRI group has been based on microencapsulated formulations of pheromones stabilized by these types of compounds. The capsules are 1–5 μm in diameter, forming a stable suspension in water by means of conventional anti-settling agents (Hall *et al.* 1981). These formulations can be sprayed as such or diluted with water, the small capsules adhere well to foliage, and no extra stickers are required. Several exposure experiments carried out during heavy rainfall have shown that the formulation is extremely rainfast once it has dried out after spraying (unpublished results).

Pyrethroid insecticides microencapsulated by interfacial polycondensation techniques can be stabilized against photochemical degradation by ultraviolet screeners of the hydroxybenzophenone type in the core and by similar types of screeners reacted with the capsule wall (Barber *et al.*, 1977). The latter are typically hydroxybenzophenones possessing amino groups to react with free isocyanate groups in the capsule wall, or carboxyl groups to react with free amino groups. Such compounds had little effect on microencapsulated pheromone formulations when the stabilities of pheromone and formulation were measured by the above techniques (unpublished results).

9.3.3.3 *Formulation of different pheromone types*

9.3.3.3.1 *Monounsaturated acetates*
Acetate esters of monounsaturated 12-, 14-, 16- and 18-carbon alcohols have been formulated in the standard ICI/ODNRI microcapsules. This formulation

TABLE 9.6 Half-lives for microencapsulated formulations containing 2% (Z)-9-tetradecenyl acetate (Z9-14 : Ac) and 2% tetradecyl acetate (14 : Ac) tested in Egypt

Formulation:	Unstabilized		Stabilizer A[1]		Stabilizer B[2]		Stabilizer A and B	
					Half-life (days)			
	Z9-14 : Ac	14 : Ac	Z9-14 : Ac	14 : Ac	Z9-14 : Ac	14 : Ac	Z9-14 : Ac	14 : Ac
Exposed	<1	3.5	2.4	3.0	2.8	4.6	7.2	8.9
Shielded	7.2	>20	>20	>20	>20	>20	>20	>20
(% remaining after 20 days)		(70)	(75)	(85)	(65)	(85)	(90)	(90)

[1] Stabilizer A is N,N'-dimethyl-N-phenyl-N'-(2-octyl)-p-phenylene-diamine.
[2] Stabilizer B is Waxoline Black.

provides good protection against photochemical degradation of the core materials, as indicated by the similarity in rates of loss of the unsaturated compound and of the corresponding, photochemically stable, saturated acetate (Table 9.6). Photochemical degradation of the capsule wall is reduced, relative to that in the unstabilized formulation, but it is still significant, as indicated by the difference in the rates of loss of the saturated acetate from samples exposed and shielded from direct sunlight (e.g. Table 9.6).

Exposure of samples of a formulation to sunlight on horizontal filter papers is probably exposure to the most extreme conditions. Significant amounts of a formulation sprayed onto trees or plants will be deposited on the lower foliage and branches and on the undersurfaces of leaves, especially when ULV techniques are used. This formulation will be protected to a greater or lesser extent from direct sunlight, and the effect of this on a microencapsulated pheromone formulation was examined in an experiment carried out on sugarcane in Mauritius. The standard ICI/ODNRI microencapsulated formulation of (Z)-13-octadecenyl acetate, for use against the sugarcane borer, *Chilo sacchariphagus* Boyer (Nesbitt *et al.*, 1980), was sprayed onto filter papers which were then pinned to the upper and lower surfaces of leaves at the top of the canes and also to leaves midway up the canes. Analysis of residual pheromone at intervals illustrated the increased persistence of formulation which penetrated the crop canopy (Table 9.7).

9.3.3.3.2 Non-conjugated, diunsaturated acetates

Microencapsulated formulations of 7,11-hexadecadienyl acetate (Z,E : Z,Z 1 : 1) have been used extensively to control pink bollworm, *Pectinophora gossypiella* Saunders. Half-life of the formulation sprayed onto filter papers and fully exposed to sunlight in Egypt is typically approximately 7 days.

Methylene-interrupted dienes are notoriously susceptible to oxidation, but these compounds are well protected in the ICI/ODNRI microencapsulated formulation. The half-life of microencapsulated (Z,E)-9-12-tetradecadienyl acetate on filter papers exposed to sunlight in London was 14.5 days, compared with 22 days for tetradecyl acetate.

TABLE 9.7 Persistence of (Z)-13-octadecenyl acetate in the ICI/ODNRI microencapsulated formulation sprayed onto filter papers attached to sugarcane leaves in Mauritius

	Half-life (days)	
	Cane top	Mid-cane
Upper leaf surface	5.5	13.5
Lower leaf surface	28	>28
		(60% left after 28 days)

9.3.3.3.3 Conjugated, diunsaturated acetates

Studies with (Z,E)-9,11-tetradecadienyl acetate, the major component of the pheromone of the Egyptian cotton leafworm, *Spodoptera littoralis* Boisd., have shown that isomerization of the conjugated double bonds on exposure to sunlight is actually accelerated by phenylene diamines. In the ICI/ODNRI microencapsulated formulation sprayed onto filter paper and exposed to sunlight, (Z,E)-9,11-tetradecadienyl acetate is converted to the thermodynamic equilibrium mixture $(E,E:Z,E:E,Z:Z,Z$ 65:16:16:3) within 2–3 days. Nevertheless, the conjugated diene system is considerably stabilized against photochemical degradation, and the resulting persistence of the Z,E isomer was better than with any other stabilizing system investigated.

Geometrical isomerization of conjugated dienes can be prevented if the microencapsulated formulation is applied in larger droplets, but the release rate is greatly reduced. However, this would seem to be a promising approach in attempts to increase the stability of microencapsulated formulations.

9.3.3.3.4 Aldehydes

Aldehydes cannot be encapsulated satisfactorily without some loss of active ingredient by interfacial polycondensation processes involving aliphatic polyamines, since the latter react with the aldehydes during the encapsulation process and during storage. However, aldehydes can be encapsulated without detectable loss by the *in situ* interfacial polycondensation method (9.2.1.2.2).

The aldehyde functionality provides an additional focus for photochemically induced degradation, and a stabilized, microencapsulated formulation of (Z)-11-hexadecenal and tetradecyl acetate sprayed onto filter papers and fully exposed to sunlight in London, UK, showed half-lives of 1.5 days and 6 days, respectively, for the two components, with 20% of the aldehyde remaining after 6 days. Shaver and Lopez (1982) reported half-lives of 1.2 days and 1.7 days, respectively, for (Z)-9-tetradecenal and (Z)-11-hexadecenal in a similar formulation exposed on filter papers, with 11% and 16%, respectively, remaining after 7 days. Wiesner and Silk (1982) studied the behaviour of various formulations of (E)-11-tetradecenal in a wind tunnel, attempting to simulate field conditions with 8 h light at $25 \pm 1\,°C$ and $2.4 \pm 0.1\,m\,s^{-1}$ windspeed and 16 h dark at $14 \pm 1\,°C$ and $0.55\,m\,s^{-1}$ windspeed. The ICI/ODNRI microencapsulated formulation had a half-life of approx. 20 days and a plastic laminate formulation a half-life of approximately 10 days.

Encapsulation of aldehyde 'propheromones' which release the aldehyde on irradiation (Liu *et al.*, 1984) may be a useful approach with exceptionally labile aldehydic pheromones.

9.3.3.3.5 Alcohols

Alcohols cannot be encapsulated as such by existing interfacial methods since they react rapidly with the electrophilic monomers. 2-Tetrahydropyranyl ethers of alcohols have been encapsulated, but the alcohols could not be regenerated by incorporation of acids into the core material or addition of acid to the aqueous suspension (unpublished results).

9.3.3.4 Field studies of aerial concentrations

Taylor (1982) reported studies of the concentrations of (Z)-9-tetradecenyl formate above corn fields sprayed with the ICI/ODNRI microencapsulated formulations at 300 g ha^{-1} or a 3 mm square plastic laminate formulation applied at 285 g ha^{-1}. Persistence studies showed half-lives of 13.5 days and 5.3 days, respectively, for the two formulations. The aerial concentration above the field treated with the microencapsulated formulation showed a maximum after 20 days, probably as a result of environmental factors affecting pheromone dispersion rather than as a direct consequence of changes in pheromone release from the formulation. The atmospheric concentration of pheromone was still significant after 31 days. The plastic laminate formulation gave a maximum aerial concentration after 8 days, but this fell progressively to zero after 31 days.

9.3.3.5 Field applications

Published results of field trials with pheromones formulated in microcapsules formed by interfacial processes are summarized in Section 9.5.2.

In the Gypsy Moth Programme, cross-linked polyurea microcapsule formulations of disparlure did not perform as well as gelatin-based microcapsules (Schwalbe *et al.*, 1979). Similarly, early studies with unstabilized polyamide or polyurea microencapsulated formulations of unsaturated acetate pheromones were not particularly successful. Taschenberg and Roelofs (1976) found that applications of a polyamide microencapsulated formulation of (Z)-9-tetradecenyl acetate at 25 g ha^{-1} every 5–7 days were necessary to suppress trap catches of grape berry moth, *Paralobesia viteana* Clem., whereas a single application of a hollow-fibre formulation at 3.7 g ha^{-1} suppressed trap catches for over 10 weeks. Marks (1978) and Marks *et al.* (1981) reported reduction of mating of red bollworm, *Diparopsis castanea* Hamps., with a microencapsulated formulation of the pheromone 'inhibitor', 9-dodecenyl acetate (*E* : *Z* 80 : 20), sprayed every 5 or 6 days in a field cage, but the same formulation had little effect in the open field.

Stabilized microencapsulated formulations of gossyplure have been used with great success to control pink bollworm, *Pectinophora gossypiella* Saunders, in Egypt on areas up to 5000 ha and in Pakistan and S. America. Th formulation has been applied at 10 g ha^{-1} by fixed-wing aircraft or helicopter, and optimum spraying conditions have been defined (Johnstone, 1982). Control of *P. gossypiella*, as measured by boll infestation, yield and quality of cotton, and other parameters, was at least as good as with conventional insecticides or hollow fibre and plastic laminate pheromone formulations (Critchley *et al.*, 1983, 1984, 1985). In Mauritius, trials with a stabilized microencapsulated formulation of (Z)-13-octadecenyl acetate showed over 90% suppression of trap catches of sugarcane borer, *Chilo sacchariphagus* Bojer, for up to 4 weeks at application rates of 20 g ha^{-1} and 40 g ha^{-1} (unpublished results). The acetate is only one component of the pheromone of this species, which is a 7 : 1 mixture of (Z)-13-octadecenyl acetate and (Z)-13-octadecen-1-ol (Nesbitt *et al.*, 1980). A stabilized micro-encapsulated formulation of (Z,E)-9,11-tetradecadienyl acetate suppressed trap catches of Egyptian cotton leafworm, *Spodoptera littoralis* Boisd., for up to 13 days, but only at the high application rate of 40 g ha^{-1} (Hall *et al.*, 1982).

Stabilized microencapsulated formulations of (Z)-11-hexadecenal were shown to suppress trap catches and mating of *Heliothis virescens* F. and *H. zea* Boddie (Shaver and Lopez, 1982). The formulations were ineffective after approximately 7 days, but only a low application rate of 2.5 g ha^{-1} was tested. Microencapsulated aldehydes have also been used in experiments with artificially maintained populations of rice stem borer, *Chilo suppressalis* Wlk., in 16 m^2 field cages on rice in the Philippines. These were treated with a single application of stabilized microencapsulated formulations of the synthetic pheromone (Z)-11-hexadecenal + (Z)-13-octadecenal or of pheromone 'mimics' (Z)-9-tetradecenyl formate + (Z)-11-hexadecenyl formate. The aldehyde mixture was applied at 10 g ha^{-1} and the formate mixture at 100 g ha^{-1} and 10 g ha^{-1}, and, after 30 days, the mean cumulative numbers of fertile eggs laid in the three treatments and untreated control cages were 75, 300, 500 and 1550, respectively. Subsequent open-field experiments have been less successful, because of the short persistence of the formulation and possibly because of the difficulties in spraying the formulation so that it adheres to rice in paddy (personal communications: P. S. Beevor (ODNRI) and V. A. Dyck (International Rice Research Institute).

9.3.4 Properties of microcapsules formed by other processes

Laboratory studies on the Shin-Etsu formulation of (Z)-11-tetradecenyl acetate in hydroxypropylmethylcellulose acetate succinate microcapsules (Section 9.2.1.1.2) showed that only 5% of the pheromone was lost from the formulation after 60 days in a wind tunnel at 30 °C and 0.5 m s^{-1} windspeed (Shin-Etsu Chemical Co. Ltd, 1983).

The Nitto formulation of polysulphone microcapsules 2–20 μm in diameter containing 40% (Z)-9-dodecenyl acetate (Section 9.2.2.1) released 75% of the pheromone after one month at 30 °C in the laboratory (Nitto Electrical Industry Co. Ltd, 1984). A Nitto formulation containing 36% gossyplure showed linear release of virtually all the core material over 30 days at 40 °C. In field trials, the latter formulation gave 94–97% trap catch suppression of *P. gossypiella* for 15 days at application rates of 2.5–10 g ha^{-1} (Smith *et al.*, 1983). Flint and Merkle (1984) tested microencapsulated formulations of gossyplure and of (Z,Z)-7,11-hexadecadienyl acetate alone. An application rate of 12.3 g ha^{-1} was necessary to cause trap catch reductions of 90% for 12 days and, at high populations, this rate was only effective for less than 6 days. There were no major differences between the effects of spray (0.01 μl droplets) and large droplet (50 μl) application methods, although the large droplet applications seemed to be more effective at lower application rates (2.5 g ha^{-1}).

Laboratory studies with the Zoecon oil/resin formulation (Section 9.2.2.1) of gossyplure showed that the release rate was decreased by raising the resin/oil ratio from 1.0 to 1.5, by lowering the initial pheromone loading from 750 p.p.m. to 500 p.p.m. or by cross-linking the resin. Field studies with a formulation containing the pheromone and permethrin in 1 : 10 ratio gave effective reduction in trap catch of *P. gossypiella* moths for 8 days at an application rate of 3.7 g ha^{-1} of pheromone (Evans, 1984).

Lee *et al.* (1981, 1982) reported the results of field experiments against the rice stem borer, *Chilo suppressalis* Wlk., with the pheromone (Z)-11-hexadecenal + (Z)-13-octadecenal microencapsulated in a urea/formaldehyde polymer (Section 9.2.1.2.3) with 2,6-di-*tert*-butyl-4-methlyphenol (BHT) as antioxidant. During the first generation, catches in traps baited with a virgin female moth were suppressed for 9 days with the formulation applied at 2.5 g ha^{-1}. During the second generation flight, over 90% reduction in trap catches for 18 nights was claimed with application rates as low as 0.83 g ha^{-1}, but catches were very low (Lee *et al.*, 1981). In subsequent experiments, the formulation was shown to be effective in suppressing trap catches for up to 10 days (Lee *et al.*, 1982). A similar microencapsulated formulation of the pheromone 'mimics', (Z)-9-tetradecenyl formate, and (Z)-11-hexadecenyl formate, was claimed to reduce trap catches of *C. suppressalis* for 5 days when applied at a rate of 30 mg per trap, but again, catches were too low to obtain significant results (Goh and Lee, 1984).

9.4 GENERAL CONSIDERATIONS AND CONCLUSIONS

At the present time, it is probably fair to report that there are four main types of formulation commercially available for use with insect sex pheromones in mating disruption, and, in principle, for use with other chemicals for modification of insect behaviour in plant protection. These formulations are microcapsules, hollow fibres, plastic laminate flakes and twist-ties and all have been used in major pest control programmes.

9.4.1 Operational factors

The four types of formulation are commercially available, and it must be assumed that large-scale manufacture is possible, although relative production costs are proprietary information of the companies involved. Microencapsulated formulations have generally been prepared by well-established processes which can be carried out on a large scale (Section 9.9.2).

Microencapsulated, hollow-fibre and plastic laminate formulations can be applied from the ground or from the air. Microencapsulated formulations can be applied with conventional ground-spraying equipment including hand-operated, knapsack sprayers (e.g. Campion *et al.*, 1976), motorized mist blowers (Granett and Doane, 1975) and hand-held ULV sprayers (e.g. Marks, 1978). The formulations can be diluted with water or, in ULV applications, used direct (e.g. Marks *et al.*, 1981). The hollow-fibre and plastic laminate flake formulations can be applied to crops by hand either premixed with the appropriate sticker or as successive applications of sticker and fibre or flake, although such methods are only feasible where cheap labour is readily available. The twist-tie formulations can only be applied by hand, but they do not require a sticker and the frequency of application is less than with the other formulations, greatly reducing labour costs.

Microencapsulated pheromone formulations have been sprayed aerially from fixed-wing aircraft (e.g. Plimmer, 1982; Critchley *et al.*, 1983) and from helicopters (Critchley *et al.*, 1984, 1985) with conventional spray equipment. In hot climates, where evaporation of water from the spray droplets is rapid, it is necessary to dilute the formulation, to fly as low as possible and to spray during the cooler times of the day. Thus Johnstone (1982) showed that at least 70% recovery could be achieved during application of a microencapsulated formulation of gossyplure in Egypt with an application rate of $25 \, l \, ha^{-1}$, droplet size $250 \, \mu m$ v.m.d. and application only when the temperature was less than or equal to $32 \, °C$ and the relative humidity greater than 50%. Under such conditions, fixed-wing aircraft are capable of treating 40 ha per flight at $10 \, g \, a.i. \, ha^{-1}$ but it is anticipated that this area can be doubled by halving the dilution of the formulation (L. J. McVeigh, personal communication (ODNRI)). Increasing the droplet size will also make it possible to carry out spraying from a height similar to that used for non-aqueous formulations.

The stickers used with hollow-fibre and plastic laminate formulations are typically polybutenes and aqueous emulsions of polybutenes or latex, respectively (e.g. Kydonieus *et al.*, 1982), and specialized spraying equipment is required for these. However, evaporation is not such a problem, and application can, in principle, be carried out from a greater height under more extreme conditions of temperature and humidity (Kydonieus *et al.*, 1982). In Egypt during 1984, a hollow-fibre formulation of gossyplure was applied at a rate of 2.85 g a.i. ha^{-1} in 37.5 g ha^{-1} of formulation mixed with 575 g ha^{-1} of Biotac polybutene sticker. Approximately 120 ha could be treated during one flight by fixed-wing aircraft fitted with the special application pods. Similarly, a plastic laminate formulation of gossyplure was applied at 6.9 g a.i. ha^{-1} in 150 g ha^{-1} of formulation mixed with 150 g ha^{-1} of Pherotac sticker, an aqueous emulsion of polybutenes. As with the hollow-fibre formulation, approximately 120 ha could be treated during one flight (L. J. McVeigh (ODNRI), personal communication).

The fact that microencapsulated pheromone formulations are applied as aqueous suspensions with conventional spray equipment means that they can often be tank-mixed and applied together with other materials, giving increases in convenience and economy. In Egypt, a microencapsulated formulation of gossyplure has been mixed with foliar fertilizers, and could be applied with formulations of *Bacillus thuringiensis* or insect growth regulators used against *Spodoptera littoralis*. In Pakistan, it may be necessary to combine the pheromone formulation used to control *P. gossypiella* with insecticides to control the other bollworm pests, *Earias vittella* and *E. insulana*.

Concern has been expressed over possible hazardous effects of micro-encapsulated formulations on spray operators and on honeybees. Significant inhalation of individual microcapsules by spray operators would seem to be unlikely because the spray droplets will generally contain agglomerations of many microcapsules in the suspending medium. However, 'the use of micro-encapsulated methyl parathion has probably caused more concern among beekeepers and agricultural scientists than any other formulation used previously' (Waller *et al.*, 1984) because of the similarity in size of typical micro-capsules and entomophilous pollen, and similar problems might be anticipated for microencapsulated pheromone formulations. The numerous studies of uptake of methyl parathion by honeybees have been summarized by Waller *et al.* (1984), and these authors conducted a definitive experiment which showed less uptake of the insecticide after application of the microencapsulated formulation than after application of an emulsifiable concentrate. It would thus seem that significant uptake of microencapsulated pheromone by honeybees is unlikely, presumably again because the microcapsules are sprayed as agglomerates and adhere to the foliage as such. It is also unlikely that the pheromone itself would have any deleterious effects on the hive, although the microcapsules may cause some mechanical or chemical effects.

9.4.2 Design factors

As described in Section 9.3, much of the work on the properties of pheromone formulations has been concerned with regulation of the release rate of the pheromone, and only more recently have the problems of degradation of the pheromone and formulation been considered.

With hollow-fibre formulations, there is no obvious restriction on the chemical nature of the compound to be formulated, and the release rate can be controlled by altering the diameter of the fibre, the nature of the fibre wall and the content of active ingredient. Protection of aldehydic pheromones against degradation has been reported using fibres with carbon black incorporated into the fibre wall and a hindered-phenol-type antioxidant mixed with the aldehyde (Weatherston, 1981). Similarly with plastic laminate formulations, there would seem to be no restrictions on the type of compound that can be formulated, and the release rate can be altered by changing the composition of the external laminate material, the size of the flakes and the active ingredient content. No studies of pheromone degradation have been reported, although increased release of pheromone from hollow fibres due to degradation of the Celcon fibre wall in sunlight has been reported by Weatherston *et al.* (1985).

With gelatin-based microencapsulated formulations and some of the newer types of microencapsulated formulations there would seem to be no restrictions on the type of compounds that can be encapsulated. With microcapsules produced by interfacial processes, compounds that react with one of the monomers, for example unprotected alcohols, cannot be encapsulated. Release rates of the core materials can be altered by the numerous design features described in Section 9.3, but degradation of the pheromone and/or of the capsule walls is always a potential problem because of the extremely small size of the capsules and the corresponding thinness of the capsule walls. Thus, even with a core : wall ratio of 1 : 1, a capsule 10 μm in diameter has walls only 1 μm thick.

Disparlure (*cis*-7,8-epoxy-2-methyloctadecane) is not particularly susceptible to degradation in sunlight and oxygen, but in studies with this pheromone in gelatin-based microcapsules, disappearance of pheromone from formulations exposed to sunlight was much more rapid than from those protected from sunlight (Table 9.3; Bierl *et al.*, 1976), suggesting that degradation of the capsule walls was taking place. Limited studies with gelatin-based microencapsulated formulations of (Z)-9-tetradecenyl acetate indicated that, in the absence of any stabilizers, degradation of both pheromone and capsule wall is rapid on exposure to sunlight (unpublished results). In the ICI/ODNRI stabilized formulation formed by interfacial polycondensation, degradation of unsaturated pheromones is greatly reduced, but there is still significant degradation of the capsule wall on exposure to sunlight. Indeed, although it is generally assumed that release of pheromone in the absence of sunlight occurs by diffusion of the contents through the capsule wall, release by spontaneous degradation of the thin capsule wall cannot be excluded, since, as shown in Table 9.1, most capsule design characteristics will have a similar effect on release by diffusion and release by capsule degradation. Actual diffusion of a relatively non-polar pheromone through a highly polar material such as gelatin, polyamide or polyurea would be expected to be extremely slow, even through the thin walls of microcapsules.

Microencapsulated pheromone formulations have many extremely useful operational and design features. However, it is clear that future developmental work should aim to produce formulations that:

(i) can be used with any type of compound;

(ii) provide at least the level of protection for the pheromone which is available in current formulations, with scope for improvements in protection of aldehydes and pheromones with conjugated, multiple unsaturation;

(iii) have capsule walls more resistant to environmental degradation, particularly in the presence of sunlight.

9.5 PUBLISHED REPORTS OF FIELD TRIALS WITH MICROENCAPSULATED PHEROMONES

9.5.1 Gelatin-based microcapsules

9.5.1.1 Gypsy moth, Lymantria dispar (L.)

NCR microcapsules; 300–400 μm diameter; 1% disparlure in xylene; Rhoplex B15/hydroxy-ethylcellulose/1% aqueous KOH sticker	Aerial application to 16 ha plots at 1.8 g ha^{-1}; >90% reduction in capture of released males after 44 days	Beroza et al. (1973)
NCR microcapsules; 25% plastic coated; 100–300 μm diameter; 2.2% disparlure in xylene; UCAR 680 latex/hydroxyethyl cellulose/1% KOH sticker	Aerial application to 16 ha plots at 2.59 g ha^{-1}; reduction in capture of released males after 55 days	Beroza et al. (1973)
	Aerial application to 60 km^2 at 5 g ha^{-1}; trap catch suppression for at least 4 weeks; mating of tethered female moths less than 10% for 2 weeks	Beroza et al. (1974)
	Mistblower application to 1 ha plots at 18 g ha^{-1}; trap catches suppressed for up to 2 months; mating of tethered females completely suppressed	Granett and Doane (1975)

Aerial application to 2.6 km² (A) or 6 km² (B) plot.

Application rate		% Reduction (6 weeks)	
		Mating	Trap catch
5 g ha⁻¹	(A)	47	55
10+10 g ha⁻¹	(A)	94	88
20 g ha⁻¹	(A)	97	83
20 g ha⁻¹	(B)	97	99
+two insecticide sprays			

NCR microcapsules; 25% plastic coated; 25–200 μm diameter; 2.2% disparlure in 3 : 1 xylene/amyl acetate; UCAR 680/hydroxyethyl-cellulose/1% aqueous KOH sticker

Beroza et al. (1975b)

Aerial application to 16 ha plots at 2.5 g ha⁻¹, 5 g ha⁻¹ and 15 g ha⁻¹
Up to 90% reduction in mating of tethered female moths at 15 g ha⁻¹

NCR microcapsules; 17.6% plastic coated; 100–250 μm diameter; 2.2% disparlure in xylene; UCAR 680/hydroxyethylcellulose/KOH sticker

Cameron et al. (1974)
Schwalbe et al. (1974)

Aerial application to 16 ha plots at 20 g ha⁻¹ in Massachusetts (MA, high population) and Maryland (MD, low population)

Formulation	% Mating suppression (7 weeks)	
	MA	MD
76-NCR-2	83	97
76-NCR-4	71	92
76-NCR-10	63	94
75-NCR-2	68	82

NCR microcapsules; 25% plastic coated; 50–400 μm diameter; 2.2% disparlure in 3 : 1 xylene/amyl acetate; Rhoplex B15 sticker (75-NCR-2)

NCR microcapsules; 25% plastic coated; 50–250 μm diameter; 2.2% disparlure in 3 : 1 xylene/amyl acetate; RA 1654 sticker (76-NCR-2)

NCR microcapsules; 25% plastic coated; 50–250 μm diameter; 11% disparlure in 3 : 1 xylene/amyl acetate; RA 1645 sticker (76-NCR-10)

3 parts 76-NCR-2 + 1 part 76-NCR-10 (76-NCR-4)

Schwalbe et al. (1979)

NCR microcapsules; 25% plastic coated; 20–60 μm diameter; 2.2% disparlure in 3 : 1 xylene/amyl acetate; RA 1645 sticker (78-NCR-2)	Aerial application of 75-NCR-2 to 15 400 ha at 20 g ha^{-1}; 68% reduction in mating of tethered female moths, 61% reduction in trap catches	Plimmer (1982)

Formulation	Application rate	% Reduction: Trap catch	% Reduction: Mating
78-NCR-2	50 g ha^{-1}	86	97
76-NCR-2	50 g ha^{-1}	88	95
78-NCR-2	5 g ha_{-1}	34	76

	Comparison of 78-NCR-2 microcapsules with hollow fibre and plastic laminate flake formulations on 16 ha plots	Plimmer *et al.* (1982)

Formulation	Application rate (g ha^{-1})	% Reduction (4 weeks): Trap catch	% Reduction (4 weeks): Mating
78-NCR-2	2	44.6	33.6
Plastic laminate	2	42.9	38.4
Hollow fibre	2	46.2	40.3
Hollow fibre	10	86.0	61.1
78-NCR-2	20	63.4	62.2
Plastic laminate	20	75.7	88.3
Hollow fibre	20	91.6	90.9

(*Continued*)

NCR microcapsules; 25% plastic coated; 100 μm diameter; 2.2% (Z)-2-methyl-7-octadecene in xylene; UCAR 680 sticker

(i) Aerial application on 16 ha plots at 61.7 g ha^{-1}; 80% reduction in trap catch of released males on 17th day
(ii) Aerial application on 16 ha plots at 15 g ha^{-1}; 29.6% mating suppression over 11 days (cf 79.5% mating suppression with disparlure at 15 g ha^{-1})

Cameron et al. (1975)

9.5.1.2 European cornborer (ECB), Ostrinia nubilalis (Hübner), Redbanded leafroller (RBLR), Argyrotaenia velutinana

NCR microcapsules; 50–200 μm diameter; 2.2% (E)-11-tetradecenyl acetate (E11-14:Ac) or 11-tetradecynyl acetate (11yn-14:Ac) in xylene; UCAR 680 sticker

(i) Spray circles 0–10 m diameter with hand sprayer at 90 g ha^{-1} (E)-11-tetradecenyl acetate; trap catches of both species suppressed for at least 9 days

(ii) Hand application at 33.2 g ha^{-1}, 66.4 g ha^{-1} and 132.8 g ha^{-1}

	% Trap catch reduction	
Compound	ECB (9 days)	RBLR (21 days)
E11-14:Ac	84–100	95
11yn-14:Ac	74–91	96

Klun et al. (1975)

9.5.1.3 Summerfruit tortrix moth, Adoxophyes orana (F.v.R.)

NCR microcapsules; 25% plastic-coated; 50–200 μm diameter; 2.2% 9 : 1 (E)-9-tetradecenyl acetate/(E)-11-tetradecenyl acetate in xylene; UCAR 680 sticker

Ground application on 0.2 ha plot at 40 g ha^{-1}; three sprays at 2-week intervals Trap catch reduction over 7 weeks 94% (synthetic pheromone), 95.5% (virgin female); reduction in larvae in next generation

Minks et al. (1976)

9.5.1.4 Oriental fruit moth, *Grapholitha molesta* (Busck.)

NCR microcapsules; uncoated or 25% plastic coated; 50–250 μm diameter; 2.2% 8-dodecenyl acetate (Z : E 94 : 6) (A) or 1-dodecanol (B) in xylene; UCAR 680 sticker

Hand application of 3 : 1 mixture of (B)+(A) to 0.8 ha plots at 20 g total per hectare

Gentry *et al.* (1974)

	% Trap catch reduction of released males	
Week	Uncoated	Coated
1	73	93
2	79	97
3	9	50
4	8	33

Hand application of formulation (A) only to 0.8 ha plots

Gentry *et al.* (1975)

	% Trap catch reduction (native/released males)					
	Week					
Application rate	1	2	3	4	5	6
5 g ha^{-1}	92/96	61/47	65/40	0/25	50/0	33/20
20 g ha^{-1}	94/98	95/98	73/77	100/86	57/78	33/30

NCR microcapsules; 25% plastic coated; 50–250 μm diameter; 8-dodecenyl acetate (Z : E 94 : 6) in 3 : 1 xylene/amyl acetate at 2.2% (C) or 11% (D); Rhoplex B15 sticker

Two successive hand applications to 0.8 ha plots

Gentry *et al.* (1980, 1982)

Formulation	Application rate (g ha^{-1})	% Trap catch reduction
1976		
C	20	92% (9 weeks)
D	20	95% (9 weeks)
Plastic laminate	0.82	78% (9 weeks)
		89% (15 weeks)
		98% (15 weeks)
		99% (15 weeks)

(Continued)

1977
D	20	77% (12 weeks)	0% (16 weeks)
3 : 1 C+D	20	97% (12 weeks)	58% (16 weeks)
Plastic laminate	2.3	100% (12 weeks)	87% (16 weeks)

Campion *et al.* (1981)

9.5.1.5 Egyptian cotton leafworm, *Spodoptera littoralis* (Boisd.)

Gelatin-based microcapsules; thin, medium or thick plastic coated; 50–100 μm diameter; 0.4% (Z)-9-tetradecenyl acetate in 3 : 1 xylene/amyl acetate

Hand application to 0.7–6.2 ha plots at 100 g ha^{-1}. Trap catch reduction not more than 69% over 5 days

9.5.1.6 Peachtree borer, *Synanthedon exitiosa* Say (PTB); Lesser peachtree borer, *S. pictipes* Grote and Robinson (LPTB)

NCR microcapsules; 25% plastic coated; 50–250 μm diameter; (Z,Z)-3,13-octadecadienyl acetate in 3 : 1 xylene/amyl acetate at 4% (A) or 10% (B); RA 1645 sticker

Hand application to 0.8 plots

Gentry *et al.* (1982)

Formulation	Application rate (g ha^{-1})	% Trap catch reduction (32 weeks) PTB	LPTB
A	19.8	57	81
B	19.8	57	64
hollow fibre	2.7	93	69

9.5.1.7 Spruce budworm, *Choristoneura fumiferana* (Clemens)

NCR microcapsules; 10–300 μm diameter; 2.2% (E)-11-tetradecenal in xylene; Rhoplex B15 sticker

Aerial application to 12 ha plot at 7.4 g ha^{-1}; 95.7% reduction in trap catches over 18 days

Sanders (1976)
Sanders and Seabrook (1982)

9.5.1.8 Codling moth, *Laspeyresia pomonella* (L.)

NCR microcapsules; 50–250 μm diameter; 2.2% (E,E)-8,10-dodecadien-1-ol+5.5% di-*tert*-pentylhydroquinone in xylene

Hand application to orchards at 7.4 g ha^{-1} (i) 93–95% trap catch reduction over 7 days; 65% reduction on day 8 (ii) 93% trap catch reduction in first week; 43% reduction in second week

Rothschild (1982)

9.5.1.9 Cabbage looper, *Trichoplusia ni* (Hbn.)

NCR microcapsules containing (Z)-7-dodecenyl acetate; 10 different formulations tested; best results with 2.2% solution in xylene, high level of plastic coating, 5% UOP 688 as antiozonant

Hand application, 900–4900 m^2 plots; 32–36 g ha^{-1}; 80–90% trap catch reduction for 24 days

McLaughlin *et al.* (1975)

9.5.2 Microcapsules formed by interfacial processes

9.5.2.1 Gypsy moth, *Lymantria dispar* (L.)

Pennwalt nylon microcapsules, 37.4% slurry; 0.36% disparlure, 39% water, 19.9% of 1% hydroxyethylcellulose, 2% UCAR 680 sticker, 1.7% of 1% KOH; 2.82 g disparlure in 2.11 ha^{-1}

Aerial application to 16 ha plots at 2.82 g ha^{-1}; reduction in capture of released male moths after 55 days

Beroza *et al.* (1973)

Pennwalt nylon microcapsules, 64% slurry; 0.33% disparlure, 15% water, 2% UCAR 680, 19% of 1% hydroxyethylcellulose, 0.2% of 10% KOH; 4.4 g disparlure in 2.11 ha^{-1}

Aerial application to 16 ha plots at 4.40 g ha^{-1}; reduction in capture of released male moths after 55 days

Beroza *et al.* (1973)

(Continued)

234

Stauffer cross-linked polyurea microcapsules; 10–40 μm diameter; disparlure in xylene, 3 parts 2.2% +1 part 11%; core : wall 10 : 1; 20 g disparlure in 2.1 l ha⁻¹

Aerial application to 16 ha plots at 20 g ha⁻¹; disruption of mating of female moths measured over 7-week season in Massachusetts (MA; high population) and Maryland (MD; low population)

Formulation	% Mating disruption	
	MA	MD
Stauffer microcapsules	46	58
gelatin microcapsules	63–83	82–97
hollow fibres	51	

Schwalbe et al. (1979)

9.5.2.2 Cabbage looper, *Trichoplusia ni* (Hbn.)

Pennwalt nylon microcapsules of varying porosity, containing (Z)-7-dodecenyl acetate

Ground application to plots 900–4900 m² at 11–30 g ha⁻¹; best results give 80% trap catch reduction over 24 days with high-porosity capsules

McLaughlin et al. (1975)

9.5.2.3 Redbanded leafroller, *Argyrotaenia velutinana*

Pennwalt polyamide microcapsules; 30–50 μm diameter; 10.4% of 11-tetradecenyl acetate (Z : E 89 : 11); 0.01% Triton B-1956 spreader-sticker

Hand application at 22 g ha⁻¹

	% Trap catch reduction	
	Expt 1 (35 days)	Expt 2 (42 days)
1 application	75	91
Applications at 5–7 day intervals	86	99

Carde et al. (1975)

Pennwalt polyamide microcapsules; 30–50 µm diameter; 14.6% of 11-tetradecenyl acetate (Z : E 89 : 11) or 19% dodecyl acetate

Taschenberg and Roelofs (1978)

Hand application to 0.27 ha plots

Treatment	% Trap catch reduction
Expt 1 (8 days)	
25 g ha⁻¹ pheromone	89
25 g ha⁻¹ pheromone + 25 g ha⁻¹ dodecyl acetate	90
Expt 2 (7 days)	
37 g ha⁻¹ pheromone	82
18.5 g ha⁻¹ pheromone + 18.5 g ha⁻¹ dodecyl acetate	77
37 g ha⁻¹ dodecyl acetate	7

9.5.2.4 Grape berry moth, *Paralobesia viteana* (Clem.)

Pennwalt polyamide microcapsules; 30–50 µm diameter; 12–15% of 9-dodecenyl acetate (Z : E 96 : 4)

Tractor application to 0.4 ha plots at 25 g ha⁻¹; 20 applications at 5–7 day intervals; Total catch of male moths: 99 in untreated, 5 in treated plots; single application at 25 g ha⁻¹ gave little disruption after 7–8 days; cf. hollow-fibre formulation suppresses trap catches for over 10 weeks at 3.7 g ha⁻¹

Taschenberg and Roelofs (1976)

9.5.2.5 Red bollworm, *Diparopsis castanea* (Hamps.)

ICI polyurea microcapsules; median diameter 3.5 µm; 2% 9-dodecenyl acetate (E : Z 80 : 20)

(i) Hand application with ULV sprayer in 0.1 ha field cage

Rate (g ha⁻¹)	Interval (days)	Total (days)	% Reduction Mating	Oviposi-tion	Boll infesta-tion
105.3	6	22	66.2	87.4	75.8
38.7	6	24	62.2	71.8	56.9
15.9	6	28	27.9	39.0	
26.34	6	8	61.9		
37.9	12	26	19.2	67.6	

Marks (1978); Beevor and Campion (1979)

(Continued)

(ii) Hand application to 0.5 ha open field plot; 38.7 g ha^{-1} gave 75% reduction in oviposition; 15.6 g ha^{-1} gave no reduction in oviposition

Marks *et al.* (1981)

(i) Hand application with ULV sprayer in 0.1 ha field cage: 30 g ha^{-1} decreasing by 10% at 4-day intervals; mean reduction in mating (1981) 60.2% at different population densities
(ii) Hand application with ULV sprayer to open field plots 0.4–1.2 ha at 30 g ha^{-1} and 60 g ha^{-1} at 7-day and 14-day intervals; no reduction in oviposition or larval infestation, some reductions in trap catches and egg fertility

9.5.2.6 Egyptian cotton leafworm, *Spodoptera littoralis* (Boisd.)

ICI polyurea microcapsules; average diameter 2 μm; 2% (Z)-9-tetradecenyl acetate core : wall 1 : 0.6.
Four other variations with thicker wall and/or polyamide cross-linking

(i) Hand application with knapsack sprayers at 70 l ha^{-1} on 300–3000 m^2 plots of lucerne; best results at 100 g ha^{-1}

Day 1-5	% Trap catch reduction 6–10	11–15	16–20	21–25
98	87	71	85	53

Campion *et al.* (1976)

(ii) Samples of formulation placed inside trap with synthetic attractant source gave >99% trap catch reduction for over 40 days

ICI polyurea microcapsules; average diameter 2 μm; 4% (Z)-9-tetradecenyl acetate core : wall 1 : 0.6	Hand application with knapsack sprayers to 50 ha plot				Beevor and Campion (1979)

Application rate (g ha^{-1})	% Larval reduction after 2 weeks
5	0
20	49
40	53

ICI polyurea microcapsules; six variations on previous entry (Beevor and Campion, 1979), for example with polyamide cross-linking, larger capsules (20 μm) and addition of aqueous UV screener (2-hydroxy-4-methoxy-benzophenone-5-sulphonic acid, sodium salt)	Hand application with knapsack to plots of lucerne, 0.7–6.2 ha; less than 76% trap catch reduction over 5 days	Campion et al. (1981)

ICI polyurea microcapsules; as above with (Z)-9-tetradecenyl acetate replaced by (Z,E)-9,11-tetradecadienyl acetate	Hand application with knapsack sprayer to plots of lucerne, 0.7–6.2 ha

Application rate (g ha^{-1})	% Trap catch reduction			
	Day 1–5	6–10	11–15	16–20
1	64	51	30	0
5	85	36	44	25
10	94	78	70	0
20	98	57	57	32

Campion et al. (1981)

(Continued)

ICI polyurea microcapsules; 1–5 μm diameter; 2% (Z,E)-9,11-tetradecadienyl acetate, 2% tetradecyl acetate with stabilizers

Hand application with knapsack sprayer to 0.01 ha plots of cotton; over 92% reduction in trap catches for 13 days at 40 g ha^{-1}

Hall et al. (1982)

9.5.2.7 Tobacco budworm, *Heliothis virescens* (F.); corn earworm *Heliothis zea* (Boddie)

ICI polyurea microcapsules; average diameter 2 μm; 2% aldehyde, with stabilizers

Ground application to 0.47 ha plots of cotton at 2.5 g ha^{-1}

Shaver and Lopez (1982)

Aldehyde	% Reduction trap catch/mating	
	Night 1	Night 7
Heliothis virescens		
(Z)-9-Tetradecenal	82.4/100	30.2/00
(Z)-11-Hexadecenal	91.9/100	59.6/70
Heliothis zea		
(Z)-9-Tetradecenal	67.6/26	23.4/00
(Z)-11-Hexadecenal	99.6/72	57.4/75

9.5.2.8 Pink bollworm, *Pectinophora gossypiella* (Saunders)

ICI polyurea microcapsules; 1–5 μm diameter; 2% 7,11-hexadecadienyl acetate $(Z,E:Z,Z\ 1:1)$ with stabilizers	Hand application with knapsack sprayer to 0.01 ha plots of cotton; over 98% trap catch reduction for 16 days at $5\,g\,ha^{-1}$; over 98% trap catch reduction for $\geqslant 28$ days at $20\,g\,ha^{-1}$	Hall *et al.* (1982)
	Egypt 1981: Aerial application with fixed-wing aircraft to 50 ha plots; 5 applications at $10\,g\,ha^{-1}$; boll infestations, yield and quality of cotton no different from plots with conventional treatment of 4 insecticide applications	Critchley *et al.* (1983)
	Egypt 1982. Aerial application with helicopters to 50 ha plots; 5 or 6 applications at $10\,g\,ha^{-1}$; boll infestations, yield and quality of cotton no different from plots treated with hollow fibre or plastic laminate flake pheromone formulations, or with 3–4 insecticide applications	Critchley *et al.* (1985)
	Egypt 1983: Aerial application with helicopters: (i) 250 ha; 5 applications of pheromone at $10\,g\,ha^{-1}$ compared with 160 ha treated with 4 applications of insecticide; (ii) 5 plots of 40 ha each treated with different regimes of pheromone and insecticides. In (i) boll infestation maintained below 10% in pheromone-treated plot, but infestation above 10% on 3 out of 8 sampling occasions in insecticide-treated plot. In (ii) replacement of earlier insecticide sprays with pheromone gave low boll infestation, but use of pheromone later or reduction in application rate from $10\,g\,ha^{-1}$ to $7.5\,g\,ha^{-1}$ less successful	Critchley *et al.* (1984)

(*Continued*)

Egypt 1984. Helicopter application on 2000 ha; 4 applications at $10\,g\,ha^{-1}$. Boll infestations similar and yields higher than in insecticide-treated fields.	Hosny (1986)
Egypt 1985. Helicopter and fixed-wing application on 4950 ha; 4 applications at $10\,g\,ha^{-1}$. No significant differences between infestations and yields in plots treated with pheromones and those treated with insecticides, although more than 8 times more beneficial arthropods in pheromone-treated fields.	
Pakistan 1984–8. Mist-blower application on 10 ha plots; 2–3 applications at $10\,g\,ha^{-1}$ in combination with insecticides against sucking pests and *Earias* spp. Reduction of at least 3 insecticide sprays compared with fields treated with insecticides only and no decrease in yield; numbers of beneficial arthropods increased in pheromone-treated plots, and fewer insecticide sprays against mites required.	B. R. Critchley (unpublished) (ODNRI)

9.5.3 Other microcapsules

9.5.3.1 Pink bollworm, *Pectinophora gossypiella* (Saunders)

Nitto microcapsules (Section 9.2.2.1) marketed as 'Biolure' by Bend Research Inc.; 36% 7,11-hexadecadienyl acetate (Z,E : Z,Z 1 : 1)	Two aerial applications to cotton at 15-day intervals; 2.5 g ha⁻¹, 3.75 g ha⁻¹, 5 g ha⁻¹ and 10 g h⁻¹; 94–97% trap catch reduction after 1st application, 90% after 2nd application; trap catches increase after 16 days	Smith *et al.* (1983)
'Biolure' microcapsules; 7,11-hexadecadienyl acetate, Z,E : Z,Z 1 : 1 or Z,Z only	0.04 ha cotton plots (i) Application rate 3.7 g ha⁻¹ as 50 µl droplets; trap catch reduction 92% and 81%, mating suppression 100% and 91% respectively for ZZ and ZE/ZZ. (ii) Formulations applied as 50 µl droplets or 0.01 µm spray; application rate 12.3 g ha⁻¹ gave at least 90% trap catch reduction for 12 days at low population, 6 days at high population; large droplets more effective at lower rates	Flint and Merkle (1984)
Zoecon oil/water-soluble acrylic resin capsules; 7,11-hexadecadienyl acetate (Z,E : Z,Z 1 : 1) and permethrin	210 ha cotton treated at 3.7 g ha⁻¹; 2 applications at 10-day intervals; trap catch suppression for 8 days	Evans (1984)

9.5.3.2 Rice stemborer, *Chilo suppressalis* (Wlk.)

Microcapsules doubly encapsulated with urea/formaldehyde copolymer; 4.5 : 1 (Z)-11-hexadecenal + (Z)-13-octadecenal; 2,6-di-*tert*-butyl-4-methylphenol (BHT) as antioxidant	(i) 12 ha rice treated at 2.5 g ha⁻¹; catch in female-baited traps suppressed for 9 days (ii) Over 90% trap catch reduction for 18 days at application rates of 0.83–5 g ha⁻¹ on 20 m² plots	Lee *et al.* (1981)
	Trap catch reduction for 7–10 days	Lee *et al.* (1982)
As previous entry (Lee *et al.* 1981, 1982), with 4.5 : 1 (Z)-9-tetradecenyl formate + (Z)-11-hexadecenyl formate and BHT as antioxidant	Trap catch reduction for 5 days at application rate '30 mg per trap'	Goh and Lee (1984)

9.6 REFERENCES

Bakan, J. A. (1975). Microencapsulation of pesticides and other agricultural materials. In *Proceedings 1975 International Controlled Release Pesticide Symposium*, F. W. Harris (Ed.), Wright State University, Dayton, Ohio, pp. 76–94.

Bakan, J. A. (1976). Controlled release pesticides via microencapsulation. In *Proceedings 1976 International Controlled Release Pesticide Symposium*, N. F. Cardarelli (Ed.), University of Akron, Ohio, pp. 1.1–1.32.

Bakan, J. A. (1980). Microencapsulation using coacervation/phase separation techniques. In *Controlled Release Technology: Methods, Theory, and Applications* (Vol. II), A. F. Kydonieus (Ed.), CRC Press, Boca Raton, Florida, pp. 83–105.

Baker, R. W. and Lonsdale, H. K. (1974). Controlled release: mechanism and rates. In *Controlled Release of Biologically Active Agents*, A. C. Tanquary and R. E. Lacey (Eds), Plenum Press, New York, pp. 15–71.

Barber, L. L., Lucas, A. J. and Wen, R. Y. (1977). US Patent 4,056,610 (to Minnesota Mining and Manufacturing Co.).

Bartell, R. J. (1982). Mechanisms of communication disruption by pheromone in the control of Lepidoptera: a review. *Physiol. Entomol.*, **7**, 353–364.

Baum, G., Bachmann, R. and Sliwka, W. (1971). German Patent 2,119,933 (to BASF), *Chem. Abstr.*, **78**, 31019c (1973).

Baum, G., Bachmann, R., Ludsteck, D. and Sliwka, W. (1972). German Patent 2,237,503 (to BASF), *Chem. Abstr.*, **81**, 64853k (1974).

Beevor, P. S. and Campion, D. G. (1979). The field use of 'inhibitory' components of lepidopterous sex pheromones and pheromone mimics. In *Chemical Ecology: ·Odour Communication in Animals*, F. J. Ritter (Ed.), Elsevier, Amsterdam, pp. 313–325.

Berger, B. B., Miller, C. D., Langer, W. and Langer, G. (1965). Electrostatic encapsulation. US Patent 3,208,951.

Beroza, M., Stevens, L. J., Bierl, B. A., Philips, F. M. and Tardif, J. G. R. (1973). Pre- and postseason field tests with disparlure, the sex pheromone of the gypsy moth, to prevent mating. *Environ. Entomol.*, **2**, 1051–1057.

Beroza, M., Hood, C. S., Trefrey, D., Leonard, D. E., Knipling, E. F., Klassen, W. and Stevens, L. J. (1974). Large field trial with microencapsulated sex pheromone to prevent mating of the gypsy moth. *J. Econ. Entomol.*, **67**, 659–664.

Beroza, M., Bierl, B. A., James, P. and DeVilbiss, D. (1975a). Measuring emission rates of pheromones from their formulations. *J. Econ. Entomol.*, **68**, 369–372.

Beroza, M., Hood, C. S., Trefrey, D., Leonard, D. E., Knipling, E. F. and Klassen, W. (1975b). Field trials with disparlure in Massachusetts to suppress mating of the gypsy moth. *Environ. Entomol.*, **4**, 705–711.

Bierl, B. A., DeVilbiss, E. D. and Plimmer, J. R. (1976). Use of pheromones in insect control programs: slow release formulations. In *Controlled Release Polymeric Formulations*, D. R. Paul and F. W. Harris (Eds), ACS Symposium Series 33, American Chemical Society, Washington, DC, pp. 265–272.

Bruce, W. A. and Lum, P. T. M. (1981). Insect pheromone protection from degradation by ultraviolet radiation. *J. Georgia Entomol. Soc.*, **16**, 227–231.

Bruce, W. A., Lum, P. T., Su, H. C. F., Bry, R. and Davis, R. (1979). US Patent Application 927.792 (to USDA). *Chem. Abstr.*, **91**, 70038w (1979).

Brynko, C. (1961). US Patent 2,969,330 (to NCR Co. Ltd), *Chem. Abstr.*, **55**, 109919d (1961).

Brynko, C. and Scarpelli, J. (1961). Patent 2,969, 331 (to NCR Co. Ltd), *Chem. Abstr.*, **55**, 10919f (1961).

Brynko, C. and Scarpelli, J. A. (1965). US Patent 3,190,837 (to NCR Co. Ltd), *Chem. Abstr.*, **63**, 6362h (1965).

Cameron, E. A., Schwalbe, C. P., Beroza, M. and Knipling, E. F. (1974). Disruption of gypsy moth mating with microencapsulated disparlure. *Science*, **183**, 972–973.

Cameron, E. A., Schwalbe, C. P., Stevens, L. J. and Beroza, M. (1975). Field tests of the olefin precursor of disparlure for suppression of mating in the gypsy moth. *J. Econ. Entomol.*, **68**, 158–160.

Campion, D. G., McVeigh, L. J., Murlis, J., Hall, D. R., Lester, R., Nesbitt, B. F. and Marrs, G. J. (1976). Communication disruption of adult Egyptian cotton leafworm, *Spodoptera littoralis* (Boisd.) (Lepidoptera, Noctuidae) in Crete using synthetic pheromones applied by microencapsulation and dispenser techniques. *Bull. Ent. Res.*, **66**, 335–344.

Campion, D. G., Lester, R. and Nesbitt, B. F. (1978). Controlled release of pheromones. *Pestic. Sci.*, **9**, 434–440.

Campion, D. G., McVeigh, L. J., Hunter-Jones, P., Hall, D. R., Lester, R., Nesbitt, B. F., Marrs, G. J. and Alder, M. R. (1981). Evaluation of microencapsulated formulations of pheromone components of the Egyptian cotton leafworm in Crete. In *Management of Insect Pests with Semiochemicals: Concepts and Practice*, E. R. Mitchell (Ed.), Plenum Press, New York, pp. 253–265.

Cardé, R. T., Trammel, K. and Roelofs, W. L. (1975). Distruption of sex attraction of the red banded leafroller (*Argyrotaenia velutinana*) with microencapsulated pheromone components. *Environ. Entomol.*, **4**, 448–450.

Caro, J. H., Bierl, B. A., Freeman, H. P., Glotfelty, D. E. and Turner, B. C. (1977). Disparlure: volatilization rates of two microencapsulated formulations from a grass field. *Environ. Entomol.*, **6**, 877–881.

Caro, J. H., Freeman, H. P., Brower, D. L. and Bierl-Leonhardt, B. A. (1981). Comparative distribution and persistence of disparlure in woodland air after aerial application of three controlled release formulations. *J. Chem. Ecol.*, **7**, 867–880.

Critchley, B. R., Campion, D. G., McVeigh, L. J., Hunter-Jones, P., Hall, D. R., Cork, A., Nesbitt, B. F., Marrs, G. J., Jutsum, A. R., Hosny, M. M. and Nasr, El-Sayed, A. (1983). Control of pink bollworm, *Pectinophora gossypiella* (Saunders) (Lepidoptera: Gelechiidae), in Egypt by mating disruption using an aerially applied microencapsulated pheromone formulation. *Bull Ent. Res.*, **73**, 289–299.

Critchley, B. R., Campion, D. G., McVeigh, E. M., McVeigh, L. J., Jutsum, A. R., Gordon, R. F. S., Marrs, G. J., Nasr, E. S. A. and Hosny, M. M. (1984). Microencapsulated pheromones in cotton pest management. In *1984 British Crop Protection Conference. Pests and Diseases, Conference Proceedings, Brighton*, November 19–22, 1984 (Vol. 1), British Crop Protection Council, Croydon, UK, pp. 241–245.

Critchley, B. R., Campion, D. G., McVeigh, L. J., McVeigh, E. M., Cavanagh, G. G., Hosny, M. M., Nasr, El-Sayed A., Khidr, A. A. and Negib, M. (1985). Control of pink bollworm, *Pectinophora gossypiella* (Saunders) (Lepidoptera: Gelechiidae), in Egypt by mating disruption using hollow fibre, laminate flake and microencapsulated formulations of synthetic pheromone. *Bull. Ent. Res.*, **75**, 329–345.

DeSavigny, C. B. (1976). US Patent 3.959.464 (to Pennwalt Co.).

DeSavigny, C. B. and Ivy, E. E. (1974). Microencapsulated pesticides. In *Microencapsulation: Processes and Applications*, J. E. Vandegaer (Ed.), Plenum Press, New York, pp. 89–94.

Evans, W. H. (1984). Development of an aqueous-based controlled release pheromone-pesticide system. In *Advances in Pesticide Formulation Technology*, H. B. Scher (Ed.), ACS Symposium Series 254, American Chemical Society, Washington, DC, pp. 151–162.

Finch, C. A. (1985). Polymers for microcapsule walls. *Chem. Ind.*, 752–756.

Flint, H. M. and Merkle, J. R. (1984). Studies on disruption of sexual communication in the pink bollworm, *Pectinophora gossypiella* (Saunders) (Lepidoptera: Gelechiidae), with microencapsulated gossyplure or its component Z,Z-isomer. *Bull. Ent. Res.,* **74**, 25–32.

Gentry, C. R., Beroza, M., Blythe, J. L. and Bierl, B. A. (1974). Efficacy trials with the pheromone of the oriental fruit moth and data on the lesser appleworm. *J. Econ. Entomol.,* **67**, 607–609.

Gentry, C. R., Beroza, M., Blythe, J. L. and Bierl, B. A. (1975). Captures of the oriental fruit moth, the pecan bud moth, and the lesser appleworm in Georgia field trials with isomeric blends of 8-dodecenyl acetate and air-permeation trials with the oriental fruit moth pheromone. *Environ. Entomol.,* **4**, 822–824.

Gentry, C. R., Bierl-Leonhardt, B. A., Blythe, J. L. and Plimmer, J. R. (1980). Air permeation tests with Orfralure for reduction in trap catch of oriental fruit moths. *J. Chem. Ecol.,* **6**, 185–192.

Gentry, C. R., Bierl-Leonhardt, B. A., Blythe, J. L., McLaughlin, J. R. and Plimmer, J. R. (1981). Air permeation tests with (Z,Z)-3,13-octadecadien-1-ol acetate for reduction in trap catch of peachtree and lesser peachtree borer moths. *J. Chem. Ecol.,* **7**, 575–582.

Gentry, C. R., Yonce, C. E. and Bierl-Leonhardt, B. A. (1982). Oriental fruit moth: mating disruption with pheromone. In *Insect Suppression with Controlled Release Pheromone Systems* (Vol. II), A. F. Kydonieus and M. Beroza (Eds), CRC Press Inc., Boca Raton, Florida, pp. 107–115.

Granett, J. and Doane, C. C. (1975). Reduction of gypsy moth male mating potential in dense populations by mistblower sprays of microencapsulated disparlure. *J. Econ. Entomol.,* **68**, 435–437.

Green, B. K. (1955). US Patent 2,712,507 (to NCR Co. Ltd), *Chem. Abstr.,* **50**, 12479a (1956).

Green, B. K. (1960). US Patent reissue 24,899 (to NCR Co. Ltd), *Chem. Abstr.,* **55**, 2956c (1961).

Green, B. K. and Schleicher, L. (1956). US Patent 2,730,456 (to NCR Co. Ltd), *Chem. Abstr.,* **50**, 5951b (1956).

Green, B. K. and Schleicher, L. (1957). US Patent 2,730,457 (to NCR Co. Ltd), *Chem. Abstr.,* **50**, 5951d (1956).

Goh, H. G. and Lee, J. O. (1984). Mating inhibition of striped rice borer (*Chilo suppressalis* W.) by pheromone mimics. *Korean J. Entomol.,* **14**, 9–12.

Gutcho, M. H. (1972). *Capsule Technology and Microencapsulation,* Noyes Data Corporation, Park Ridge, New Jersey.

Gutcho, M. H. (1976). *Microcapsules and Microencapsulation Techniques,* Noyes Data Corporation, Park Ridge, New Jersey.

Gutcho, M. H. (1979). *Microcapsules and other Capsules. Advances since 1975.* Noyes Data Corporation, Park Ridge, New Jersey.

Hall, D. R., Marrs, G. J., Nesbitt, B. F. and Lester, R. (1981). Stabilised compositions containing behaviour-modifying compounds. *PCT International,* WO 81 02,505. *Chem. Abstr.,* **96**, 64215p (1982).

Hall, D. R., Nesbitt, B. F., Marrs, G. J., Green, A. St J., Campion, D. G. and Critchley, B. R. (1982). Development of microencapsulated pheromone formulations. In *Insect Pheromone Technology: Chemistry and Applications,* B. A. Leonhardt and M. Beroza (Eds), ACS Symposium Series 190, American Chemical Society, Washington, DC, pp. 131–143.

Harbort, L. (1966). German Patent 1,519,930 (to Gunther Wagner Pelikan Werke), *Chem. Abstr.,* **71**, 62441u (1969).

Heinrich, R., Frensch, H. and Albrecht, K. (1979). German Patent 2,757,017 (to Hoechst AG), *Chem. Abstr.*, **91**, 108634c (1979).

Heinrich, R., Frensch, H. and Albrecht, K. (1981). German Patent 3,020,781 (to Hoechst AG), *Chem. Abstr.*, **96**, 86652f (1982).

Herbig, J. A. (1968). Microencapsulation. In *Encyclopaedia of Polymer Science and Technology* (Vol. 8), Interscience, New York, pp. 719–736.

Hosny, M. M. (1986). The role of pheromones in the management of the pink bollworm infestation in Egyptian cotton fields. In *Proceedings Parasitis* **86**, Omni-Expo, Geneva.

Johnstone, D. R. (1982). Factors affecting aerial applications of a microencapsulated pheromone formulation for the control of *Pectinophora gossypiella* (Saunders) by communication disruption on cotton in Egypt. *Miscellaneous Report No. 56*, Centre for Overseas Pest Research, London, 11 pp.

Klun, J. A., Chapman, O. L., Mattes, K. C. and Beroza, M. (1975). European corn borer and redbanded leafroller: disruption of reproduction behavior. *Environ. Entomol.*, **4**, 871–876.

Koestler, R. C. (1980). Microencapsulation by interfacial polymerization techniques—agricultural applications. In *Controlled Release Technologies: Methods, Theory, and Applications* (Vol. II), A. F. Kydonieus (Ed.), CRC Press, Boca Raton, Florida, pp. 117–132.

Kydonieus, A. F., Gillespie, J. M., Barry, M. W., Welch, J., Henneberry, T. J. and Leonhardt, B. A. (1982). Formulations and equipment for large volume pheromone applications by aircraft. In *Insect Pheromone Technology: Chemistry and Applications*, B. A. Leonhardt and M. Beroza (Eds), ACS Symposium Series 190, American Chemical Society, Washington DC, pp. 175–191.

Langer, G. and Yamate, G. (1964). US Patent 3,159,874.

Langer, G. and Yamate, G. (1966). US Patent 3,294,704. *Chem. Abstr.*, **66**, 60180f (1967).

Lee, J. O., Park, J. S., Goh, H. G., Kim, J. H. and Jun, J. G. (1981). Field study on mating confusion of synthetic sex pheromone in the striped rice borer, *Chilo suppressalis (Lepidoptera: Pyralidae). Korean J. Plant Protection*, **20**, 25–30.

Lee, J. O., Goh, H. G., Kim, Y. H., Kim, J. H. and Park, C. H. (1982). Evaluation of microencapsulated formulation of pheromone as a control agent for the striped rice borer, *Chilo suppressalis* (Lepidoptera: Pyralidae). *Korean J. Entomol.*, **12**, 25–28.

Liu, X., Macaulay, E. D. M. and Pickett, J. A. (1984). Propheromones that release pheromonal carbonyl compounds in light. *J. Chem. Ecol.*, **10**, 809–822.

Macaulay, N. (1962). US Patent 3,016,308 (to Moore Business Forms).

Marks, R. J. (1978). Mating disruption of the red bollworm of cotton, *Diparopsis castanea* Hampson (Lepidoptera: Noctuidae) by ultra-low-volume spraying with a microencapsulated inhibitor of mating. *Bull. Ent. Res.*, **68**, 11–29.

Marks, R. J., Hall, D. R., Lester, R., Nesbitt, B. F. and Lambert, M. R. K. (1981). Further studies in mating disruption of the red bollworm, *Diparopsis castanea* Hampson (Lepidoptera: Noctuidae), with a microencapsulated mating inhibitor. *Bull. Ent. Res.*, **71**, 403–418.

Matson, G. W. (1962). Canadian Patent 742,643 (to Minnesota Mining and Manufacturing Co.).

McLaughlin, J. R., Mitchell, E. R. and Tumlinson, J. H. (1975). Evaluation of some formulations for dispensing insect pheromones in field and orchard crops. In *Proceedings 1975 International Controlled Release Pesticide Symposium*, F. W. Harris (Ed.), Dayton, Ohio, pp. 209–215.

Meghir, S. (1984). Microencapsulation of insecticides by interfacial polycondensation: the benefits and problems. *Pestic. Sci.*, **15**, 265–267.

Minks, A. K., Voerman, S. and Klun, J. A. (1976). Disruption of pheromone communication with microencapsulated antipheromones against *Adoxophyes orana*. *Ent. Exp. Appl.*, **20**, 163–169.

Nitto Electric Industry Co. Ltd (1984). Microencapsulated liquid pesticide. Japanese Patent JP 59 95,928, *Chem. Abstr.*, **101**, 186163 (1984).

Nixon, J. R. (Ed.) (1976). *Microencapsulation*. Marcel Dekker, New York.

Nesbitt, B. F., Beevor, P. S., Hall, D. R., Lester, R. and Williams, J. R. (1980). Components of the sex pheromone of the female sugarcane borer, *Chilo sacchariphagus* (Bojer) (Lepidoptera: Pyralidae): identification and initial field trials. *J. Chem. Ecol.*, **6**, 385–394.

Patwardhan, S. A. and Das, K. G. (1983). Microencapsulation. In *Controlled Release Technology: Bioengineering Aspects*, K. G. Das (Ed.), Wiley, New York, pp. 121–141.

Pelah, Z. and Marcus, A. (1982). Israel Patent IL 56,709 (to Ben Gurion University of the Negev), *Chem. Abstr.*, **99**, 39535x (1983).

Plimmer, J. R. (1982). Disruption of mating in the gypsy moth. In *Insect Suppression with Controlled Release Pheromone Systems* (Vol. II), A. F. Kydonieus and M. Beroza (Eds), CRC Press Inc., Boca Raton, Florida, pp. 135–154.

Plimmer, J. R., Bierl, B. A., DeVilbiss, E. D. and Smith, B, L. (1976). Evaluation of controlled-release formulations of insect pheromones for mating disruption. In *Proceedings 1976 International Controlled Release Pesticide Symposium*, N. F. Cardarelli (Ed.), Akron, Ohio, pp. 3.29–3.39.

Plimmer, J. R., Bierl, B. A., Webb, R. E. and Schwalbe, C. P. (1977). Controlled release of pheromone in the gypsy moth program. In *Controlled Release of Pesticides*, H. B. Scher (Ed.), ACS Symposium Series 53, American Chemical Society, Washington, DC, pp. 168–183.

Plimmer, J. R., Caro, J. H. and Freeman, H. P. (1978). Distribution and dissipation of aerially-applied disparlure under a woodland canopy. *J. Econ. Entomol.*, **71**, 155–157.

Plimmer, J. R., Leonhardt, B. A., Webb, R. E. and Schwalbe, C. P. (1982). Management of the gypsy moth with its sex attractant pheromone. In *Insect Pheromone Technology: Chemistry and Applications*, B. A. Leonhardt and M. Beroza (Eds), ACS Symposium Series 190, American Chemical Society, Washington, DC, pp. 231–242.

Rodson, M. and Scher, H. B. (1984). Encapsulation process. US Patent 4,448,929 (to Stauffer Chemical Co.), *Chem. Abstr.*, **101**, 395862 (1984).

Roth, W., Heinrich, R., Knauf, W., Bestmann, H. J., Brosche, T. and Vostrowsky, O. (1981). Formulierung von Pheromonen — Möglichkeiten und Schwierigkeiten. *Mitt. Dtsch. Ges. Allg. Angew. Entomol.*, **2**, 279–288.

Rothschild, G. H. L. (1982). Suppression of mating in codling moths with synthetic sex pheromone and other compounds. In *Insect Suppression with Controlled Release Pheromone Systems* (Vol. II), A. F. Kydonieus and M. Beroza (Eds), CRC Press Inc., Boca Raton, Florida, pp. 117–134.

Ruus, H. (1969). US Patent 3,429,827 (to Moore Business Forms), *Chem Abstr.*, **70**, 107685z (1969).

Sanders, C. J. (1976). Disruption of sex attraction in the Eastern spruce budworm. *Environ. Entomol.*, **5**, 868–872.

Sanders, C. J. (1982). Disruption of male spruce budworm orientation to calling females in a wind tunnel by synthetic pheromone. *J. Chem. Ecol.*, **8**, 493–506.

Sanders, C. J. and Seabrook, W. D. (1982). Disruption of mating in the spruce budworm, *Choristoneura fumiferana* (Clemens). In *Insect Suppression with Controlled Release Pheromone Systems* (Vol. II), A. F. Kydonieus and M. Beroza (Eds), CRC Press, Boca Raton, Florida, pp. 175–183.

Scher, H. B. (1975). British Patent 1,371,179 (to Stauffer Chemical Co.).

Scher, H. B. (1977a). Microencapsulated pesticides. In *Controlled Release Pesticides,* H. B. Scher (Ed.), ACS Symposium Series 53, American Chemical Society, Washington, DC, pp. 126–144.

Scher, H. B. (1977b). US Patent 4,046,741 (to Stauffer Chemical Co.).

Scher, H. B. (1979). US Patent 4,140,516 (to Stauffer Chemical Co.).

Scher, H. B. (1981). US Patent 4,285,720 (to Stauffer Chemical Co.).

Scher, H. B. (1983). Microencapsulation of pesticides by interfacial polymerization: process and performance considerations. In *Pesticide Chemistry: Human Welfare and the Environment* (Vol. 4), J. Miyamoto and P. C. Kearney (Eds), Pergamon Press, Oxford, pp. 295–300.

Schwalbe, C. P., Cameron, E. A., Hall, D. J., Richarson, J. V., Beroza, M. and Stevens, L. J. (1974). Field tests of microencapsulated disparlure for suppression of mating among wild and laboratory-reared gypsy moths. *Environ. Entomol.,* **3,** 589–592.

Schwalbe, C. P., Paszek, E. C., Webb, R. E., Bierl-Leonhardt, B. A., Plimmer, J. R., McComb, C. W. and Dull, C. W. (1979). Field evaluation of controlled release formulations of disparlure for gypsy moth mating disruption. *J. Econ. Entomol.,* **72,** 322–326.

Shaver, T. N. and Lopez, J. D. (1982). Effect of pheromone components on mating and trap catches of *Heliothis* spp. in small plots. *Southwestern Entomologist,* **7,** 181–187.

Shin-Etsu Chemical Co. Ltd (1983). Japanese Patent 58,183,601. *Chem. Abstr,* **100,** 116487b (1984).

Sliwka, W. (1975). Microencapsulation. *Angew. Chem. Internat. Edn.,* **14,** 539–550.

Smith, K. L., Myers, D. A. and Hyori, T. (1983). Control of pink bollworm with Biolure microencapsulated gossyplure. *10th International Symposium on Controlled Release of Bioactive Materials,* San Francisco, July 24–27, pp. 212–214.

Somerville, G. R. (1962). US Patent 3,015,128 (to Southwestern Research Institute).

Somerville, G. R. (1967). US Patent 3,310,612 (to Southwestern Research Institute), *Chem. Abstr.,* **66,** 1015632Y (1967).

Somerville, G. R. (1968). US Patent 3,389,194 (to Southwestern Research Institute).

Somerville, G. R. and Goodwin, J. T. (1980). Microencapsulation using physical methods. In *Controlled Release Technologies: Methods, Theory, and Applications* (Vol. II), A. F. Kydonieus (Ed.), CRC Press Inc., Boca Raton, Florida, pp. 155–164.

Taschenberg, E. F. and Roelofs, W. L. (1976). Pheromone communication disruption of the grape berry moth with microencapsulated and hollow fibre systems. *Environ. Entomol.,* **5,** 688–681.

Taschenberg, E. F. and Roelofs, W. L. (1978). Male redbanded leafroller moth orientation disruption in vineyards. *Environ. Entomol.,* **7,** 103–106.

Taylor, A. W. (1982). Field measurement of pheromone vapour distribution. In *Insect Pheromone Technology: Chemistry and Applications,* B. A. Leonhardt and M. Beroza (Eds), ACS Symposium Series 190, American Chemical Society, Washington, DC, pp. 193–207.

Thies, C. (1975). Physicochemical aspects of microencapsulation. *Polymer-plast. Technol. Eng.,* **5,** 23–53.

Vandegaer, J. E. (1971). US Patent 3,577,515 (to Pennwalt Corp.).

Vandegaer, J. E. (Ed.) (1974). *Microencapsulation: Processes and Applications,* Plenum Press, New York.

Vandegaer, J. E. and Meier, F. G. (1969). US Patent 3,464,926 (to Pennwalt Corp.), *Chem. Abstr.,* **71,** 113854r (1969).

Waller, G. D., Erickson, B. J., Harvey, J. and Archer, T. L. (1984). Comparison of honeybee (Hymenoptera: Apidae) losses from two formulations of methyl parathion applied to sunflowers. *J. Econ. Entomol.*, **77**, 230–233.

Weatherston, J. (1981). Device for disseminating a pheromone. US Patent Applic. 230,759, *Chem. Abstr.*, **98**, 121397s (1983).

Weatherston, I., Miller, D. and Lavoie-Dornik, J. (1985). Commercial hollow-fiber pheromone formulations: the degrading effect of sunlight on Celcon fibers causing increased release rate of the active ingredient. *J. Chem. Ecol.*, **11**, 1631–1644.

Wiesner, C. J. and Silk, P. J. (1982). Monitoring the performance of Eastern spruce budworm formulations. In *Insect Pheromone Technology: Chemistry and Applications*, B. A. Leonhardt and M. Beroza (Eds), ACS Symposium Series 190, American Chemical Society, Washington, DC, pp. 209–218.

Yonce, C. E. and Gentry, C. R. (1982). Disruption of mating of peachtree borer. In *Insect Suppression with Controlled Release Pheromone Systems* (Vol. II), A. F. Kydonieus and M. Beroza (Eds), CRC Press, Boca Raton, Florida, pp. 99–106.

Insect Pheromones in Plant Protection
Edited by A. R. Jutsum and R. F. S. Gordon
©1989 John Wiley & Sons Ltd

10

Alternative Dispensers for Trapping and Disruption[1]

I. Weatherston

Département de Biologie, Université Laval, Québec G1K 7P4, Canada

10.1 INTRODUCTION

The use and acceptance of behaviour-modifying chemicals in the commercial arena depends on how they can be integrated, with current methods, into plant protection strategies. This in turn partially depends on the continued development of economic synthetic routes to the chemicals themselves, together with the evolution of efficacious and cost-effective controlled-release devices. The previous three chapters describe those systems which, to date, have had the widest use within the agrochemical industry. There are, however, many other controlled-release devices such as those undergoing development, and those used by researchers in their studies on the chemical elucidation, behavioural aspects, and preliminary field tests of semiochemicals. This chapter focuses on these alternative dispensers, which, for ease of presentation are discussed under four arbitrarily chosen headings: (a) commercial dispensers; (b) rubber septa; (c) polyvinyl chloride rods; and (d) miscellaneous devices.

10.2 COMMERCIAL DISPENSERS

One of the difficulties encountered in discussing commercial dispensers is the reluctance of companies to reveal details of their systems in regard to such factors as methods of preparation and loading, mode of release and cost data, etc. Claims of proprietary information and submitted patent

[1]Contribution no. 418, Department of Biology, Laval University.

applications were the reasons cited by these companies which have pheromone-dispensing systems in restricted commercial use at this time, hence this section is limited to a discussion of the systems of Bend Research Inc., Bend, Oregon, (now Consep Membranes Inc.) and Shin-Etsu Chemical Co. Ltd, Tokyo, Japan.

Shin-Etsu Chemical Co. Ltd, a leader in the production of acetylenic intermediates for the polymer, fragrance and pharmaceutical industries, has over the last ten or so years also produced insect sex pheromones. More recently they have developed two types of controlled-release dispensers marketed under the trade name Hamakicon and for use in the disruption technique. Very little information is available on the one dispenser referred to as the 'suspending particle' which would appear to be a finely ground polymeric material impregnated with the active ingredient. The loading per particle is variable from 0.1 to 2.0 mg, with a longevity span of 10–60 days, the formulation is sprayable with conventional equipment, gives many point sources and is environmentally degradable (Anon., 1984).

The other system is composed of thin polyethylene tubes which may be loaded with 30–300 mg per tube of active material, with release periods from 30 to 200 days and which is hand applied. This device when loaded with (Z)-11-tetradecenyl acetate (90 mg per tube) is registered in Japan as a pesticide against both the tea tortrix (Homona magnanima Niet.) and the smaller tea tortrix (Adoxophyes spp). The company claim that controlling two species simultaneously, with a component common to their pheromones, is a breakthrough in pest management using pheromones. In tests carried out in 1983 at Miyazaki an average of 69.8% reduction over controls in the larval population of both tortrix species was achieved using the above formulation on a 0.1 ha plot. There are five generations of insects from March through October and three applications, each of 300 tubes/plot, took place on March 25, July 7 and August 18, with one insecticide spray being applied on August 15. No cost analysis is available and at the rate of 3000 tubes/ha this system is very labour-intensive and costly. Data are also available (Anon. 1984) on the use of this system in the control of the peach fruit moth (Carposina niponensis Walsingham), the peach-tree borer (Synanthedon exitiosa Say) and the oriental fruit moth (Cydia molesta L.). With (Z)-13-eicosen-10-one as active ingredient (loading not specified) and $1500 + 800$ tubes ha^{-1} in conjunction with three insecticide sprays, fruit damage in two plots was reduced by 81 and 72% over a conventionally treated control plot (1.7 and 2.5% fruit damage compared to 8.8% in the control). In conjunction with seven insecticide sprays damage reduction in one plot was 100% while in the other was an anomalous 41% (0 and 1.0% fruit damage compared to 1.7% in the control plot).

The system has been used in Australia against the oriental fruit moth on peaches. Two applications of 1000 tubes ha^{-1} (75 mg per tube) charged with (Z)-8-dodecenyl acetate (93%), (E)-8-dodecenyl acetate (6%) and

(Z)-8-dodecenol (1%) gave season-long protection. Treated orchards received no insecticide, with control orchards receiving azinphosmethyl (four applications) and cyhexatin (two applications). Trap catch was reduced 98.6% and this resulted in 0.15% damaged fruit and 1.80% damaged new growth as compared to 0.35% and 2.1%, respectively, in the control orchards.

Data cited from these tests indicate that insect control and crop protection are being achieved; however, they are incomplete since release rates, acceptable damage levels and comparative costs of pheromone, pheromone plus insecticide and insecticide treatments are not given. Only in the Miyazaki data is the area treated given, and at 0.1 ha it is too small for a reliable, unreplicated test.

Field trials in Egypt, Pakistan and the USA of this type of dispenser ('PB-Rope') used against pink bollworm, *Pectinophora gossypiella*, were described in Chapter 5.

BioLure controlled-release insect pheromone systems have been developed and marketed by Bend Research Inc., which over the last ten years has been involved in the research and development of products based on controlled-release technology. There are two types of BioLure controlled-release pheromone devices: a microencapsulated form for aerial broadcast application use in the mating disruption control technique (see Chapter 9); and a membrane-coated reservoir device which can be used for mating disruption control by hand placement, but which has a wider application as a trap bait for detection and monitoring use.

This dispenser is a membrane dispenser type which consists of three parts: an impermeable backing, a microporous polymeric pheromone-containing reservoir, and an enveloping rate-controlling membrane through which the pheromone diffuses. As long as the pheromone concentration remains constant within the enveloped reservoir, Fick's law will apply and the pheromone will diffuse through the membrane at a constant rate. As reported by Smith *et al.* (1983) the pheromone release rate is the product of the available surface area (A), the diffusivity of the pheromone in the membrane (D) and the solubility of this pheromone in the membrane (S) divided by the thickness of the membrane (l) (i.e. rate = ADS/l). By the use of various polymeric membranes of different thickness, different available areas and different permeabilities, desired release rates can be achieved with all types of semiochemicals. The longevity of the formulation can be easily adjusted since it is dependent only on the initial loading and the release rate. By varying the polymeric membrane to reflect the permeability and adjusting the initial load, it has been shown (Smith *et al.*, 1983) that (Z)-7-dodecenyl acetate can be released at 4 μg day^{-1} over 50 days; 25 μg day^{-1} over 35 days; 100 μg day^{-1} over 38 days or 450 μg day^{-1} over 9 days.

In field tests, disc dispensers loaded with (Z)-9-tetradecenyl formate, releasing at a rate of 230 μg day^{-1} achieved 98.6–100% disruption of *Heliothis zea* Boddie males over 40 days in Florida sweet corn, as measured by female baited trap catch (Smith *et al.*, 1983).

The stated advantages of this device (Anon., 1983) are ease of manufacture, zero-order release, and widely variable release rates with concomitant latitudes of longevity.

10.3 RUBBER SEPTA

Of all the pheromone-release devices, rubber septa (sleeve stoppers) are the most widely used as trapping baits by researchers, since they can be obtained from several supply houses, are easily loaded with pheromone and are convenient to handle. Several companies, including International Pheromones in Britain, Raylo-Terrochem in Canada and Zoecon in the United States, retail septa pheromone lures. The efficacy of these devices is attested to by the fact that many people consider them to be the 'industry standard' of pheromone lures against which experimental and new commercial lures are measured. A quick perusal of the literature indicates that a detailed account of the field use of rubber septa is impracticable in this overview.

In many cases the septa are used as they are received from the supplier without any pretreatment. They are loaded by placing the desired amount of pheromone dissolved in a suitable solvent, usually hexane or dichloromethane, in the 'cup' end of the septum; the solvent aids in the penetration of the pheromone into the rubber. Once the solvent has evaporated the septa may be stored in capped vials or hermetically sealed foil envelopes, preferably at refrigerator temperatures, until use. In the preparation of gossyplure-impregnated septa for monitoring the pink bollworm, *Pectinophora gossypiella* Saunders, Flint *et al.* (1978) prepared the septa by dissolving gossyplure in 100 μl dichloromethane and placed the solution into the septum; once the material was absorbed into the rubber and the solvent had evaporated a further portion of 100 μl solvent was added to the septum to aid further in the penetration of the pheromone into the rubber. In this manner septa were prepared loaded at nine concentrations from 11.2 μg to 36.5 mg per septum. The effect of using different solvents in the loading process was studied in the case of (E,E)-8,10-dodecadienol, the pheromone of the codling moth, *Cydia* (*Laspeyresia*) *pomonella* L., by Maitlen *et al.* (1976). These authors tested hexane, dichloromethane, chloroform and dimethylformamide and found that the percentage loss of the pheromone after 24 h was 57.8, 8.2, 11.2 and 4.8%, respectively. One can postulate that the volatility of the solvent affects the depth of penetration of the pheromone into the rubber, the solubility of the pheromone in the solvent may be affected as the solution penetrates into the rubber, and the type of solvent will have an effect on the degree of swelling of the rubber. Such factors would lead to a different distribution of the pheromone within the rubber, causing the differences observed by Maitlen *et al.* (1976). These differences would be greatest over the first day or so since the release from the rubber follows first-order kinetics (i.e. the release rate is dependent on the amount of material remaining in the device). Dimethylformamide had the disadvantage that its odour still persisted after 24 h, whereas none of the other solvents tested exhibited this possible complication. The authors (Maitlen *et al.*, 1976) concluded that dichloromethane was the best solvent, although it should be noted that after the first 24 h the release rates from similarly charged septa loaded in hexane or dichloromethane were the same. It was also shown that chemical degradation was not a significant factor in the loss of the diunsaturated alcohol from the septa, even on storage for more than 10 weeks at 4 °C.

Although the use of non-pretreated septa does not appear to affect the field efficacy of lures containing acetate, alcohol, epoxide or ketone pheromones, Steck *et al.* (1979) found that traps baited with some types of rubber septa when loaded with aldehydic pheromones or blends containing aldehydic components either failed to catch males of the target species or caught males of other species whose pheromone blend did not contain aldehydic components. Extraction of such septa with hot alcohol resulted in the removal of more than 20 compounds from the rubber. Septa treated in this manner before loading were subsequently efficacious in the field when used with aldehyde-containing pheromones. Steck *et al.* (1979) isolated and characterized 1,2-dianilinoethane in the ethanolic extract; this compound, an antiozonant, is known to react rapidly with aldehydes to form imidazolidines. Not all commercially available septa contain 1,2-dianilinoethane; Steck *et al.* (1979) recommend that septa be pre-extracted or dianilinoethane-free septa be obtained from the manufacturer for use with aldehyde-containing pheromones. In the development of a lure for the tobacco budworm, *Heliothis virescens*, utilizing the two major components of the pheromone, (Z)-11-hexadecenal and (Z)-9-tetradecenal, Flint *et al.* (1979) compared in the field non-extracted and extracted (hexane : dichloromethane, 1 : 1) septa with septa to which urea had been added to stabilize the aldehydes, and found that extraction and/or stabilization with urea offered no advantage over untreated septa for periods of up to 5 weeks. Untreated septa baited with the two aldehydes captured as many male moths as did live females.

Methods for the determination of release rates of pheromones from laboratory and field-aged controlled-release devices has been reviewed (Weatherston *et al.*, 1981). Until very recently the majority of release rate determinations from rubber septa were carried out by analysing the pheromone residue in the septa at various time intervals. The method of extraction of pheromones from septa is exemplified by the report of Maitlen *et al.* (1976) in regard to (E,E)-8-10-dodecadienol. Septa were extracted with a 1 : 1 mixture of hexane and dichloromethane (25 ml) over 15 min with mechanical shaking. The extract was then chromatographed through a hexane pre-washed alumina column (15 g), eluting the pheromone with an 85 : 15 hexane : dichloromethane solution (125 ml). This removed all coloured and non-pheromonal materials extracted from the rubber. The eluate was taken to dryness at 30 °C on a rotary evaporator and the residue redissolved in a small amount of hexane for gas chromatographic analysis. Protocol controls indicated that the recoveries from septa loaded with 0.05, 0.10, 0.25, 0.50, 1.0, 2.0 and 4.0 mg (E,E)-8-10-dodecadienol were 92, 100, 109, 94, 104, 107 and 107%, respectively. Essentially the same method, with the solvent mixture volume being increased to 50 ml, the extraction time being increased to 1 h and the alumina column chromatography being omitted, has been reported for the extraction of gossyplure (Flint et al., 1978), more than 20 aliphatic acetates (C_6–C_{18}) (Butler and McDonough, 1979), 15 aliphatic alcohols (C_8–C_{18}), five monounsaturated acetates (C_{12} and C_{14}) and disparlure (Butler and McDonough, 1981). To extract [^{14}C]gossyplure from septa Golub et al. (1983) allowed the septa to sit in a toluene-based liquid scintillation cocktail for several days prior to counting.

Since the release of such compounds emitting from a rubber septum is expected to follow first-order kinetics (McDonough, 1978) the release rate (R) may be calculated from the equation

$$R = Mt_{1/2} \ln 2$$

where M is the amount of the compound residual in the septum after a given time, and $t_{1/2}$ is the half-life of the formulation.

M may be calculated from the equation

$$M = M_0 \exp(-t/t_{1/2} \ln 2)$$

where M_0 is the initial amount of the compound present in the septum, and t is the time taken for M_0 to decrease to M. The mathematical relationships and the derivation of these equations have been elegantly described by McDonough (1978).

By using the residue analysis method and applying the equations above Butler and McDonough (1979) determined within 95% confidence limits the half-lives of 19 acetates and a formate. The average half-life varied from 0.48 days for hexyl acetate to almost 24 years for (Z,Z)-3,13-octadecadienyl acetate.

The half-life is a parameter thermodynamically equivalent to the gas chromatographic retention time since both are a measure of volatility, hence the logarithm of the half-life should be linearly related to the carbon number of homologous series provided the rubber substrate behaves as an ideal liquid, as is the case with most gas chromatographic liquid phases. For n-alkyl acetates Butler and McDonough (1979) found such a linear relationship for carbon numbers 10–15. Deviation from linearity for carbon numbers less than 10 and greater than 15 is postulated as being due to the presence of polymer cross-links in the rubber, with the creation of molecular-sized cages from which the small molecules are able to diffuse in and out as if the cages did not exist. For molecules exhibiting linearity, they diffuse in and out of the cages with a degree of difficulty and because of this at a slower rate than can be accounted for by their increasing molecular weight alone. The half-life of hexadecyl acetate is less than that of pentadecyl acetate (481 days as compared to 1353 days). This may be explained by the inability of a molecule of greater than critical size to penetrate into the cage structures, and hence to diffuse through the rest of the rubber faster and be released at a faster rate. It would appear that this postulation, although most probably valid, is incomplete.

The cage structure postulation can also be used to explain the small differences between the half-lives of saturated, monounsaturated and diunsaturated acetates of 12 and less carbon number. The larger effect observed with diunsaturated C_{16} acetates is explained by the fact that some conformations, depending on the position and geometry of the double bonds, will impart to the molecule a much smaller effective size, thus increasing the likelihood of facile diffusion in and out of the cages.

Subsequent studies of *n*-alkanols (Butler and McDonough, 1981) gave analogous results with half-lives increasing from 0.90 days for 1-octanol to 3 years for 1-heptadecanol and then decreasing to 1.67 years for 1-octadecanol. In this series linearity of the relationship between the logarithm of the half-life and carbon number is valid between carbon numbers 10 and 17. This paper also reports that the position of the double bond in tetradecenyl acetates has a significant and greater effect on the half-life ((Z)-7- and (Z)-9-tetradecenyl acetates having half-lives of 154 and 199 days, respectively) than on the half-lives of dodecenyl acetates.

To measure the amount of pheromone emitting from a controlled-release device over a specified time period and then measuring the residue at the end of this period, measuring the residue is a superior method for determining release rates. The effluent vapour may be adsorbed on a substrate from which it can be desorbed and quantified. Provided break-through did not take place and chemical degradation did not occur then mass balance should be achieved, and the rate calculated from both the effluent vapour analysis and the residue analysis will be the same. Several adsorbents have been used for this purpose (Weatherston *et al.*, 1981 and references therein). The release rate of [14C]gossyplure from septa was determined by Golub *et al.* (1983) using the mini-airflow method, in which the pheromone is adsorbed on glass beads. A mass balance of 94.2–100% was obtained, with the mean release rate over 21 days being 0.53–0.87 μg day^{-1}. A silanized silica gel was employed as adsorbent by McDonough (1983) and McDonough and Butler (1983) for the determination of the half-lives of various pheromones by the collection of effluent vapour. Based on previous work (Butler and McDonough, 1979, 1981) the following equation was used to calculate the half-life:

$$t_{1/2} = \frac{P \ln 2}{R}$$

where R is the instantaneous release rate and P is the amount of pheromone in the septum at time t. Although R cannot be measured it can be approximated by measuring the amount released in a discrete time period. Mass balance determinations carried out for four alcohols of carbon number 10, 12, 14 and 16 resulted in values of between 93 and 103%. A comparison of the half-lives of six unsaturated acetates and two unsaturated alcohols determined by the effluent collection method was made with those obtained by the residue method and the results were found to be in good agreement. One advantage of the effluent method is the ease and speed by which half-lives can be determined: half-lives of over 1000 days can be determined in a day. McDonough and Butler (1983) reported half-lives for a further 17 pheromone compounds. They obtained data indicating that Z and E isomerism at the same position in both 12-carbon alcohols and their acetates does not significantly affect the half-lives. With a carbon number of 16, geometrical isomerism of the acetates has a significant effect ($t_{1/2}$ for (E)-7-hexadecenyl acetate = 1168 ± 65 days; $t_{1/2}$ for the (Z)-7- isomer = 794 ± 24 days). In both C-16 alcohols and acetates with E configuration double bonds, positional isomerism affected the half-life more than in Z-configured compounds. In the dodecenols there is no significant difference in the half-lives of (E)-7 and (E)-8 compounds, but the difference is significant between (E)-8 and (E)-9, and also between (Z)-8 and (Z)-9, the numerical values being equivalent.

10.4 POLYVINYL CHLORIDE RODS

A solid polyvinyl chloride slow-release pheromone formulation, usually in the form of cylindrical pellets or rods, has been used as trap bait for monitoring several insects, including *Trichoplusia ni* Hübner (Fitzgerald *et al.*, 1973), *Rhyacionia buoliana* D. & S. (Daterman, 1974), *Orgyia pseudotsugata* McDunnough (Livingston and Daterman, 1977; Daterman, 1978), *Eucosma sonomana* Kearfott (Sartwell *et al.*, 1980a), *Choristoneura fumiferana* Clemens (Sanders and Weatherston, 1976; Sanders 1981) and certain bark beetles (Daterman, 1982), and also in disruption studies (Sartwell *et al.*, 1980b).

This solid plastic formulation, which is similar to insecticide-polymer slow-release formulations reported by Lloyd and Matthysse (1966, 1970), was first used with pheromone as active ingredient by Fitzgerald *et al.* (1973). The method of preparation is outlined in the report of Daterman (1974). Low-temperature fusing resin (vinyl chloride monomer), plasticizer (di-2-ethylhexyl phthalate), 49 parts of each, together with 2 parts of an antioxidant (Advastab BC 109), are thoroughly mixed by stirring and air bubbles removed by vacuum evacuation. The pheromone is then added as a weight percentage, the mixture again thoroughly stirred and air bubbles removed by vacuum evaporation in a rotary evaporator. The viscous mixture is then either poured or drawn into glass tubes used as fusing moulds, sealing the ends by touching a hot plate, and the plastisol fused at 145 °C for 2–5 min or until it becomes translucent. The product is stored at 0 °C in the moulds, wrapped in aluminium foil. For use, the mould is broken and the plastic rod slid out and cut into the desired lengths. A 40 : 60 ratio of resin to plasticizer was used by Sanders and Weatherston (1976) and Sanders (1981) since it produced a less viscous mixture, which facilitated the removal of air bubbles.

The release rate of the pheromone can be easily determined by weight loss studies. The weight loss from pheromone-loaded pellets is compared to the weight loss from unimpregnated pellets over time. This method was used by Fitzgerald *et al.* (1973) to obtain the release rate at 26 ± 2.5 °C from polyvinyl chloride releasers 7 cm long $\times$ 1 cm in diameter loaded with 1, 5, 10 and 20% of (Z)-7-dodecenyl acetate, over a period of 270 days. After a period of 25 days during which the release rate of the pheromone decreased rapidly (e.g. the 10% loading decreased from $350\,\mu g\,h^{-1}$ to $90\,\mu g\,h^{-1}$), the active material released slowly until the experiment was terminated at 270 days, at which time the rate was $30\,\mu g\,h^{-1}$. The release curves for the other loadings were similar, with the release rate at the midpoint of the experiment (135 days) being 6, 25, 55 and $93\,\mu g\,h^{-1}$, respectively, for the 1, 5, 10 and 20% loadings. To determine if the loss of plasticizer and/or stabilizer from the pellets varied the pheromone concentration and hence the release rate, the authors analysed the residual concentration of the pheromone gas chromatographically and compared it to the data from weight loss studies. The residual concentration of (Z)-7-dodecenyl acetate, after 270 days, obtained from the chromatographic analysis of the 5 and 10% loadings was 0.60% and 1.0%, respectively, compared to 0.81% and 1.1% from the corrected weight loss.

Working with (E)-9-dodecenyl acetate, a pheromone component of the European pine shoot moth, Daterman (1974) determined the release rate at 24.4 ± 0.5 °C for pellets 5 mm long and 3 mm in diameter and found that it was between 0.30 and $1.98\,\mu g\,h^{-1}$ over a period of 50 days, the rate closely approximating the equation $\log y = 4.1 - 0.047x$, where y = release rate in $ng\,min^{-1}$ and x = duration in days. He also showed that using a 5% loading of the pheromone in different-sized pieces of PVC the most effective formulations in a field trapping test released at between 0.3 and $3.36\,\mu g\,h^{-1}$, and that lures (5 mm long $\times$ 3 mm in diameter) similarly loaded would emit the pheromone at optimum levels during a 5–6-week period even at 27–29.5 °C, the upper limits of temperature normally experienced during the flight period of the European pine shoot moth in Oregon.

Sanders (1981) developed a PVC lure incorporating a blend of (E)- and (Z)-11-tetradecenal in the ratio 95–97% E and 5–3% Z isomers for use in monitoring the spruce budworm. His objective was to develop a lure which would release the attractant at a uniform rate over the 6-week flight period, would protect the aldehyde from degradation and would be reproducible from year to year. He determined the release rate from rods 10 mm long and 2.5, 4.0 and 10 mm in diameter, at both 10 °C and 21 °C, the temperature limits likely to be experienced during evening flight periods of the budworm. He found that the 10 mm diameter pellet gave the most uniform rate at the higher temperature over the 180-day experiment, releasing at $130\,\mu g\,day^{-1}$ on day 10, $70\,\mu g\,day^{-1}$ on day 40 and $35\,\mu g\,day^{-1}$ on day 100. At the lower temperature the rates from three different sized pellets were slower but more uniform. By varying the active ingredient concentration and the pellet dimensions, Sanders (1981) developed a lure 10 mm long $\times$ 4 mm in diameter loaded at a concentration of 0.03% pheromone, which was released at a rate of $0.1–1.0\,\mu g\,day^{-1}$ and approximated the attractiveness of a female spruce budworm.

To protect the aldehydes from degradation Sanders (1981) tested Advastab BC 109, UOP 688 and BHT (butylated hydroxytoluene) at concentrations between 2 and 10% and concluded that the Advastab used at a low concentration (2%) was effective. Colouring the formulation by incorporation (0.1% by weight) of fluorescent dyes (orange, light green, pink, dark green or yellow) into the plastisol did not affect the catch of spruce budworm males, and made easy the handling of different formulations. Working with *Heliothis virescens*, the tobacco budworm, a species which also utilizes aldehydic pheromone components, Hendricks (1983) has shown that the colour of the plastic lure, whether in rod or surrogate form (made in a mould shaped like the wings and body of the moth) did affect the number of male moths captured. Field testing various pheromone blends in PVC lures coloured red, green, black, orange and yellow, Hendricks found that black lures caught more male moths than those of the other four colours, in part because the black pigment (carbon black) afforded the aldehydes protection against ultraviolet-catalysed oxidative decomposition. One difference between the field tests of Sanders (1981) and Hendricks (1983) is that Sanders used Pherocon wing traps with the bait affixed to the upper interior surface, thereby being protected from sunlight, while Hendricks used cone traps (Hartstack *et al.*, 1979, 1980), where the bait is offered little or no protection from direct sunlight.

Polyvinyl chloride devices containing a 4 : 1 ratio of (Z/E)-9-dodecenyl acetate have been used in disruption trials, as exemplified by those of Sartwell *et al.* (1980b). Using an application rate of 3.5 g active ingredient per hectare they obtained an 85% reduction in the population of the western pine shoot borer in the test plots, and an 83% reduction in plant damage. The PVC formulation used consisted of pellets (50 mm × 3 mm) containing 35 mg attractant (10% loading). At a rate of 100 releasers per hectare spaced at 10 m intervals and 400 releasers per hectare spaced at 5 m intervals this gave application rates of $3.5 \, \text{g ha}^{-1}$ and $14.0 \, \text{g ha}^{-1}$ of active ingredient, respectively. Release rates were determined at 2-week intervals by weight loss measurements and gas chromatographic analysis. Although both methods gave similar results with laboratory-aged formulation, with field-aged material weight.loss measurements indicated a mean release rate of $8.8 \, \mu\text{g h}^{-1}$ whereas gas chromatography gave a mean rate of $15.2 \, \mu\text{g h}^{-1}$. This is explained by the deterioration of the pheromone by sunlight prior to release, since no sunscreen was used in the formulation.

The proponents of this type of formulation point out the flexibility in formulation design, and that release rates can be controlled by varying the length and/or diameter of the pellet, initial loading and rigidity of the plastic. However, the major drawback for general use (i.e. trapping and disruption) is that at the desired release rate and use time period the amount of material utilized is but a small percentage of the initial loading—in other words, suboptimal utilization of an expensive active ingredient. As detailed by Fitzgerald (1973), even after 270 days at $24.4 \pm 0.5 \, °\text{C}$, formulations loaded at 1, 5, 10 and 20% still contained 31, 38, 35 and 45% of their initial loading of (Z)-7-dodecenyl acetate. Like many other release devices the use of PVC pellets for control through disruption is severely limited by the lack of application equipment, application usually being by hand placement.

10.5 MISCELLANEOUS DEVICES

As noted by Weatherston *et al.* (1981) the use of pheromone traps, whether it be for quarantine, monitoring or mass trapping purposes, has employed as dispensers of the active materials such substrates as polyethylene vials and vial caps, glass capillary tubes, dental wicks, cigarette filters and tubing made of polyethylene and rubber. This area has been reviewed by Daterman, (1982). In strategies aimed at controlling insect pests with semiochemicals by interfering with their natural olfactorally dependent orientation or mating behaviour, the variety of substrates used as release devices has been equally varied, and included rice seed, rubber discs, polyethylene vial caps, corn cobs, cork fragments, string, metal planchets, polyethylene and rubber tubing.

A compilation of the various substrates used as controlled-release devices for behaviour modifying chemicals is given in Table 10.1. Although not an exhaustive list, it is representative of the materials used and serves to highlight particular needs of specific cases, whether they be exceptionally fast or slow release rates, environmental acceptability, application problems, the use of extenders and preservatives, protection from external debilitating factors, special modes of action for the release of the semiochemical, the use of the attract and kill technique or the use of the bioirritant technique, etc.

10.5.1 Cigarette filters

The four components of the *Anthonomus grandis* Boheman pheromone grandlure, namely (+)-*cis*-2-isopropenyl-l-methylcyclobutanc cthanol (23.4%), (Z)-3,3-dimethyl-$\Delta^{1,\beta}$-cyclohexane ethanol (17.3%), (Z)-3,3-dimethyl-$\Delta^{1,\alpha}$-cyclohexane ethanal (29.65%) and the corresponding E isomer (29.65%), are very volatile compounds. According to Bull *et al.* (1973), when 100 μg grandlure were allowed to volatilize from 1-dram glass vials $\simeq 35\%$ of the alcohols and $\simeq 65\%$ of aldehydes had evaporated after 100 min at 25 ± 2 °C; at 37 ± 2 °C the respective percentages evaporated after 60 min were 80 and 100%. In order to compensate for the slower release of the alcohol components and extend the field viability of the pheromone, the percentage ratio of the components was changed to 33.3 : 24.6 : 21.05 : 21.05, respectively, and cigarette filters were injected with 0.5 ml of a solution of grandlure (25 mg) in a mixture of methanol (34.5%), water (12.8%), polyethylene glycol (20.4%) and glycerol (32.3%). The loaded filters were placed in foil-wrapped 1-dram glass vials (11 mm aperture), and in weathering tests were shown to release 60–70% of the pheromone in 8 days at a near linear rate. In attractancy tests the accumulated number of weevils captured is stated to follow closely the release curves, which show 90% of the grandlure released over 14 days. However, examination of the data reveals that 3000 out of a total of approximately 8000 weevils were trapped on days 3 and 4, indicating that the release rate was optimized during this time ($\simeq 2$ mg day^{-1}); by days 12–14 the mean rate had dropped to less than 0.4 mg day^{-1}, with less than 200 weevils being captured. The purpose of the glass vial is to act as a physical barrier and to aid in the control of the release rate, while the foil gives some measure of protection to the labile aldehydes from light-catalysed oxidative decomposition. Since the release data were obtained only by residue analysis no insight is offered into any decomposition which might have taken place.

With subsequent changes in trap design the cigarette filter in a glass vial proved no longer efficacious in trapping boll weevils because of reduced release rates caused by lure placement. After field testing five cigarette filter formulations, McKibben *et al.* (1980) found that a commercially available polyester-coated filter was the most efficacious of the formulations tested. During 1978 and 1979 half a million such polyester filters were used as the standard dispenser for the Boll Weevil Eradication Trial in North Carolina, realizing substantial savings in both material and labour costs over previous dispensers.

Cellulose acetate cigarette filters in glass vials have also been used in the development of a monitoring system for *Heliothis virescens* F. incorporating the two-component $((Z)$-11-hexadecenal : (Z)-9-tetradecenal) virelure attractant (Hendricks *et al.*, 1977; Shaver *et al.*, 1981). Various substances such as cottonseed oil, wheat-germ oil, polyethylene glycol-600 distearate and polyethylene glycol-6000 distearate were tested as extenders, with d-α-tocopherol being used as an antioxidant. The best of such formulations contained 10.5 μl virelure (20 : 1 (Z)-11-hexadecenal : (Z)-9-tetradecenal) in 0.4 ml methylene chloride containing 5% (v/v) of a 9 : 1 mixture of wheat-germ oil and d-α-tocopherol (Hendricks *et al.*, 1977). Even with the additives, trapping data indicated that over seven nights the pheromone release rate was significantly lower on nights 3–7 than it was on the first two, as reflected by trap catch. Changes in component ratio which would be a factor in reduced catches was ruled out when residue analysis of the filters after 5 days showed that the ratio was within 7% of the original value.

Shaver *et al.* (1981) indicated that over 5 days in the field the release rate of virelure changed from 36.2 μg h^{-1} over the first 2 days to 40.9 μg h^{-1} over the next 2 days, and to 32.1 μg h^{-1} on the fifth day. Such variation may be accounted for by changes in temperature. However, the ratios of the components in the atmosphere for the same time periods were 8 : 1, 21 : 1 and 7 : 1, which are much different from the initial 15.7 : 1 (Z)-11-hexadecenal : (Z)-9-tetradecenal ratio. This illustrates yet again that the ratio of components loaded onto a device often bears no relation to their ratio in the emitting vapour, a phenomenon which may be caused by several factors, including possible decomposition, differences in diffusivity and the solubility rates of the individual components travelling through and out of the device.

The field performance of the cigarette filter devices for monitoring *Heliothis virescens* was much inferior to that of commercially available plastic laminates loaded with the same components. In general the use of cigarette filters will probably be restricted to research purposes, either where candidate pheromones are being field bioassayed over short time intervals or where exceedingly high release rates are required, say, over a 24 h period.

10.5.2 Polyethylene vial caps

Amongst the first release devices to be used for the dissemination of lepidopteran sex pheromones was the polyethylene vial cap. This is a hollow enclosure with a top, normally made of higher density polyethylene,

which snaps into place. To study the release characteristics of the caps Kuhr
et al. (1972) loaded them with 10 μl neat radiolabelled saturated analogues
of pheromone acetates and placed them in a hood at room temperature with
an air draft of 0.54 ms^{-1}. Residue analysis at various time intervals over
80 days by scintillation counting yielded half-lives of 7, 11 and 20 days
for dodecyl, tridecyl and tetradecyl acetates, respectively, while even after
80 days the amount of hexadecyl acetate had only been reduced by 15%.
Varying the amount of dodecyl acetate from 10 μl to 30 μl and 100 μl increased
the half-life from 7 to 16 and 39 days. Loading this compound in petroleum
ether (10 μl in 25 μl solvent) also increased the half-life to 12.5 days. In an
attempt further to extend the longevity by retarding the release rate, glass
wool, dental wick, cotton, boiling chips, filter paper, rubber foam, rubber
strips, aluminium foil and florisil were each added to caps prior to loading
with 10 μl dodecyl acetate; only florisil significantly reduced the emission
rate, only 0.5 μl being released in 33 days.

The release curves obtained when 20 μl of a 1 : 1 mixture of [^{14}C] dodecyl
acetate : cis-11-tetradecyl acetate or 20 μl of a 1 : 1 mixture of dodecyl
acetate : [^{14}C] tetradecyl acetate were field exposed showed that both
labelled compounds released linearly over 60 days, the dodecyl acetate
having a half-life of 41 days and the tetradecyl acetate one of 56 days
according to the authors (Kuhr et al., 1972). The slower release pattern,
compared with that obtained in the laboratory, was probably caused by two
factors: (a) the lower temperatures, to which the caps were exposed,
especially during night time; and (b) the use of a total of 20 μl acetates per
cap rather than the 10 μl used in the laboratory experiment. A third
contributing cause is that the caps were loaded with a mixture of two
components of different vapour pressures and diffusivities, causing a
depression of the release rate over that exhibited by the pure compounds.

Disruption trials on a small scale (0.04 ha) against Rhyacionia frustrana Comstock
have been carried out using 2-dram polyethylene vial caps (Berisford and
Hedden, 1978). The caps were loaded with 1 mg neat (E)-9-dodecenyl acetate,
the major component of the sex pheromone of the European pine shoot moth,
Rhyacionia buoliana D. & S., and deployed at the rate of 1 and 2 g ha^{-1} (i.e. 1000
and 2000 caps ha^{-1}) to disrupt the attraction of Rhyacionia frustrana males to
traps baited with live females. At the higher application rate the trap catch was
reduced by 71.9%, whereas a 57.1% reduction was observed at a rate of 1 g ha^{-1}.

Vial caps have also been shown to be effective in the release of apple
volatiles used as bait for trapping male and female apple maggot flies
(Reissig et al., 1982). Of the seven apple volatiles identified as being attractive
to Rhagoletis pomonella Walsh (Fein et al., 1982) six — hexyl acetate, butyl
2-methylbutanoate, propyl hexanoate, hexyl propanoate, butyl hexanoate
and hexyl butanoate (36 : 7 : 12 : 5 : 29 : 11) — were tested at three rates — 50,
100 and 300 mg — (Reissig et al., 1982) in conjunction with sticky red spheres

(Prokopy, 1977). The spheres baited with apple volatiles caught significantly more male and female flies than did the visually attractive spheres without volatiles. The two lower dosages were more effective than the 300 mg loading. The longevity of the caps is at least 2 weeks; however, they have been superseded by small polyethylene bottles containing 700 mg of the mixture, which lasts throughout the season (May–September) (Prokopy, personal communication).

Polyethylene vial caps are still amongst the release devices most widely used by university and government researchers for field bioassays of newly elucidated pheromones. For disruption studies they are only suitable for very small field tests, such as the one described above, because of the need for hand placement of the devices.

10.5.3 Miscellaneous capsules

From Table 10.1 it can be determined that polyethylene vials and capsules, usually centrifuge tubes or Beem capsules, have also had a global use as pheromone dispensers for both monitoring and trapping studies. From the examples given in the table, two are discussed in detail.

n-Dipropyl disulphide, a known plant-derived oviposition stimulant and attractant for male and female onion fly, *Hylemya (Delia) antiqua* Meigen has been dispensed from Beem capsules (Dindonis and Miller, 1981). For characterization of the release as a function of time, and also temperature, the capsules (200 μl capacity) were filled with neat dipropyl disulphide, maintained at the desired temperature and the release rate determined gravimetrically by residue analysis. At $33 \pm 1\,^{\circ}$C, and after a latency period of about 2 h, the disulphide was emitted at a constant rate ($\bar{X} = 0.9\,\mathrm{mg\,h^{-1}}$) for approximately 8 days, by which time no visible reservoir remained. It was also determined that release rate increased exponentially as a function of temperature; however, within the temperatures encountered in the field (17–41 $^{\circ}$C) there was a seven-fold increase in the release rate, from $0.29\,\mathrm{mg\,h^{-1}}$ to $2.0\,\mathrm{mg\,h^{-1}}$.

In field attractancy tests, traps baited with 10, 100 and $10 \times 100\,\mu$l dipropyl disulphide caught significant numbers of both male and female flies over the controls, but there was no statistical difference in trap catches among the three loadings. Increasing the amount of attractant from 0.1 μl to 100 μl in decade steps increased the trap catches but at the $10 \times 100\,\mu$l loading the trap catch plateaued. The authors comment that the greater amount of attractant emitted, the greater the active space within which flies will respond; however, a threshold may be reached which will negate the long-distance effect by a close-range repellency or arrestment of attracted flies. Although Dindonis and Miller do not recommend the *n*-dipropyl disulphide/Beem capsule system as an effective monitoring system, it is the chemical aspect which they consider must be improved since they report that polyethylene capsules appear to be ideally suited for dispensing highly volatile semiochemicals. Over an extended time, a constant release rate with only relatively small temperature-caused deviations is possible by varying the size of the reservoir and the surface area of the capsule exposed to the atmosphere.

10.5.4 Polyethylene tubes

In disruption trials against the oriental fruit moth in Australian peach orchards, Rothschild (1979) experimented with a release device consisting of three closed polyethylene microcentrifuge tubes wired together, each tube containing 50 μl (Z)-8-dodecenyl acetate. One such device was attached to every tree (242) in 2×0.6 ha plots, and the catch of male moths in these treated plots compared with the catch in two similarly sized control plots. Trap catch reduction of 98–99% was achieved over an 89-day period, the release rate at the temperatures during the flight period exceeding the 5 mg h^{-1} ha^{-1} necessary to cause trap catch reduction. In another trial where the observed release rate was less than 5 mg h^{-1} ha^{-1} the trap catch was reduced by only 38%.

Laboratory release rate measurements were carried out at 15±1 °C, 20±1.5 °C and 29±3 °C, with a wind speed of about 0.5 m s^{-1} and measured by weight loss. At 15 °C after a latency period of 2 days the dodecenyl acetate was released at a constant rate over 55 days ($\bar{X}$ rate/device = 22.7 μg h^{-1}); however, only 20% of the initial loading had been emitted. At 21 °C, over 55 days ≈30% of the initial charge had been released at a mean rate of 33.0 μg h^{-1}, while at 29 °C the release rate during the first 20 days was greater than that over the remaining 35 days, with 56% of the load being utilized over the whole time period. The mean release rate at this temperature was 63.6 μg h^{-1} per device. Rothschild (1979) further showed that at this highest temperature the release rate was almost directly related to the surface area of the dispenser still in contact with the pheromone, and that weight loss in the field was consistent with laboratory data based on the daily maximum temperatures.

Rothschild (1979) also experimented with polyethylene tubing dispensers (20 cm × 0.7 mm i.d. and 1.4 mm o.d.) containing 60–70 μl pheromone and sealed at both ends. This type of device did not give a constant release at 29 °C, 76% of the dodecyl acetate being released in the first 10 days. At 15 and 20 °C it took 30 and 47 days to release the same amount of material. The decreased longevity caused by the rapid release was overcome by leaving the tubing unsealed and placing both ends into a stoppered glass vial (45 × 10 mm) containing 1 g of the active material. However, these modified devices were not efficacious in reducing trap catches, this in part being caused by water condensation in the reservoirs.

Amongst several other authors reporting the use of polyethylene tubing Charmillot *et al.* (1981) carried out disruption trials against *Adoxophyes orana* (FvR) and achieved high trap catch reduction of males, and complete mating inhibition of tethered females with one polyethylene tubing dispenser per 400 m². The devices were made by filling several metres of capillary tubing (0.5–0.9 mm i.d. and 1.0–2.0 mm o.d.) with a 9 : 1 mixture of (Z)-9-tetradecenyl acetate and (Z)-11-tetradecenyl acetate containing 0.1% butylated hydroxytoluene as antioxidant and a little β-carotene for colour. After 2–4 days any excess material not absorbed into the walls was expelled. The tubing was then cut to desired lengths, the loading being between 25 and 100 mg h^{-1}. The release rate was reported as a function of degree-days above a 10 °C threshold, with the amount released determined either gravimetrically or by gas chromatographic analysis of the residue. That oxidative degradation of the attractant was of little importance was indicated by there always being <5% decomposition products in the residual pheromone as evidenced by gas chromatography. Data given for tubing 1.7 mm o.d. and 0.8 mm i.d. gave a half-life, for *Adoxophyes orana* pheromone, of 3.2 days at 20 °C, 6.4 days at 15 °C or 16 days at 12 °C. A half-life of 15 degree-days was obtained with 1.3 mm o.d. and 0.5 mm i.d. tubing, i.e. 50% of the charge is released in 3 days at 15 °C or 1.5 days at 20 °C.

10.5.5 Dental roll

Dental roll has also been widely used to disseminate pheromones (see Table 10.1 for representative examples) but the paper of interest is that of Beroza *et al.* (1971), dealing with the activity of natural and synthetic gypsy moth attractants, since it is one of the first reports concerning the prolongation of longevity by experimenting with keepers and extenders. The need for a keeper or extender in the use of a pheromone release device such as dental roll, from which pheromones volatilize very quickly, is to increase the field life of the formulation by retarding the release rate and/or decreasing the susceptibility of the attractant to decomposition. Preliminary work in this area with filter paper wicks and gypsy moth extracts was followed by the use of racemic disparlure on dental roll (Beroza *et al.*, 1971). Two types of keepers were tested: involatile materials which would remain in the device; and compounds of similar volatility to disparlure which would co-emit with the attractant. Based on the fact that the activity of crude extracts persisted longer than purified material, Beroza *et al.* (1971) selected the triglycerides trioctanoin and triolein to be tested as involatile keepers since they believed that fats in the crude extract were responsible for the increased longevity. Furthermore, liquid fats were preferred because the attractant could more likely be renewed at the surface of a liquid than when held in a solid keeper. Trioctanoin proved to be an excellent keeper and much superior to triolein.

TABLE 10.1 Controlled-release devices for behaviour-modifying chemicals

Dispenser type	Insect		Purpose	Chemical loading	Release rate	Notes	Reference
	Species	Order					
Cigarette filter	*Heliothis virescens* (F.)	Lepidoptera	Trapping	1.25–50 mg	—	Cellulose acetate filters (30×8 mm) were loaded with virelure ((Z)-11-hexadecenal/(Z)-9-tetradecenal) in ratios of 16/1 and 32/1 in a 90:9:1 v/v mixture of solvent: wheat germ oil: d-α-tocopherol. The filter was contained in a glass vial (30×9 mm).	Shaver et al., 1981
	Heliothis virescens (F.)	Lepidoptera	Trapping	10.5, 10.6 and 21 μl	—	Cellulose acetate filters (30×8 mm) were loaded with virelure (20/1) in hexane or dichloromethane containing 5 or 10% oil additives.	Hendricks et al., 1977
	Anthonomus grandis (Boheman)	Coleoptera	Trapping	25 mg	—	The filters in foil wrapped 1 dram glass vials (11 mm aperture) released 60–70% of their grandlure loading in 8 days at near linear rate.	Bull et al., 1973
	Anthonomus grandis (Boheman)	Coleoptera	Trapping	8 mg	—	The following formulations were tested: (a) cellulose acetate filter (30×8 mm), (b) filter wrapped in aluminium foil, (c) filter in a glass vial, (d) filter inside a 30×8 mm piece of polypropylene tubing, (e) filter inside a 50×8 mm polypropylene tube, (f) cellophane coated filter, and (g) polyester coated filter.	McKibben et al., 1980

Fibre-glass strips	*Musca domestica* (L.)	Diptera	Killing Station	100 mg	2.87 mg day^{-1}	The 2.5×4.5 cm strips were coated with muscalure and permethrin (90.75 mg) from an acetone solution.	Carlson and Leibold, 1981
Acrylic yarn	*Musca domestica* (L.)	Diptera	Killing Station	100 mg	1.02 mg day^{-1}	One metre lengths of the yarn were soaked in a muscalure/permethrin solution to give the same initial loading as the fibre-glass strips.	Carlson and Leibold, 1981
Sulphonated polystyrene resins	*Costelytra zealandica*	Coleoptera	Trapping	—	—	Two types of resin were used to dispense phenol, the sex attractant of the beetle: (a) 8% cross-linked sulphonated polystyrene beads treated at 0 °C with aqueous benzene diazonium chloride, and (b) the cross-linked sulphonated polystyrene beads treated with aqueous phenol.	Lauren, 1979
Rice seed	*Dendroctonus frontalis* (Zimm.)	Coleoptera	Disruption	1.95 and 19.5 mg/1 g rice	—	Frontalure (Frontalin+α-pinene) at each loading was aerially applied at 23 kg of rice seed per hectare but failed to disrupt the process by which host trees are colonized by *D. frontalis*.	Vité et al., 1976
Polyvinyl chloride capsules	*Heliothis virescens* (F.)	Lepidoptera	Trapping	10 or 20 mg	—	The capsules were constructed from two bullet shaped, moulded PVC wire tip protectors to give dispensers 17 mm long, 4.2 mm outside diameter with a wall thickness of 0.73 mm, and loaded with various *H. virescens* pheromone blends.	Hendricks, 1982

(Continued)

Dispenser type	Insect		Purpose	Pheromone loading	Release rate	Notes	Reference
	Species	Order					
Silicone rubber discs	Cydia molesta (L.)	Lepidoptera	Disruption	≈50 mg	variable	The extremely variable release rate may be partially explained by changes occurring in the disc as the rubber cures.	Rothschild, 1979
Leather	Synanthedon pictipes (Grote and Robinson)	Lepidoptera	Trapping	500 μg	<0.0014 μg h^{-1}	These devices were made from a piece of pet collar (12×12×3 mm), pieces of shoe lace (30×4×1 mm) and (30×3×2 mm) and loaded by applying a hexane solution of (E,Z)-3,13-octadecadienyl acetate to the surface.	Sharp and Cross, 1980
				500 μg	<0.19 μg h^{-1}	A piece of chamois (12×12×1 mm) loaded as above.	
Polyethylene vial caps	—	Lepidoptera	Trapping	10–100 μl	—	Saturated analogues of acetate pheromones ^{14}C labelled in the acetate moiety were used to determine release rates both in the laboratory and in the field.	Kuhr et al., 1972
	Rhyacionia frustrana (Comstock)	Lepidoptera	Disruption	1 mg/cap	—	The vial caps (2 dram) were loaded with neat (E-9-dodecenyl acetate, and applied to small plots (0.04 ha) at rates of 1 and 2 g ha^{-1}. At the higher rate a 71.9% reduction in trap catch to female baited traps was achieved.	Berisford and Hedden, 1978
	Rhagoletis pomonella (Walsh)	Diptera	Trapping	50 and 100 mg	—	Apple volatiles (6 components) dispensed from vial caps in conjunction with red spheres significantly increased the catch of male and female flies over that obtained from the use of the visually attractive spheres alone.	Reissig et al., 1982

Dispenser	Species	Use	Loading	Release rate	Remarks	Reference
	Carposina niponensis (Walsingham)	Lepidoptera	Trapping	0.1–0.3 mg 1.0–3.0 mg	—	The caps were loaded with a 20 : 1 mixture of (Z)-7-eicosen-11-one and (Z)-7-nonadecen-11-one.	Shirasaki et al., 1979
Polyethylene capsules and vials	*Heliothis armigera* (Hb.)	Lepidoptera	Trapping	2.08 mg	—	Vials (36×16 mm with 1.5 mm wall thickness) were loaded with a hexane solution of various pheromone blends containing 10% butylated hydroxytoluene as antioxidant. The dispensers were ineffective as trap baits.	Kehat et al., 1980
	Synanthedon pictipes (G. & R.)	Lepidoptera	Trapping	500 µg	<0.0053 µg h^{-1}	A 1 ml vial was loaded with a hexane solution of the pheromone, and closed after the evaporation of the solvent.	Sharp and Cross, 1980
	Spodoptera littoralis (Boisd.)	Lepidoptera	Disruption and trapping	0.5–2.0 mg	—	This laboratory study determined release curves (% remaining *vs* time) for (Z,E)-9,11-tetradecadienyl acetate (long range attractant for *S. littoralis*), (Z)-9-tetradecenyl acetate (inhibitor for *S. littoralis*), and (Z)-7-tetradecenal (sex attractant for *P. citri*) from 4 ml vials.	Campion et al., 1978
	Prays citri (Millière)	Lepidoptera	Trapping		—		
	Hylemya antiqua (Meigen)	Diptera	Trapping	0.1–1000 µl	—	n-dipropyl-disulphide, an oviposition stimulant and attractant was dispensed from capsules. There was no statistical difference between the catches at loadings of 10–1000 µl.	Dindonis and Miller, 1981

(Continued)

Dispenser type	Insect Species	Order	Purpose	Pheromone loading	Release rate	Notes	Reference
	Staphylinochrous whytei	Lepidoptera	Trapping	0.2–1.25 mg	—	Various combinations of the E and Z isomers of 11-tetradecenol and its acetate, and an equal amount of butylated hydroxytoluene in 500 μl of hexane were dispensed from vials.	Hall and Read, 1979
	Doratopteryx plumigera	Lepidoptera	Trapping	1.0 mg	—		
	Cydia molesta (L.)	Lepidoptera	Disruption	150 mg	22.7 μg h^{-1} (15°) 33.0 μg h^{-1} (20°) 63.6 μg h^{-1} (29°)	Three centrifuge tubes wired together constitute a device, the release rates given are the means over 55 days for Z-8-dodecenyl acetate at the stated temperatures, measured in the laboratory with a windspeed of 0.5 m s^{-1}. With these dispensers 98–99.5% trap catch reduction was obtained in peach orchards over a period of 3 months.	Rothschild, 1979
	Diparopsis castanea (Hamps.)	Lepidoptera	Trapping	1 mg	—	85–87% (E)- and 13–15% (Z)-9-11-dodecadienyl acetate in a 4:1 ratio with 11-dodecenyl acetate were placed in a sealed 0.5 ml vial. Field longevity of this device was 5–6 weeks.	Marks, 1977, 1978
	Carposina niponensis (H.-S.)	Lepidoptera	Trapping	0.1–0.3 mg 1.0–3.0 mg	—	A 20:1 mixture of (Z)-7-eicosen-11-one and (Z)-7-nonadecen-11-one were loaded into the capsules.	Shirasaki et al., 1979

Polyethylene tubing	*Adoxophyes orana* (F.v.R)	Lepidoptera	Disruption	25–100 mg m^{-1}	Several metres of capillary tubing (1.0–2.0 mm o.d. and 0.5–0.9 i.d.) were filled with a 9 : 1 mixture of (Z)-9-tetradecenyl acetate and (Z)-11-tetradecenyl acetate containing 0.1% butylated hydroxytoluene and a little β-carotene. After 2–4 days the excess material not absorbed into the walls was expelled. One dispenser per 400 m^2 resulted in high trap catch reduction of feral males, and complete inhibition of mating in tethered females.	Charmillot et al., 1981
	Cryptophlebia leucotreta (Meyrick)	Lepidoptera	Trapping	1, 5, 10 and 20 mg	The tubing (30×12 mm) was filled with neat pheromone (1 : 1 (E,Z)-8-dodecenyl acetate) or a hexane solution of the attractant.	Angelini et al., 1980
	Cydia molesta (L.)	Lepidoptera	Disruption	≈65 mg	Release rates of both (Z)-8-dodecenyl acetate (29 °C) and dodecyl acetate (15, 20 and 29 °C) were measured in the laboratory.	Rothschild, 1979
Rubber tubing	*Adoxophyes orana* (F.v.R)	Lepidoptera	Disruption		Tubing (4 mm o.d. ×2 mm i.d.) was impregnated with a hexane solution of a 1 : 1 mixture of (E,E)-8,10-dodecadien-1-ol and a 9 : 1 mixture of (Z)-9-, and (Z)-11-tetradecenyl acetates.	Charmillot et al., 1981
	Cydia pomonella (L.)	Lepidoptera	Disruption			
	Pectinophora gossypiella (Saunders)	Lepidoptera	Trapping		Pieces of red rubber tubing are placed in a hexane solution of gossyplure for 2 h at 23 °C allowing the pheromone to be absorbed into the rubber.	Huber and Hoffman, 1979

(Continued)

Dispenser type	Insect Species	Order	Purpose	Pheromone loading	Release rate	Notes	Reference
Cellulose acetate discs	*Anthonomus grandis* (Boh.)	Coleoptera	Trapping	2–25 mg	—	Cellulose acetate discs were prepared using various plasticizers, and loaded with grandlure or its 2 alcohol components. Laboratory and field studies were undertaken to determine release rate, longevity and attractancy of the devices.	Bull *et al.*, 1973
Dental roll	*Heliothis armigera* (Hb.)	Lepidoptera	Trapping	104 µg	—	Dental roll impregnated with a 100 : 4 mixture of (Z)-11, and (Z)-9-hexadecenal	Kehat *et al.*, 1980
	Prionoxystus robinae (Peck)	Lepidoptera	Trapping	500 µg	—	Dental roll impregnated with a 60/40 mixture of (Z,E) and (E,E)-3,5-tetradecadienyl acetates in hexane. Various preservatives were tested as to their effect on longevity and attractancy.	Solomon *et al.*, 1978
	Lymantria dispar (L.)	Lepidoptera	Trapping	1–200 µg 100 µg 50 µg–500 µg	—	(+),(–)(±) disparlure (±) disparlure (+), (–) (±) disparlure — dispensed in 100 µl of petroleum ether on to cotton wicks (1×1 cm).	Cardé *et al.* 1977a Cardé *et al.*, 1977b Cardé *et al.*, 1978
	Heliothis armigera (Hb.)	Lepidoptera	Trapping	0.5–5.0 mg	—	Dental roll (2 cm long) was impregnated with either (Z)-11-hexadecenal or (Z)-11-tetradecenal in hexane.	Gothilf *et al.*, 1979
	Lymantria dispar (L.)	Lepidoptera	Trapping	5–50 mg	—	The use of 'keepers' to increase the longevity of disparlure impregnated on dental roll (12 mm×6 mm) was studied in the laboratory and the field.	Beroza *et al.*, 1971

				Amount	Release rate	Description	Reference
Polyurethane foam	Synanthedon pictipes (G. & R.)	Lepidoptera	Trapping	500 µg	15–25 µg h^{-1}	Loaded in the same manner as the leather devices (see above) the $(10 \times 10 \times 10$ mm) foam cubes were differentiated by two pore sizes: (a) 1.0 mm and (b) <0.5 mm.	Sharp and Cross, 1980
Glass capillaries, tubes & beakers	Ips acuminatus	Coleoptera	Trapping		I & II 30 µg h^{-1} III 20 µg h^{-1} IV 400 µg h^{-1}	The four components of the pheromone, ipsenol (I), ipsdienol (II), trans-verbenol (III) and 2-methyl-3-buten-2-ol (IV) were dispensed from separate capillaries (0.8 mm i.d.) contained in a polyethylene vial $(50 \times 9$ mm).	Bakke, 1978
	Trypodendron lineatum (Oliver)	Coleoptera	Trapping			Lineatin was dispensed from 40×0.9 mm capillaries; α-pinene : ethanol $(1:4)$ from 50×0.9 mm capillaries. Although lineatin dispensed in this way was attractive, the co-release of the α-pinene : ethanol mixture increased the attractancy by a factor of 2.	Klimetzek et al., 1980
	Ips cembrae (Meer.)	Coleoptera	Trapping	10–40 µl		The synthetic compounds field tested, ipsenol, ipsdienol and 3-methyl-3-buten-1-ol were dispensed from separate capillaries contained in polyethylene capsules or from 50×10 mm glass tubes with 2 mm diameter holes in their plastic stoppers.	Stoakley et al., 1978
	Various	Lepidoptera	Screening	1 mg		The various acetates and alcohols tested were dispensed either singly or as binary mixtures from 11 glass vials containing 1 ml of sunflowerseed oil.	Toth et al., 1979
	Sitotroga cerealla (Oliver)	Lepidoptera	Disruption	100 µl	0.009 µg h^{-1}	These glass devices were 5 ml beakers with an inside bottom area of 229 mm^2.	Vick et al., 1978

Whereas the disparlure/trioctanoin-baited lures had a field longevity of at least 3 months, the disparlure/triolein combination performed very poorly. It is suggested that the epoxide group of disparlure reacts with the allylic hydrogens of the double bonds of the triolein. Similarly in the case of co-emitting keepers, unsaturated compounds such as methyl oleate and methyl linoleate performed poorly. Although not noted by Beroza *et al.* (1971), unsaturated compounds, in addition to reacting with disparlure, would also be subject to oxidative decomposition (see Shani and Klug, 1980); however, this could be prevented by the addition of an antioxidant. Further discussion of pheromone decomposition is to be found in Chapter 8. Of the several volatile keepers tested, e.g. hexadecyl acetate, 1-octadecanol, methyl palmitate, hexadecylamine, eicosane, methyl octadecyl ether, etc., only 1-octadecanol adversely affected the attractancy of disparlure; methyl palmitate, which has a volatility only slightly greater than disparlure, consistently performed well.

10.5.6 Miscellaneous carriers

Two other devices worthy of mention because of novel aspects have been reported by Lauren (1979) and Carlson and Leibold (1981). In the former case there was a requirement for the controlled release of phenol, the sex attractant of the grass grub beetle. The method of choice was to prepare resins which would yield phenol on hydrolysis. Details of the preparation of the resins are given in Table 9.1. The diazonium sulphonate fomulation (500 and 1000 mg) was effective in the field for up to 11 days; the direct phenol formulation retained its attractancy for 5 days. No release data are reported but the release from the diazonium sulphonate is dependent on the hydrolysis of the azo- link, and in both formulations hydrogen bonding between the phenol moiety and the sulphonic acid groups would regulate the release rates. It is suggested that field release is modified by the structure of the polymeric carrier. These baits are used in conjunction with water traps, and are reported to be easily prepared, convenient to handle and have good inter-batch reproducibility.

Carlson and Leibold (1981) developed pheromone-toxicant devices (PTDs) for houseflies, incorporating a mixture of muscalure and permethrin. After preliminary work on release devices for muscalure alone they tested two PTDs. Fibre-glass strips (4.5×2.5 cm) were each coated with muscalure (100 mg) and permethrin (90.76 mg) from an acetone solution. Significant losses of both materials occurred during preparation, since on day 0 of longevity tests only 48 mg muscalure and 52.2 mg permethrin were recovered. These devices release the pheromone at the rate of 2.87 mg day^{-1} and the toxicant at 0.69 mg day^{-1}. The acrylic yarn device (see Table 9.1) had an even greater initial loss (78.1 mg muscalure and 74.7 mg permethrin) but released the pheromone at 1.02 mg day^{-1} and the permethrin at the rate of 0.35 mg day^{-1}. Tests in swine and sheep barns indicated that the PTDs could increase the efficiency of pesticide usage by both decreasing the amount used and restricting the application. However, several problems, e.g. the large initial loss of active ingredients, must be overcome.

10.5.7 Acknowledgements

I wish to thank Drs Gary Daterman, Les McDonough, Chris Sanders and Kelly Smith for supplying data used in the preparation of this overview, and for helpful discussions. Mme Michèle Carignan is sincerely thanked for her assistance in the preparation of the manuscript.

10.6 REFERENCES

Angelini, A., Descoins, C., Le Rumeur, C. and Lhoste, J. (1980). Nouveaux résultats obtenus avec un attractif sexuel de *Cryptophlebia leucotreta* Meyr (Lepidoptera) *Cot. Fib. Trop.*, **XXXV**, 277–281.
Anonymous (1983). Bend research controlled release technology, *Techbrief*, 4 pp.
Anonymous (1984). *Pheromones*, Shin-Etsu Chemical Co. Ltd, Tokyo, 10 pp.
Bakke, A. (1978). Aggregation pheromone components of the bark beetle *Ips acuminatus Oikos*, **31**, 184–188.
Berisford, C. W. and Hedden, R. L. (1978). Suppression of *Rhyacionia frustrana* response to live females by the sex pheromone of *R. buoliana. Environ. Entomol.*, **7**, 532–533.
Beroza, M., Bierl, B. A., Tardif, J. G. R., Cook, D. A. and Pasek, E. C. (1971). Activity and persistence of synthetic and natural sex attractants of the gypsy moth in laboratory and field trials, *J. Econ. Entomol.*, **64**, 1499–1508.
Bull, D. L., Coppedge, J. R., Hardee, D. D. and Graves, T. M. (1973). Formulations for controlling the release of synthetic pheromone (grandlure) of the boll weevil. 1. Analytical studies. *Environ. Entomol.*, **2**, 829–835.
Butler, L. I. and McDonough, L. M. (1979). Insect sex pheromones: evaporation rates of acetates from natural rubber septa. *J. Chem. Ecol.*, **5**, 825–837.
Butler, L. I. and McDonough, L. M. (1981). Insect sex pheromones: evaporation rates for alcohols and acetates from natural rubber septa. *J. Chem. Ecol.*, **7**, 627–633.
Campion, D. G., Lester, R. and Nesbitt, B. F. (1978). Controlled release of pheromones. *Pestic. Sci.*, **9**, 434–440.
Cardé, R. T., Doane, C. C., Granett, J., Hill, A. S., Kochansky, J. and Roelofs, W. L. (1977a). Attractancy of racemic disparlure and certain analogues to male gypsy moths and the effect of trap placement, *Environ. Entomol.*, **6**, 765–766.
Cardé, R. T., Doane, C. C., Baker, T. C., Iwaki, S. and Marumo, S. (1977b). Attractancy of optically active pheromone for male gypsy moths. *Environ. Entomol.*, **6**, 768–772.
Cardé, R. T., Doane, C. C. and Farnum, D. G. (1978). Attractancy to male gypsy moths of (+) disparlure synthesized by different procedures. *Environ. Entomol.*, **7**, 815–816.
Carlson, D. A. and Leibold, C. M. (1981). Field trials of pheromone-toxicant devices containing muscalure for houseflies (Diptera: Muscidae). *J. Med. Entomol.*, **18**, 73–77.

Charmillot, P-J., Scribante, A., Pont, V., Deriaz, D. and Fournier, C. (1981). Technique de confusion contre la tordeuse de la pelure *Adoxophyes orana* F.v.R. (Lep., Tortricidae): I. Influence de la diffusion d'attractif sexuel sur le comportement. *Bull. Soc. Entomol. Suisse*, **54**, 173–190.

Daterman, G. E. (1974). Synthetic pheromone for detection survey of European pine shoot moth, *U.S. For. Serv. Res. Pap. PNW*, **180**, 12 pp.

Daterman, G. E. (1978). Monitoring and early detection. In *The Douglas-Fir Tussock moth: a synthesis*, M. H. Brookes, R. W. Stark and R. W. Campbell (Eds), *U.S. For. Serv. Tech. Bull.*, **1585**, 99.

Daterman, G. E. (1982). Monitoring insects with pheromones: Trapping objectives and bait formulations. In *Insect Suppression with Controlled Release Pheromone Systems*, A. F. Kydonieus and M. Beroza (Eds), CRC Press, Boca Raton, Florida, pp. 195–212.

Dindonis, L. L. and Miller, J. R. (1981). Onion fly trap catch as affected by release rates of *n*-dipropyl disulfide from polyethylene enclosures. *J. Chem. Ecol.*, **7**, 411–418.

Fein, B. L., Reissig, W. H. and Roelofs, W. L. (1982). Identification of apple volatiles attractive to the apple maggot *Rhagoletis pomonella*. *J. Chem. Ecol.*, **8**, 1473–1487.

Fitzgerald, T. D., St Clair, A. D., Daterman, G. E. and Smith, R. G. (1973). Slow release plastic formulation of the cabbage looper pheromone *cis*-7-dodecenyl acetate: release rate and biological activity. *Environ. Entomol.*, **2**, 609–610.

Flint, H. M., Butler, L., McDonough, L. M., Smith, R. L. and Forey, D. E. (1978). Pink bollworm: response to various emission rates of gossyplure in the field. *Environ. Entomol.*, **7**, 57–61.

Flint, H. M., McDonough, L. M., Salter, S. S. and Walters, S. (1979). Rubber septa: a long lasting substrate for (Z)-11-hexadecenal and (Z)-9-tetradecenal, the primary components of the sex pheromone of the tobacco budworm. *J. Econ. Entomol.*, **72**, 798–800.

Golub, M., Weatherston, J. and Benn, M. H. (1983). Measurement of release rates of gossyplure from controlled release formulations by mini-airflow method. *J. Chem. Ecol.*, **9**, 323–333.

Gothilf, S., Kehat, M., Dunkelblum, E. and Jacobson, M. (1979). Efficacy of (Z)-11-hexadecenal and (Z)-9-tetradecenal as sex attractants for *Heliothis armigera* on two different dispensers. *J. Econ. Entomol.*, **72**, 718–720.

Hall, D. S. and Read, J. S. (1979). Sex attractants for two zyagaenid moths, *J. Ent. Soc. Sth. Afr.*, **42**, 115–119.

Hartstack, A. W., Witz, J. A. and Buck, D. (1979). Moth traps for the tobacco budworm. *J. Econ. Entomol.*, **72**, 519–522.

Hartstack, A. W., Lopez, J. D., Klun, J. A., Witz, J. A., Shaver, T. N. and Plimmer, J. R. (1980). New trap designs and pheromone bait formulations for *Heliothis, Beltwide Cotton Prod. Res. Conf. Proc.*, St. Louis, MO, pp. 132–136.

Hendricks, D. E., Hartstack, A. W. and Shaver, T. N. (1977). Effect of formulations and dispensers on the attractiveness of virelure to the tobacco budworm. *J. Chem. Ecol.*, **3**, 497–506.

Hendricks, D. E. (1982). Polyvinyl chloride capsules: A new substrate for dispensing tobacco budworm (Lepidoptera: Noctuidae) sex pheromone bait formulations. *Environ. Entomol.*, **11**, 1005–1010.

Hendricks, D. E. (1983). Formulations and field bioassay of tobacco budworm sex pheromone components impregnated into PVC plastic dispensers, in surrogate and rod forms, *Proc. 10th Intl. Symp. Controlled Release Bioactive Materials*, San Francisco, pp. 231–233.

Huber, R. T. and Hoffman, M. P. (1979). Development and evaluation of an oil trap for use in pink bollworm mass trapping and monitoring programs. *J. Econ. Entomol.*, **72**, 695–697.

Kehat, M., Gothilf, S., Dunkelblum, E. and Greenberg, S. (1980). Field evaluation of female sex pheromone components of the cotton bollworm *Heliothis armigera*. *Ent. Exp. Appl.*, **27**, 188–193.

Klimetzek, von D., Vité, J. P. and Mori, K. (1980). Zur Wirkung und Formulierung des Populationslockstoffes des Nutzholzborkenkäfers *Trypodendron* (= *Xyloterus*) *lineatum* Z. *Angew. Entomol.*, **89**, 57–63.

Kuhr, R. J., Comeau, A. and Roelofs, W. L. (1972). Measuring release rates of pheromone analogues and synergists from polyethylene caps, *Environ. Entomol.*, **1**, 625–627.

Lauren, D. R. (1979). Controlled release formulations for phenols: use as sex attractant lures for the grass grub beetle. *Environ. Entomol.*, **8**, 914–916.

Livingston, R. L. and Daterman, G. E. (1977). Surveying for Douglas-fir tussock moth with pheromone. *Bull. Entomol. Soc. Am.*, **23**, 172.

Lloyd, J. E. and Matthysse, J. G. (1966). Polymer-insecticide systems for use as livestock feed additives. *J. Econ. Entomol.*, **59**, 363–367.

Lloyd, J. E. and Matthysse, J. G. (1970). Polyvinyl chloride-insecticide pellets fed to cattle to control face fly larvae in the manure. *J. Econ. Entomol.*, **63**, 1271–1281.

Maitlen, J. C., McDonough, L. N., Moffitt, H. R. and George, D. A. (1976). Codling moth sex pheromone: baits for mass trapping and population survey. *Environ. Entomol.*, **5**, 199–202.

Marks, R. J. (1977). Assessment of the use of sex pheromone traps to time chemical control of red bollworm *Diparopsis castanea* Hampson (Lepidoptera: Noctuidae) in Malawi. *Bull. Ent. Res.*, **67**, 575–587.

Marks, R. J. (1978). The influence of pheromone trap design and placement on catch of the red bollworm of cotton, *Diparopsis castanea* Hampson (Lepidoptera: Noctuidae), *Bull. Ent. Res.*, **68**, 31–45.

McDonough, L. M. (1978). Insect sex pheromones: Importance and determination of half life in evaluating formulations, SEA-ARS Research Report, Western series 1, 20 pp.

McDonough, L. M. (1983). Evaporation rates of insect sex pheromones from a natural rubber matrix, *Proc. 10th Intl. Symp. Controlled Release Bioactive Materials*, San Francisco, pp. 206–208.

McDonough, L. M. and Butler, L. I. (1983). Insect sex pheromones: determination of half lives from formulations by collection of emitted vapor. *J. Chem. Ecol.*, **9**, 1491–1502.

McKibben, G. H., Johnson, W. L., Edwards, R., Kotter, E., Kearny, J. F., Davich, T. B., Lloyd, E. P. and Ganyard, M. C. (1980). A polyester-wrapped cigarette filter for dispensing grandlure, *J. Econ. Entomol.*, **73**, 250–251.

Prokopy, R. J. (1977). Attraction of *Rhagoletis* flies to red spheres of different sizes. *Can. Ent.*, **109**, 593–596.

Reissig, W. H., Fein, B. L. and Roelofs, W. L. (1982). Field tests of synthetic apple volatiles as apple maggot (Diptera: Tephritidae) attractants. *Environ. Entomol.*, **11**, 1294–1298.

Rothschild, G. H. L. (1979). A comparison of methods of dispensing synthetic pheromones for the control of oriental fruit moth *Cydia molesta* (Busck.) (Lepidoptera: Tortricidae) in Australia. *Bull. Ent. Res.*, **69**, 115–127.

Sanders, C. J. (1981). Release rates and attraction of PVC lures containing synthetic sex attractant of the spruce budworm. *Choristoneura fumiferana* (Lepidoptera: Tortricidae). *Can. Ent.*, **113**, 103–111.

Sanders, C. J. and Weatherston, I. (1976). Sex pheromone of the eastern spruce budworm (Lepidoptera: Tortricidae): Optimum blend of *trans* and *cis* 11-tetradecenal, *Can. Ent.*, **108**, 1286–1290.

Sartwell, C., Daterman, G. E., Korber, T. W., Stevens, R. E., Sower, L. L. and Medley, R. D. (1980a). Distribution and hosts of *Eucosma sonomana* in the western United States as determined with pheromone baited traps. *Ann. Entomol. Soc. Am.*, **73**, 254–256.

Sartwell, C., Daterman, G. E., Sower, L. L., Overhulser, D. L. and Koerber, T. W. (1980b). Mating disruption with synthetic sex attractant controls damage by *Eucosma sonomana* (Lepidoptera: Tortricidae: Olethreutinae) in *Pinus ponderosa* plantations. I. Manually applied polyvinyl chloride formulations. *Can. Ent.*, **112**, 159–162.

Shani, A., and Klug, J. T. (1980). Sex pheromone of Egyptian cotton leafworm (*Spodoptera littoralis*): its chemical transformations under field conditions. *J. Chem. Ecol.*, **6**, 875–881.

Sharp, J. L. and Cross, J. H. (1980). Longevity and attractivity to male lesser peachtree borer of EZ-ODDA evaporated from various dispensers. *Environ. Entomol.*, **9**, 818–820.

Shaver, T. N., Hendricks, D. E. and Hartstack, Jr., A. W. (1981). Dissipation of virelure, a synthetic pheromone of the tobacco budworm from laminated plastic and cigarette filter dispensers. *Southwestern Entomologist*, **6**, 205–210.

Shirasaki, S., Yamada, M., Sato, R., Yaginuma, K., Kumakura, M. and Tamaki, Y. (1979). Field tests on attractiveness of the synthetic pheromone of the peach fruit moth *Carposina nipponensis* Walsingham (Lepidoptera: Carposinidae). *Jap. J. Appl. Ent. Zool.* **23**, 240–245.

Smith, K. L., Baker, R. W. and Ninomiya, Y. (1983a). Development of Biolure controlled release pheromone products. In *Controlled Release Delivery Systems*, T. J. Roseman and S. Z. Mansdorf (Eds), Marcel Dekker Inc., New York, pp. 325–335.

Solomon, J. D., Dix, M. E. and Doolittle, R. E. (1978). Attractiveness of the synthetic carpenterworm sex attractant increased by isomeric mixtures and prolonged by preservatives. *Environ. Entomol.*, **7**, 39–41.

Steck, W., Bailey, B. K., Chisholm, M. D. and Underhill, E. W. (1979). 1,2-Dianilinoethane, a constituent of some red rubber septa which reacts with aldehyde components of insect sex attractants and pheromones. *Environ. Entomol.*, **8**, 732–733.

Stoakley, J. T., Bakke, A., Renwick, J. A. A. and Vité, J. P. (1978). The aggregation pheromone system of the larch bark beetle *Ips cembrae* Meer. *Z. Ang. Entomol.*, **86**, 174–177.

Toth, M., Szöcs, G., Novak, L. and Szantay, Cs. (1979). Post-season field screening of lepidopterous sex attractant candidates in Hungary. *Acta Phytopathol. Acad. Scient. Hungaricae*, **14**, 195–199.

Vick, K. W., Coffelt, J. A. and Sullivan, M. A. (1978). Disruption of pheromone communication in the Angoumois grain moth with synthetic female sex pheromone. *Environ. Entomol.*, **7**, 528–531.

Vité, J. P., Hughes, P. R. and Renwick, J. A. A. (1976). Southern pine beetle: effect of aerial pheromone saturation on orientation. *Naturwissen.* **63**, 44.

Weatherston, J., Golub, M. A., Brooks, W. W., Huang, Y. Y. and Benn, M. H. (1981). Methodology for determining the release rates of pheromones from hollow fibers. In *Management of Insect Pests with Semiochemicals, Concepts and Practice*, E. R. Mitchell (Ed.), Plenum Press, New York, pp. 425–443.

PART D

Commercialization
and the Future

Insect Pheromones in Plant Protection
Edited by A. R. Jutsum and R. F. S. Gordon
©1989 John Wiley & Sons Ltd

11

The Development
and Marketing of
a Pheromone System

G. J. Jackson

ICI Agrochemicals, Fernhurst,
Haslemere, Surrey GU27 3JE, UK

11.1 INTRODUCTION

When an agrochemical company is considering a compound for development
and eventual marketing as a pest control product, it examines the project
very carefully against a number of criteria. This detailed study is essential
because today the cost of bringing such a new product to the market can
easily exceed £20 million, and that figure does not include the opportunity
cost of other beneficial work left undone because resources were fully
occupied.

In this context, pheromones and other behaviour modifying chemicals are
no different from any other new product: the detailed appraisal is carried
out, the financial parameters are subjected to close scrutiny using computer
modelling techniques, and the whole written up into a 'development case'
which will be the basis of a 'go/no-go' decision by the company concerned.

In the early 1980s, ICI Agrochemicals developed the sex pheromone of
Pectinophora gossypiella Saunders, the pink bollworm, as a control agent for
this serious pest of cotton. The control technique used was mating disruption
by confusion of the males, as described in Chapter 5. The initial evaluation
and field development were carried out in Egypt, and the product is now on
the market in this and several other countries under the trade names Pectone
and Pectimone. I shall use this project to exemplify the salient points of how
a pheromone system can be developed.

11.2 CHARACTERISTICS CONSIDERED IN A DEVELOPMENT CASE

11.2.1 Effectiveness

The first point is that the chemical must produce the desired effect, and it must do so reliably and under field conditions.

Another critical factor is the reliability of the system from one season to another: a pheromone normally suppresses only one species in the pest complex, and management of the others must depend largely upon natural enemies or further pesticides, allied to compensatory growth by the cotton plants to replace lost bolls. Clearly it is only possible to establish the reliability of the system from several years' experience in any given location.

11.2.2 Safety

This requirement includes not only the mammalian toxicology profile but also the behaviour of the chemical in the environment. The studies needed to determine these characteristics are complex, expensive and numerous (see Chapter 12), and prior to the development decision only a limited amount of work will have been done.

Based upon this, together with any relevant information in the literature and knowledge of the group of compounds to which the candidate belongs, a judgement must be made as to the likelihood of the compound passing the stringent safety tests which will be required before sales can take place.

The more expensive of these tests can only commence after the decision to develop, but if the compound fails the tests it will be dropped, no matter how far along the development track it has progressed by then.

In the case of Pectone, the active ingredient is a natural product of a female moth which had previously been characterized, synthesized and sold in the USA in various formulations. As a result, a considerable amount of basic toxicological data was available, plus the evidence of a Californian registration showing that the compound had been cleared by the Environmental Protection Agency (EPA) for control of *P. gossypiella* by mating disruption.

The data showed an exceptionally good profile for the pheromone with acute oral and dermal LD_{50}s in excess of $15\,000\,\mathrm{mg\,kg^{-1}}$, and no longer-term effects.

11.2.3 The market

Ideally, market research should have shown that there was a large potential market for the compound, and that this particular product could be expected to take a worthwhile share of that market. At present these requirements are unlikely to be met by pheromones, and this is a major obstacle to their development; a 'small' compound cannot repay the enormous costs involved, particularly the safety tests, and show a profit within a reasonable time.

Pheromones tend to be highly specific, and sex pheromones in particular can usually only be used against a single species in practice. Their potential market is therefore inevitably smaller than, for example, that of a broad-spectrum insecticide. Furthermore, there must be doubt about the ability of a new sex pheromone product to take a good share even of such a limited market from the established, reliable, broad-spectrum insecticides.

The other side of the 'size' coin—price—is no help either: no cotton producer, whether private farmer or government organization, can be expected to pay a premium for a pheromone system of pest control as long as the established insecticides continue to give adequate results.

Clearly, any pheromone product is likely to be categorized as 'small' in a development appraisal.

11.2.4 Registerability

This springs largely from Section 11.2.2 above, but it is clearly an advantage if the product is expected to be easy to register, with no hold-ups; possibly part of the registration package may already be available. Pheromones in general can be expected to have a relatively easy registration track. They are not, however, exempt from the standard appraisal on toxicology and environmental safety; a recent EPA ruling does exempt sex pheromones when used in traps, but their use in other modes, for example broadcast application for mating disruption, requires full appraisal. In the case of Pectone in Egypt, registration was not problematic, owing to the prior registration of the active ingredient in California under EPA rules.

11.2.5 Project value

The value of the project arises from the interaction of the four previous factors, and is an important criterion in the development decision. Ideally, a project should have a low total cash outflow in the early years; it should repay these costs, i.e. turn cash-positive, in a short time; and it should generate a sound net cash inflow over a ten-year period.

This cash flow will then be compared to those of other projects competing for resource within the company, and it can happen that intrinsically worthwhile projects can be shelved because they show a less favourable financial profile than other competing projects.

As discussed earlier, a pheromone project is likely to be small and would be at risk of being rejected for this reason. The only way to overcome this difficulty is to ensure that the costs of development are kept to a minimum, so that the net cash outflow in the development years is small. This allows the project to turn cash-positive as early as possible once sales commence. The ways in which development costs can be minimized are discussed in Section 11.3.

11.3 THE DEVELOPMENT STRATEGY

Having assembled all relevant information on the project, fed this into a computer model, and produced a cash flow which is likely to be acceptable to the company, it is essential to evolve a strategy for the development of the project. In practice, this will often be part of the work which has gone into producing the cash flow.

For a pheromone, let us assume that we are faced with a project where the effectiveness, especially the reliability, is not well established; where product safety is believed to be good but still requires toxicological studies; where sales are expected to be small and limited perhaps to five or six countries, but with a straightforward registration track in those countries; and finally where the financial return on a ten-year basis is small but just about acceptable, yet is dependent upon costs being kept to a minimum in the early years of development.

The strategy to follow in this case would be low-cost development initially, in one or two countries only: a 'spearpoint' strategy. During this development much information could be gained about biological effectiveness and some about year-on-year reliability. Studies on safety, to assemble an appropriate regulatory package would, of course, have to be carried out.

Field work in the chosen country could be performed on a collaborative basis; it is one of the favourable aspects of pheromones that entomologists are likely to show considerable interest, and it is possible to set up collaborative development programmes which involve government or private scientific institutions, the agrochemical company, and the Ministry of Agriculture of the chosen development country.

If this initial development is satisfactory and leads to sales in the chosen country, then the spearpoint strategy can be widened to include other countries and, possibly, by admixture with other control agents, to include other members of the pest complexes found there.

By this strategy, early expenditure is kept low and early returns can repay the costs; risks of product failure are minimized; and, a critical factor, occupation of scarce company resources is of a level not to appear a threat to larger, more profitable projects.

11.4 IMPLEMENTING THE STRATEGY

The spearpoint strategy was the one chosen by ICI Agrochemicals for the development of its pheromone system. Implementation of this strategy is detailed below. However, this imposes an artifical order on what was in fact a complex web of interacting factors and decisions; in practice only some of the latter were made deliberately—some were imposed by circumstance.

11.4.1 Selection of target pest

P. gossypiella was chosen for a number of reasons. Its pheromone has a chemical composition which is suitable to ICI Agrochemicals' patented formulation technology (see below), and it is available commercially as a synthetic mixture of only two isomers. This pheromone was already registered by the EPA for mating disruption, and was known to be capable of reducing populations in the field.

The biology and behaviour of the pest were also critical factors, in that the female moths, once mated, do not migrate significantly. Thus one would not expect an influx of fertilized females to a pheromone-treated area.

Finally, *P. gossypiella* is a key pest in a number of countries; if the early work was successful, the spearpoint could be widened to expand the project.

11.4.2 The collaborating institution

For some years, ICI Agrochemicals had been working with the Overseas Development Natural Resources Institute (ODNRI), and this was an obvious choice for collaboration on the *P. gossypiella* project. ODNRI had considerable experience of pheromone work, and were willing to enter into a joint programme. The fact that a good mutual understanding and working relationship already existed was an added bonus.

11.4.3 The country of operations

Again, this choice was almost automatic. Egypt has a very highly developed system of agriculture, and the control of cotton pests is handled solely by the Ministry of Agriculture on a central basis. *P. gossypiella* is the key member of the pest complex, and several prominent members of the Egyptian scientific community were willing to support attempts to control it by means of pheromones.

ICI Agrochemicals has a strong operational base in Egypt, capable of handling the local administration and logistics of the programme from its office in Cairo. Excellent telephone, telex and radio communication were available. Most important, the office could co-ordinate the work of the collaborating partners: ICI, ODNRI, the Egyptian Ministry of Agriculture and the University of Ain Shams.

Finally, ODNRI had collaborated for years with the Egyptian Ministry of Agriculture and universities in a variety of programmes. A close working relationship between the parties had developed, and this would naturally be important.

Thus the selection of Egypt allowed the project to be slotted into an existing infrastructure of administration and operations, and this helped greatly in keeping costs to the minimum.

11.4.4 Registration trials

Early evaluation work had already suggested that the ICI Agrochemicals system could be effective in Egypt, and trials for registration were therefore planned. The assistance of the collaborating officials was particularly valuable in defining a trials programme that would demonstrate effectiveness on a large scale and would also count towards ultimate registration by the Ministry of Agriculture. In 1982, trials were set up on 50 ha plots. If successful, these were to be followed by larger trials in 1983 and by large-scale 'proving' trials in 1984.

At the same time two other companies, BASF and Sandoz, also entered the collaborative programme with ODNRI and the Egyptian Government, using their own formulations of pheromone. BASF used the system of plastic laminated flakes developed by Hercon in the USA, and Sandoz used the hollow fibre system of Albany International, also developed in the USA. Both these systems require special application machinery, whereas the ICI system, described in Section 11.4.5 below, can be applied through any standard arrangement of hydraulic nozzles.

As to the performance of the product, the duration of control provided by a spray of pheromone, which would determine the interval between spray rounds, was critical. Less than a 2-week interval would require five spray rounds to protect the cotton from the first susceptible stage (end of May in Egypt) until picking time. A 2–3 week interval would need only four rounds, which was current Egyptian practice.

One particular consideration was the danger of attack by *Spodoptera littoralis* (Boisd.), the cotton leafworm. Any sudden influx by this highly mobile noctuid moth would certainly overwhelm the natural enemy populations and would necessitate a blanket spray of insecticide. It was decided to circumvent this problem by locating the trials in the isolated El Fayoum area to the south of Cairo, where some 13 000 ha of cotton are grown. In El Fayoum, damaging populations of *S. littoralis* do not occur. The other main pest, apart from *P. gossypiella*, is *Earias insulana* (Boisd.), the spiny bollworm. This pest builds up its numbers as the season progresses, and causes severe losses in the late season if unchecked in a normal spray regime. In the proposed pheromone system, control of E. *insulana*, during the early and middle season at least, would have to be achieved by the natural enemies. It was hoped that by the time infestations of E. *insulana* built up to intolerable levels, the cotton would have reached the picking stage and no action would be necessary.

11.4.5 Formulation technology

In the development of pheromones as useful pest control agents, formulation is the key. The primary objective is to produce a formulation which is stable in storage and after application in the field, but which will provide slow release of the active ingredient for as long as possible following application.

There are a number of ways of achieving this, and these have been described in Part C of this book. All of these systems provide acceptable stability and slow release, but most, except the liquid formulation of microencapsulated material, require special application techniques or must be applied by hand. In some crops and circumstances this may be acceptable or even advantageous, but in others, such as most large-scale cotton production, it is a distinct disadvantage.

The system adopted by ICI Agrochemicals was microencapsulation, and the first commercial formulation developed was a 2% aqueous suspension of microcapsules. This has favourable toxicological characteristics and is non-flammable; these useful qualities result in a zero hazard classification for transport (including air freight), handling and storage.

In all agrochemical development, work on improvement of the formulation may continue for some time—even well into the product's commercial life. The improvement of pheromone formulations can be expected to make considerable strides under the impetus of commercial development.

11.4.6 Patenting

Without patent protection to assure freedom from 'pirating' of ideas and investment, no company would consider developing any agrochemical. In the case of those pheromones whose structures have already been published, for example in the academic literature, patent protection for the compound *per se* is not available.

It might be possible to secure a patent on a special process to make the chemical synthetically, but such patents do not prevent others inventing and using a different process to make it.

A more common form of patent protection for pheromones is on the formulation technology, and this is another reason why this is a key feature of pheromone development. In the case of the microcapsule formulation sold by ICI Agrochemicals, both the active pheromone and the plastic capsule wall need to be stabilized; ICI holds sole ownership of a patent on the particular method of stabilization employed. Thus, while anyone may purchase the sex pheromone of *P. gossypiella,* an alternative method of stabilization, not covered by the patent, would be needed before anyone could develop a microencapsulated formulation for sale.

11.4.7 Toxicology

Generally speaking, it is to be expected that pheromones will be of very low hazard from both the mammalian and the environmental safety point of view. They are not, after all, pesticides in the strict sense of the word, and they are not designed to be toxic to individual organisms. It is of course well recognized that, just because they are naturally occurring compounds, they cannot be deemed *ipso facto* to be safe. Indeed, as mentioned earlier, the American EPA has been careful to observe this in the case of sex pheromones.

Nevertheless, a very large margin of safety can be demonstrated. The sex pheromone of *P. gossypiella* is a good example, where studies have shown very low acute toxicity (Table 11.1).

TABLE 11.1 Toxicology profile of *P. gossypiella* sex pheromone (data for technical active ingredient).

Test	Species	Result
Acute oral LD_{50}	Rat	$> 15\,000$ mg kg^{-1}
Acute inhalation LC_{50}	Rat	> 3.3 mg l^{-1}
Skin irritation	Rat	Not an irritant
Eye irritation	Rabbit	Not an irritant
Ames mutagenicity		Negative
Acute oral LD_{50}	Mallard	$> 10\,000$ mg kg^{-1}
96 h acute LC_{50}	Trout	540 p.p.m.

This is not to suggest that all pheromones would appear in such a favourable light, but it will be the minimum requirement that a pheromone should pass 'Tier 1' of the EPA system, before proceeding to registration.

11.4.8 Production

The development of any new agrochemical involves the careful planning of production facilities from an early stage. New plant will usually be required: this may involve heavy capital expenditure which can account for a significant proportion of the total development costs.

However, a pheromone will have special problems. Pheromones are at the very frontier of agricultural pest control, and therefore the level of risk is much higher than would be the case for the development of, for example, a new pyrethroid; both the effectiveness of the product and the customers' response to it are likely to be in doubt at the early stages. These uncertainties make it difficult to commit even comparatively modest sums to new production facilities. Furthermore, because pheromones do not fit into the normal registration process, the timing and size of requirements are likely to undergo prolonged discussion and negotiation.

For example, in Egypt the cotton insecticide tenders are negotiated during November, and orders placed in December for the following year's requirements. For the 1984 pheromone trials, however, technical discussion on the appropriate way forward continued into March; the tender was eventually held in April and orders were placed on 28 April with an on-site requirement of not later than the last week of May. This posed a serious problem for ICI, where manufacture of the formulation takes place in-house, and the delay almost caused the cancellation of the project.

The problem was only overcome by taking far greater risks on ordering and holding stocks of materials, and making other preparations, than would normally have been justified by the size of the project. If the Egyptian Ministry of Agriculture had decided for any reason not to allow the project to progress to the stage of final large-scale proving trials, for which the products are actually purchased by the Ministry, then significant financial losses would have been incurred. In the event, production was able to begin within days of receiving the order, and 5000 l of formulated product were air freighted in stages to Cairo in time for spraying operations.

Again, an unorthodox view had to be taken of this, since normal transportation would be by sea at a fraction of the cost of air freight. However, this order was regarded as 'continued development', and not a commercial sale, so that the normal requirements of profitability could be disregarded.

The success of the 1984 operations, and the virtual certainty of a larger order from Egypt for 1985, permitted the establishment of better production facilities over the winter period. This removed a major bottleneck from the supply route.

11.4.9 Registration

Normally in Egypt, with the successful completion of final large-scale proving trials, a formal registration as an integral part of the cotton pest control system would be granted. However, we have already seen that pheromones do not fit the normal registration process. The pheromone programme is still regarded as separate from the main cotton pest operations, and no formal registration has yet been granted. The main reason for this is concern about year-to-year reliability of the system, as mentioned earlier, added to the fact that the work is at the frontier of pest control.

This cautious attitude is shared by the companies involved, and there is general agreement that the promotion of this new technique will best be served by a step-by-step approach which builds upon success. Any failure would be a major setback for the technique, not only in one particular country, but in affecting attitudes worldwide.

With pheromones, therefore, registration should be a gradual process of expansion, built upon increasing confidence that the systems will work reliably. Once acceptance has been gained, however, a pheromone product could be slotted into a pest control programme, although naturally the nature and extent of the niche occupied will be different in every country, crop and geographical location.

11.5 MARKETING

Once registration has been achieved, either by a formal notification or perhaps by gradual acceptance over several years, then sales can commence. The work of marketing the product, however, will have been an integral part of the later stages of development.

11.5.1 Customer attitudes and expectations

'The customer' may be one of a very wide variety of types, ranging from peasant farmers with smallholdings in some countries to, as in Egypt, a top-level committee of highly experienced and qualified government scientists headed by an Under-Secretary of State. In all cases, the customer will be used to dealing with traditional insecticides, designed to kill insects. A considerable amount of effort has to go into ensuring that the customer understands the new pheromone product and expects to get from it what it was designed to deliver.

With the pheromone project in Egypt, the task was made easier by a number of crucial factors: several prominent members of the agricultural scientific community were interested and were willing to promote the project; the centralization of responsibility in a technically qualified committee made it possible to present a reasoned case; the Ministry officials were keen to diversify their control operations and to reduce their dependence on traditional insecticides; and there was strong opinion that the amount of insecticides used on Egyptian cotton should be minimized. Thus, the general climate of thought was receptive to the idea, and initial approval was gained.

It cannot be emphasized too strongly that in the development and marketing of pheromone systems, the management of customer attitudes and expectations is a crucial part of the work. It is essential that the customer understands that what he is buying with any pest control agent is *added value to the crop*. The improvement in yield, quality or both must be sufficient to justify the cost of the control operations, and if it does so, the simple killing of insects is a crude irrelevance. The management of pest populations in such a way that they do not cause damage beyond the economic threshold of control is greatly to be preferred (Collins *et al.*, 1984).

It will be immediately clear that such an understanding is more likely to be gained by relatively sophisticated farmers and by government-controlled farming operations where pest control is under the management of qualified entomologists.

11.5.2 Market positioning

Market positioning is concerned with where the product will 'fit' into the market-place, and has two aspects: technical position and commercial position.

Most agrochemicals are capable of doing a number of particular jobs, and pheromones are no exception. Deciding which, and perhaps how many, of these jobs the product should actually do is the technical positioning of the product. For the sex pheromone in Egypt this was done in consultation between Ministry officials and the companies, but the final word lay with the Technical Recommendation Committee of the Ministry of Agriculture. It has already been explained that because of the virtual absence of *Spodoptera littoralis* from Upper and Middle Egypt, it had been decided to restrict pheromone usage to these areas for the time being. The question was, should the pheromone products be used on their own, as a season-long programme of four or five applications with no measures being taken against other pests? Or should they be used only in the early part of the season, perhaps for three applications, with a final spray of e.g. pyrethroid against *Earias* sp. at the end? Or should they be restricted to one application, and if so where should it be? There are arguments in favour of all these positions, and at the time of writing the Committee has not yet made a decision.

Under the season-long pheromone regime, beneficial parasites and predators are preserved in the cotton during the early and middle part of the season (Jackson, 1984). These beneficial organisms assist in maintaining larval populations of the various lepidopterous pests at a low level until August, when the numbers of these beneficials decline naturally. Another important point is that during the period of greatest flowering of the cotton, when bees are foraging in the crop, no pesticides will be used. This is an important consideration, for honey is an Egyptian export crop, and bee-keepers have now begun to move their hives into the pheromone-treated area so that their bees may forage in the cotton—a source which was totally closed to them before (see Chapter 5).

Technical and commercial positioning are of course closely linked, like all other aspects of a development project. This link was mentioned earlier in connection with the duration of mating suppression achieved from one application. At the time, the Egyptian government was buying a basic four spray rounds for the cotton, at a total price of, say, $95 per hectare per year. Each of the spray rounds gave approximately 2 weeks' control. For the pheromone programme to be attractive to the government, it would have to perform the same function for, at most, the same number of dollars per hectare. That would mean four rounds of pheromone for $95, or one round for about $24, and so on. To achieve this, a minimum of a 2-week interval between pheromone rounds was essential, since if the interval were shorter, five pheromone rounds would be needed. The government would still be willing to pay only $95 for the five, rendering the whole project unattractive financially to the agrochemical company.

In the event, application intervals of up to 3 weeks were achieved, giving extended suppression from a four-round pheromone programme.

The above positioning refers to a season-long, insecticide-free programme of sex pheromone usage. There are, of course, alternatives to this, in which one or more pheromone sprays could be integrated with a programme of insecticide usage. Insect growth regulator (IGR) compounds are now available which are safe to beneficials, but which effectively control gross leaf-feeding Lepidoptera such as *Spodoptera littoralis*.

The permutations are numerous, and this illustrates the flexibility of the pheromone as a component of the 'integrated pest management tool-chest'.

11.5.3 Technical support

In a sense the first sale is a critical point in the life of any new product, but it is a well-known saying in agrochemical marketing that a sale is only complete when a repeat order is won. This refers to the need for sound technical support following the sale, in order to ensure complete satisfaction of the customer's expectations. As explained earlier, the management of customer expectations, and the vital necessity of ensuring that they are fulfilled, is even more important with a pheromone than with a standard insecticide. For this reason the effort devoted to technical support of a pheromone sale must err, if anywhere, on the side of excess.

To support ICI Agrochemicals' operations in Egypt, it was judged necessary to have one senior entomologist, experienced in the pheromone work, present throughout the main 2-month application period. In addition, other staff spent some weeks in Egypt at critical times, and full logistic and administrative support was provided by the Cairo office of ICI Agrochemicals.

It must be understood that the ICI Agrochemicals staff worked in an advisory capacity alongside the Ministry's own very extensive organization. ICI Agrochemicals' main role was to act as an observer of the operations and to ensure that Ministry officials at all levels were always fully informed of the progress of events, and to consult with and, where appropriate, to advise the Ministry's entomologists upon any questions arising.

11.5.4 The expansion phase

Once the project has been successfully established on one pest in one country, the 'spearpoint' approach can be deepened to include other countries for the same pest, and possibly broadened to one or two other pests. It will be remembered that development costs overall must be kept low, so the expansion will be slow and careful. Certainly each new geographical area, even for the same crop and pest, will present an entirely new set of biological, technical and commercial problems which will have to be solved again before the system can be put on a commercial footing in that area.

11.6 CONCLUSIONS

The development and marketing of pheromones for agricultural pest control is in its infancy, and there is much to be learned before it can become a smooth management process. The idea of using sex pheromones in the confusion technique of mating disruption was first suggested by Beroza in 1960, and the fact that it has taken many years to bring this idea to the farmer in a usable form is proof enough of the difficulties involved.

Nevertheless, it does appear that progress is now being made and we can expect the development and marketing of pheromones to gain momentum over the next ten years. By then, pheromones should be a standard item in the entomological 'tool chest' for integrated pest management programmes.

11.7 REFERENCES

Beroza, M. (1960). Insect attractants are taking hold. *Agricultural Chemicals*, **15**, 37–40.
Collins, M. D., Perrin, R. M., Jutsum, A. R. and Jackson, G. J. (1984). Insecticides for the future: a package of selective compounds for the control of major crop pests. *British Crop Protection Conference—Pests and Diseases 1984* (Vol. 1), pp. 299–304.
Jackson, G. J. (1984). Present trends in pesticide development regarding safety to beneficial organisms. *British Crop Protection Conference—Pests and Diseases 1984* (Vol. 1), pp. 387–394.

Insect Pheromones in Plant Protection
Edited by A. R. Jutsum and R. F. S. Gordon
©1989 John Wiley & Sons Ltd

12

The Registration of Pheromones

N. Punja
ICI Agrochemicals, Fernhurst, Haslemere, Surrey GU27 3JE, UK

12.1 INTRODUCTION

Behaviour-modifying chemicals (BMCs) have been registered for use as pest control agents in a number of countries over the last few years. Most of these chemicals are pheromones, used *per se* or in combination with conventional insecticides as baits or in traps. Other types of BMCs include synthetic analogues of pheromones and oviposition deterrents. Because of several exceptional properties inherent to BMCs, their registrations have often been achieved more easily, within shorter time and with fewer data than those demanded of conventional pesticides.

This chapter reviews registration aspects of pheromones taking into account their special features in relation to regulatory requirements for pesticides in general. Particular reliance and emphasis is placed upon the guidelines published by the United States Environmental Protection Agency (EPA) on 'Biorational Pesticides' (1982) with subsequent amendments in the Federal Register (1983, 1984). These guidelines, which reflect statutory authority as well as policy considerations, serve as very useful reference documents. They also give a lead to many other countries which have hitherto operated on a case-by-case basis for the registration of pheromones.

12.2 REGISTRATION OF PESTICIDES

Registration of pesticides is a formal process of obtaining approval by government agencies empowered or entrusted to control and regulate the manufacture, commercial distribution and use of pesticides in agriculture, horticulture, forestry, domestic, public health and veterinary outlets.

Such regulations, often enforceable by statutes, are designed to evaluate the benefit versus risk of the proposed pesticides so as to protect the health and safety of humans, their domestic animals, wildlife and the environment.

Application for approval is made to the relevant government bodies set up to receive and evaluate the data ('registration package') generated from extensive studies in the laboratory and field. These studies normally consist of investigations in several different scientific disciplines, as exemplified below:

Biological screening to evaluate product efficacy in the glasshouse and field.

Product chemical analysis to characterize and determine the purity and physicochemical properties of the technical and end-use product.

Residue chemical analysis to determine the nature, extent and magnitude of the residues of the parent compound and its metabolites following application of the product on the relevant crops according to good agricultural practices.

Environmental studies to determine the fate of the applied product on the current and rotational crops, the surroundings (e.g. air, soil and water) and its impact on various target and non-target organisms under laboratory, simulated or actual field conditions.

Toxicological studies to determine the acute, subacute or chronic toxicological effects of the active ingredient or formulated product when administered to test animals, as well as additional *in vivo* and *in vitro* studies on relevant organisms, all of which serve as models for assessing short- and long-term toxicological effects on humans.

The pesticide is approved and granted a temporary or full registration when it is judged to be safe in its proposed use, when its benefit : risk ratio is acceptable and when it is considered that its registration is in the public interest.

Registration of BMCs follows a similar pattern, except that, in recognition of their built-in safety features, several detailed studies are often waived.

12.3 REGISTRATION OF BEHAVIOUR-MODIFYING CHEMICALS

Pheromones, as behaviour-modifying chemicals, are regarded as a special group of chemicals and *prima facie* as safer pest control agents to replace, supplement or complement conventional pesticides. Some countries (such as the USA) have made policy decisions to encourage the use of pheromones by exempting them from various time-consuming and costly registration data requirements that are obligatory for conventional pesticides.

Pheromones are considered safer and environmentally more acceptable than conventional pesticides for the following reasons.

12.3.1 Species specificity

Careful studies of target pests, their habitat, life cycle, feeding or mating habits, oviposition behaviour and interaction with other organisms have shown that pheromones are generally species-specific. They control target pests by disrupting factors such as mating or oviposition without directly affecting other species. This selectivity is especially important for the preservation of beneficial species. Conventional pesticides, on the other hand, are often non-selective and may affect many species — pests and non-pests, target as well as non-target species.

Furthermore, pheromones have non-toxic modes of action. In contrast, conventional pesticides usually rely on their property of directly killing the pest. Because many toxic modes of action are common to various species of animals and plants, conventional pesticides constitute a higher potential risk of endangering the environment than pheromones.

12.3.2 Exposure

Many pheromones occur in nature where they have been fulfilling their endowed function for a long time. When used artificially, whether in excess of their natural background levels or not, they are normally applied at very low rates. They are usually volatile chemicals and dispersed in the atmosphere or placed in baits, traps or formulated in inert dispensers such as hollow fibres, tape or encapsulated formulations. Pheromones are usually not applied directly to water. Applications of pheromones in the field under these circumstances would be expected to be less likely to deposit foreign residues in food or feed, and to result in lower levels of dietary and other hazards to humans and the environment than those expected with conventional pesticides.

12.3.3 Acute toxicity

Results of toxicological studies with pheromones show that their acute toxicities to mammals, birds, fish and plants are low compared to conventional pesticides. Thus, pheromones pose a lower risk of exposure and of adverse effect in the environment and are judged as deserving of some dispensation from the otherwise very strict rules governing registration requirements for pesticides in many countries.

However, while there is a strong presumption of safety to pheromones in favour of their registration, the burden of proof of their safety lies on the registrant, who must supply sufficient data, albeit less than those required for conventional pesticides, to discharge this burden. Until recently, pheromones in this regard had been evaluated for registration by all government agencies on a case-by-case basis, applying existing regulations where appropriate. To illustrate the registration procedure and requirements for pheromones, the USA serves as a good example for many reasons: the USA is a foremost user of pesticides, including pheromones; the subcontinent covers a wide geographic, climatic and crop-pest complex; its regulations for pesticide registration are regarded as being extensive, sophisticated and strict; and the EPA has more recently introduced new registration guidelines which include behaviour-modifying chemicals.

12.4 ENVIRONMENTAL PROTECTION AGENCY (EPA) GUIDELINES

Registration of pesticides in the USA, be it for new registration, amended registration, re-registration or experimental use, is governed by the Federal Insecticide, Fungicide and Rodenticide Act (FIFRA) 1972, as amended in 1975 and 1978. Part 158 of FIFRA covers the full range of data requirements and submission to EPA to enable regulatory evaluations and decisions pertaining to the safety of each pesticide proposed for registration or experimental use.

EPA's new guidelines on 'Biorational Pesticides' (dealing with certain behaviour-modifying chemicals defined therein as biochemical pest control agents and microbial pest control agents) were published in November 1982. The guidelines were updated in October 1984 by publication in the Federal Register under the title 'Data Requirement for Pesticide Registration: Final Rule'. EPA's approach to regulating pheromones and other behaviour-modifying chemicals reflects its intent to specify data requirements for this class of products, taking account of their special characteristics by introducing a hierarchical or tier testing scheme. The tier testing scheme is designed to ensure that only the minimum data necessary to make a scientifically sound regulatory decision are produced. Thus, the need to generate and submit extensive data may not be required for all pheromones. The testing scheme consists of three tiers. An adverse effect in the first tier triggers off the need for the next tier. The first tier measures 'maximum hazard' as the prescribed tests therein involve the most challenging exposure in terms of the treatment dose or concentration, route of administration and the age of the test animals. Negative test results, therefore, provide a high degree of confidence in predicting that no adverse effects will occur from the use of these pheromones.

Most of these pheromones are unlikely to require further testing beyond the first tier. However, it is important to note that the tier testing scheme does not eliminate the need for more severe and extensive testing when necessary. The scheme makes ample provision for the requirement to conduct the same degree of testing as conventional pesticides when necessary by tier progression.

Product chemistry data requirements for these pheromones are not tiered and are virtually identical to those for conventional pesticides.

Residue chemistry data requirements are triggered by the rate of application of the BMC and by the first tier of the toxicology tests, i.e. if the BMC is applied at rates equal to or in excess of 49 g of the active ingredient (a.i.) per hectare (0.7 ounces per acre) in a single application, or if the results of the first tier of the toxicology tests triggers off the need to go to the second tier of the toxicology test. In the absence of any adverse toxicological effect in the first tier and when less than 49 g of a.i. of the pheromone per hectare are applied, there is no need to conduct residue analysis and to establish tolerances for residues on crops.

Environmental fate data requirements are necessary to estimate the concentration of the pheromone and its degradation products occurring in various media (i.e. soil, water and air) at suitable intervals after pesticide application. Generally these data are only required if adverse effects are observed in the first tier of the environmental fate and non-target organism tests or if the pheromone is applied directly to water. When the rates of application of the pheromone are so low that it is difficult to identify the degradation products and to calculate the percentage of the parent compound remaining, the guidelines allow the use of biological monitoring techniques instead of instrumental methods of quantitative chemical analysis.

Toxicology data requirements in the first tier for acute tests (oral, dermal, inhalation, primary skin and eye irritation) with pheromones and other behaviour-modifying chemicals are the same as for conventional pesticides. The first tier also includes hypersensitivity, genotoxicity, cellular immune response, and a subchronic (90-day) feeding and dermal study with one species and a teratogenicity study with one species. The immune response test could provide valuable information as a screening procedure for carcinogenicity, since many immunomodulating agents are also carcinogens. When triggered, the tier progression leads on to chronic exposure and oncogenicity studies.

Efficacy data requirements on product performance are discretionary and generally not required unless the risk from the use of the pheromone is substantial enough for a need to assess its benefits.

Terrestrial wildlife, aquatic animal and *non-target insect and plant* data requirements are similarly tiered.

Data requirements for each tier in the tier testing scheme are illustrated in Fig. 12.1. Note that there are conditions attached for each of the listed studies, so that only the relevant studies have to be carried out within the tier.

The EPA guidelines describe criteria for classifying a behaviour-modifying chemical as 'biochemical' as distinct from a 'conventional' pesticide. The behaviour-modifying chemical is a biochemical pesticide when it is either a natural product or, if synthetic, its molecular structure is identical to the natural product, as in the case of pheromones. Where there are minor differences between stereochemical ratios, the classification depends upon

TIER I	TIER II	TIER III
Acute oral	Mammalian mutagenicity	Chronic exposure
Acute dermal	Immune response	Oncogenicity
Acute inhalation		
Primary eye irritation		
Hypersensitivity		
Hypersensitivity incidents		
Genotoxicity		
90-day feeding (one species)		
90-day dermal (one species)		
90-day inhalation (one species)		
Teratogenicity (one species)		

TIER I	TIER II	TIER III
Avian acute oral	Volatility	Terrestrial wildlife
Avian dietary	Dispenser-water leaching	Aquatic animal
Freshwater fish LC_{50}	Adsorption–desorption	Additional non-
Freshwater invertebrate LC_{50}	Octanol/water partition	target plant
Non-target plant	UV absorption	Additional non-
Non-target insect	Hydrolysis	target insect
	Aerobic soil metabolism	
	Aerobic aquatic metabolism	
	Soil photolysis	
	Aquatic hydrolysis	

Fig. 12.1 Data requirements toxicology, environmental fate and non-target organisms

whether or not there is significant difference in the toxicological properties between the isomers. The same criteria apply when the exact molecular structure of the naturally occurring compound is unknown or if the synthetic chemical is closely related, although not identical, to the naturally occurring compound. The Agency's decision to classify such a behaviour-modifying chemical as biochemical or conventional pesticide will depend upon the significance of the differences in their toxicology as demonstrated in at least the first tier testing and their mode of action. If a synthetic analogue demonstrates direct toxicity to any non-target organism in the first tier testing, the Agency may not classify the chemical as a biochemical pesticide. The Agency may classify a synthetic chemical as a biochemical pesticide and still impose certain conventional data requirements on it.

Pheromones isolated and purified from single insect species may consist of a number of structurally related organic compounds with similar properties. If the degradation products of the pheromone mixture have been characterized and they display little or no toxicity, it may not be necessary to conduct toxicity tests with each individual component of the naturally occurring mixture.

If several pheromone mixtures are combined into a single product in order to broaden the spectrum of activity against different pests, or to enhance stability, each pheromone mixture must be toxicologically tested for possible adverse effect in addition to the formulated product containing the mixtures of pheromones. Appropriate controls must also be tested to ensure that any toxicity or lethality is due to the presence of the active ingredient in the formulated product.

EPA announced in the Federal Register in 1983 an exemption from FIFRA requirements for products containing pheromone attractants. The exemption applies only to pheromones and identical or substantially similar compounds used in traps in which these compounds are the sole active ingredients. The rationale for this exemption is that since a pheromone trap is a device containing a pheromone used for the sole purpose of attracting and trapping or killing target arthropods, these traps achieve pest control by reducing the numbers of target pest organisms from their natural environment. This rationale cannot be applied, for instance, to pheromones which act by disruption of mating behaviour. Pheromone traps do not result in increased levels of pheromones over a significant area of a treated field. Thus the exemption does not apply to the use of pheromones in 'confusion techniques' which, by definition, depend on increasing the concentration of the pheromones or identical or substantially similar compounds over and above the natural background levels in the field.

12.5 CONCLUSIONS

The following pheromones have been approved (up to October 1988) for various outlets in the United States:

1 : 1 mixture of (Z,Z)- and (Z,E)-7,11-hexadecadien-1-yl acetate on cotton
(Z)-11-Tetradecen-1-ol formate on cotton
(Z)- and (E)-4-Tridecen-1-yl acetate as tape dispensers and chopped fibre
  on tomatoes to control tomato pinworm
(Z,E,E)-9,11,13-Octadecatrienoic acid methyl ester on cotton
(Z)-9-Tetradecenal and (Z)-11-hexadecenal on sweet corn and field corn
(Z)-11-Hexadecenal on artichokes
(Z,Z) and (Z,E)-3,13-Octadecadien-1-ol acetate on almonds, cherries, prunes,
  apricots, nectarines and plums.
3,7,11-Trimethyl-2,6,10-dodecatrien-1-ol and 3,7,11-trimethyl-1,6,10-dodeca-
  trien-3-ol.

(Z) and (E)-8-dodecen-1-yl and (Z)-8-dodecen-1-ol in orchards to control oriental fruit moth

(Z)-2-Iso-propenyl-1-methylcyclobutane ethanol, (Z)-3,3-dimethyl-1, β-cyclohexane ethanol, (Z)-3,3-dimethyl Δ^1 cyclohexane ethanal and (E)-3,3-dimethyl Δ^1 cyclohexane ethanal in combination with plant volatiles on cotton

(E)-10-(Z)-12-hexadecadien-1-ol to control silkworm

Exo-7-ethyl-5-methyl-6,8-dioxabicyclo-[3.2.1]-octane to control western pine beetle

8,10-Dodecadien-1-ol to control codling moth

(Z)-7,8-epoxy-α-methyloctadecane to control gypsy moth

1,5-Dimethyl-6,8-dioxybicyclo-[3.2.1]-octane to control southern pine beetle

(Z)-7-hexadecen-1-ol acetate to control pink bollworm

($-$)-α-Cubebene, ($-$)-4-methyl-5-heptanol and ($-$)-α-multistriatin to control European elm bark beetle

10-Propyl-trans-5,9-tridecadien-1-ol acetate to control pink bollworm

(Z) and (E)-9-dodecenyl acetate to control western pine shoot borer

Pheromones offer an exciting solution to many pest problems. The low exposure expected from their use has encouraged speedy registrations. Certain criteria, such as their low rates of application (less than $49 \, g \, ha^{-1}$) may not *per se* be sufficient justification for the leniency shown towards the registration of pheromones since some other highly active pesticides (for example, synthetic pyrethroid insecticides) are also applied at such low rates in the field. However, when all the criteria (such as species specificity, and low acute toxicity to target and non-target organisms) are considered together, pheromones clearly distinguish themselves from conventional pesticides by their overall safety and environmental acceptability.

Increased use of pheromones, no doubt encouraged by the relative ease of their registration, will lead to further information and possibly to the appropriate modification of data requirements for future registration in the light of this new knowledge.

12.6 REFERENCES

United States Environmental Protection Agency Pesticide Assessment Guidelines (1982). PB83-153965, EPA 540/9-82-028, November, Subdivision M: Biorational Pesticides.

United States Federal Register (1983). Vol. 48, No. 165, August 24, pages 38572–38574; Part II, Environmental Protection Agency; Products Containing Pheromone Attractants; Exemption from FIFRA Requirements.

United States Federal Register (1984). Vol. 49, No. 207, October 24, pages 42856–42905; Part II, Environmental Protection Agency, 40 CFR Part 158: Data Requirements for Pesticide Registration; Final Rule.

Insect Pheromones in Plant Protection
Edited by A. R. Jutsum and R. F. S. Gordon
Published 1989 John Wiley & Sons Ltd

13

The Adoption of Pheromones in Pest Control

M. Hebblethwaite

Overseas Development Natural Resources Institute,
Central Avenue, Chatham Maritime, Chatham, Kent ME4 4TB, UK

13.1 INTRODUCTION

To date, pheromones have been adopted in pest control only to a very limited extent in comparison with conventional pest control methods. One reason for this is the relatively recent development of pheromones and other behaviour-modifying chemicals to the stage of offering an acceptable control technique. However, with progress being made in the science and technology of pheromones, it is important to consider the social and economic issues which may influence their adoption.

In order to identify these issues, the expected mode of action of pheromones in controlling a pest population will be considered in relation to the important characteristics of national farming systems. This will be complemented by reviewing the social and economic factors which apply in general in the adoption of agricultural innovations. The extent of their applicability to pheromones in particular will be considered. Three countries will be used as examples to assess and illustrate some of the themes emerging from such an approach.

13.2 CHARACTERISTICS OF FARMING SYSTEMS

A brief listing of aspects of land use is useful at the outset as, together with pest biology and behaviour, these are important in determining the inherent presence of pests and appropriate control strategies. At the most basic level, one such aspect is the range of dominant plant species present, both crop and non-crop. For crop pests which also have non-crop species as host plants, uncultivated areas and weed populations are important elements of the farming system relevant to pest incidence. Conversely, for those pests having only crops as host plants, the pattern of cultivation can be focused upon. Other aspects of land use are the proportion of land cultivated, whether cropping patterns tend towards monocropping or mixed cropping, whether particular crops have one or several planting seasons in the agricultural year, and whether planting is consolidated into blocks of single crops or comprises small fields of various crops.

Social and economic characteristics of farming systems particularly relevant to the adoption of pest control practices can be considered in three categories: (a) whether the farming system is high or low input/output; (b) whether resources and practices are uniform or heterogeneous within the farming population; and (c) where the responsibility for decision-making is found. These categories will now be considered in relation to the adoption of pest control practices and pheromones in particular.

13.2.1 Input/output levels

Farm-level input/output levels constitute a more or less continuous spectrum between countries and sometimes also within countries. For illustrative purposes, high and low input/output systems will be considered.

High input systems, by definition, involve high levels of financial and human capital per unit area. Financial capital comprises investments and working capital for seasonal inputs, and high-input systems obviously demand the ability to purchase pest control inputs and services. Human capital comprises technological and managerial skills, and high-input systems demand the ability to assess and undertake pest control practices. This implies that new pest control practices such as the use of pheromones need to offer a clear advantage over conventional practices from the viewpoint of pest control managers, if they are to have good prospects of adoption. This is true especially when there is uncertainty over their effectiveness in controlling pests.

With high levels of financial and human capital per hectare, crop yield potential is high, increasing the returns per unit expenditure on pest control. Also, a greater degree of control over the production environment is achieved in terms of soil, water and nutrition conditions, reducing year-to-year variability.

Yield potential is highly variable where unpredictable or adverse weather conditions are not offset by high levels of financial and human capital, either because of scarcity of these resources or because the returns would be uneconomic. In such situations farmers may be less willing to invest in pest control practices (conventional or pheromones).

However, if farmers regard expenditure on pest control as a form of insurance, the aversion to risk, typical of small farmers in developing countries, may lead them to accept a lower benefit : cost ratio than would be regarded as the minimum acceptable return for other crop production inputs.

Large farm businesses cannot always be equated with high input/output systems. Thus farming systems in which large areas are farmed with proportionally little financial or human capital (land-extensive systems) may be dominated by large businesses in terms of land, but will tend to have lower yields per unit area than the converse situation of high levels of financial and human capital in proportion to land (land-intensive systems). As costs, as distinct from worthwhile expenditure on pest control, are more a function of crop area than crop yield per unit area, a large farm in a land-extensive system would be justified in spending on pest control only to a lower cost level than in a more land-intensive system.

If pheromones were to offer pest control at a lower cost per unit area than conventional practices, situations in which large farms dominate in a land-extensive system would be good candidates for the adoption of pheromone use, assuming that farm owners or managers were to have a sufficient background of education and experience.

Farming systems with high levels of input use must be backed up by input delivery agencies, and by the provision of technical information on input use and of financial information on input and output prices. Adoption of pheromones requires similar institutional capability. When this is already developed for conventional pest control practices, the use of pheromones will be assisted, provided support for both approaches to control is compatible for the agencies already involved.

In considering the adoption of a new input such as pheromones in replacement of alternatives, the good access to current price information typical of high-input systems enables farmers to make precise cost comparisons. Future costs of newly introduced pheromones may be more uncertain than those of conventional insecticides. However, this is unlikely to be a major factor at farm level, as farmers will focus largely on current costs and returns of the alternatives. Farmers' access to information on movements in relative costs of pheromones and pesticides will be an important factor in pheromone adoption in high-input systems.

13.2.2 Heterogeneity in farm resources, practices and attitudes

Farming populations within a country are never uniform. Variations exist in financial and human resources, hence in farm size, investments in land and water development, cropping patterns and intensities, and in production practices. Although less readily discernible, there is also likely to be some variation in farmers' goals and attitudes, most importantly in economic terms concerning the extent to which profit maximization is the goal and how risk aversion modifies this. Amongst peasant farmers, there is likely to be priority given to subsistence needs, and avoidance of financial investments which promise only a small net return. High opportunity costs of capital would also discourage the latter type of investment.

Variation in ease and cost of access to support services (input supply, credit and extension) is also important, especially in developing countries. Such variation is often associated not only with farm size but also with social and political status and relationships in the community.

The nature of certain innovations, for example those requiring large capital investments suited mainly to the needs and resources of large farms, leads to non-uniform adoption across the farming community. Even with practices which are divisible, hence neutral to scale, non-simultaneous adoption of agricultural innovations is universal. An S-shaped adoption pattern (number of adopters on the vertical axis and time on the horizontal axis) is frequently encountered, in developed and developing countries (Feder et al., 1982).

The terms 'early adopters', 'followers' and 'laggards' have been used to categorize farmers adopting at different stages on the S-shaped curve. While a farmer's degree of aversion to, or acceptance of, risk will influence the category into which he falls, variation in resources and institutional access will be major factors in determining if and when a farmer will adopt an innovation on offer.

Common conventional control activities at farm level comprise cultural practices (crop rotations, use of resistant varieties, destruction of crop residues etc.) and the use of pesticides. These can be adopted by an individual farmer and, with some exceptions, be effective irrespective of farm size and whether or not his neighbours are doing likewise. The adoption of pheromones cannot necessarily be regarded in the same manner.

Oviposition disruptants, insect-growth regulators and anti-feedants do not depend for their effectiveness in one plot on their simultaneous use in adjoining fields. An adoption pattern, increasing over time in the form of the S-shaped curve, would therefore be feasible. In contrast, an adoption pattern of spatially mixed early adopters, followers and laggards within a farming community would not work in the case of pheromones, as these depend for control on reducing population numbers of target pests by means of mating disruption or mass trapping. Immigration by winged adults of the pest species would counteract the efforts of early adopters.

The one exception to this infeasibility of pheromone adoption by single farmers would be if they controlled a crop area in excess of that needed for economic use of pheromones. What this area is depends on the particular pest (thus its degree of mobility), pest population pressure in the region, costs of pheromone use and crop prices. Area-wide use, generally involving single farmers, has to be the standard approach with pheromones.

13.2.3 Location of decision-making responsibilities

For most innovations in agricultural production practices, it is common for farmers to be individually responsible for taking decisions and putting these into effect, subject of course to the possibilities and constraints within which they operate. Important exceptions to this are the centrally planned and managed economies, notably the USSR, Eastern European countries, China and Cuba. Further exceptions are found in irrigation projects or less often in tree crops projects, involving contiguous smallholder-operated plots under some degree of central management.

Considering specifically the adoption of behaviour-modifying chemicals (BMCs), it is important to identify who is likely to be responsible for pest control decisions and actions. As noted above, this applies especially in the case of pheromones, for which area-wide application is necessary. Responsibility can be considered within a spectrum ranging from central government to individual farmers, with intermediate positions in the form of regional government, project authorities and local communities.

In certain countries, plant protection departments of central or regional government have this responsibility, involving aerial and/or ground spraying operations. However, well-established effective area-wide control operations under government management are rare, except possibly in the centrally managed economies.

Attempts at project-level management of pest control are more common especially in developing countries, but in these, low availability of recurrent finance and of managerial resources often renders direct intervention in agricultural production or pest control practices inefficient. Instead, a policy more in keeping with these resource limitations tends to place responsibility for field operations with farmers, supported by public or private sector service agencies in the supply of inputs, information, credit and marketing.

Community-level organization in the use of behaviour-modifying chemicals would offer an area-wide approach amongst small farmers whose average size is beneath the minimum necessary for effective and economic use of pheromones. At the risk of over-generalization, it can be said that in developing countries community-wide cooperation in agriculture is most common and most successful in service functions and in undertaking small-scale capital developments. In traditional communities, some exchanges of labour between farm families do occur in farm production activities. Where cooperation in such activities has been introduced from above, failure has been common, either because important technical or organizational factors have been overlooked, or because of insufficient social cohesion, perhaps with exploitative relations emerging within the group. Given that pheromones in a small-farmer situation will often require joint adoption amongst neighbouring farmers, the potential problems of cooperation will have to be addressed, from two directions. The minimum area for viable use of pheromones must be identified for each particular pest and particular ecological situation. Equally important, the relevant pointers must be sought from traditional and recent activities attempted jointly at rural community level.

Following the scale of increasing degrees of decentralization of decision making, the individual farmer level represents the opposite end of the spectrum to central government. At farmer level, the various forms of land tenure must be recognized as potentially affecting pest control decisions. Where farmers own all the land they farm, they incur (with their families) all on-farm expenditure and receive all farm-level income. Thus such farmers can directly relate returns to costs, and are solely responsible for decisions (subject to government restrictions and incentives).

Renting of land is common. In developed countries this is usually in the form of a fixed financial payment per unit area. The tenant would normally be responsible for choice of pest control actions, financing all costs of these and receiving all additional returns gained through pest control. The distinction between owner-operators and tenant farmers has therefore little importance in introducing pheromones in developed countries.

The tenurial situation in developing countries is usually less straightforward and may have much more relevance to pest control practices. A mixture of owner-operators, tenant-owners (renting some land and owning some) and pure tenants is common. The land-owner may decide on crops to be grown and cultural operations to be followed. The tenant may often be theoretically responsible for decision-making on use of seasonal inputs, but the land-owner may influence inputs to be used, through supplying credit or assisting in physically obtaining supplies.

Importantly, rather than fixed money rent being paid, land rent is often in the form of crop-sharing, whereby the harvest is divided between land-owner and tenant on a percentage basis, often 50 : 50. Yet the tenant normally provides all labour inputs (family labour, sometimes supplemented by hired-in labour). The shares of landlord and tenant in financing the cost of purchased seasonal inputs such as pest control materials can vary from country to country, and can be complicated by the availability and terms of credit from landlord to tenant. Under share-cropping, the economic incentive for the tenant to adopt a costly pest control practice may be seriously weakened by tenancy terms.

While a case-by-case approach is needed, it can be emphasized that in any situation involving land-owner–tenant arrangements each party's direct and indirect influence on decisions and their participation in costs and returns to BMC adoption must be fully taken into account. Ignoring this might, for example, result in a misdirected extension programme and a false assessment of the profitability and risks of adoption. The need for an area-wide approach with pheromones reinforces the need to take into full account the role of landowners in adoption, as each may influence practices on several tenanted farms.

13.3 BENEFITS OF PHEROMONES AND THEIR RELATION TO ADOPTION

13.3.1 Characteristics of pheromone benefits

A distinction can be drawn very usefully between private costs and benefits and social costs and benefits. The former accrue directly to an individual decision-maker from his own actions, while the latter are the costs and benefits of an individual's actions which pass on to society as a whole.

In the adoption of pheromones, farmers are usually the decision-makers (as qualified in Section 13.2) and society is usually equated with national boundaries (although this can be inappropriate, especially in the case of small neighbouring countries).

Decisions in pest control normally have been dominated by private costs and benefits. These comprise control costs for target pests, and control costs for non-target pests; conventionally, heavy reliance is placed upon the use of chemical pesticides for pest control. These *may* have the following costs of a largely social nature: development of resistance to pesticides (by target and non-target pests); damage to human health; and environmental pollution. Much of the advocacy of pheromones is based upon social as well as private benefits.

The target specificity of pheromones (hence their safety to non-target organisms) means that predators and parasites are not harmed. Thus a farmer using a pheromone hopes to avoid any upsurge of the target or non-target pests, and the associated avoidance of an increase in control costs is a private benefit to the farmer. If the beneficial insects are notably mobile across farm boundaries, the private benefits accruing to an adopter of pheromones would be partly transformed into social benefits. Conversely, farmers who continue with the use of pesticides damaging to mobile beneficial species would be imposing social costs upon neighbouring farmers, as well as incurring private costs themselves.

The widespread instances of resistance to pesticides entail mixed private and social costs, with the emphasis on the latter. Those farmers using a pesticide in a manner favouring the development of resistance themselves suffer reduced cost-effectiveness in control. However, the induced resistance is then present in the population of the pest species appearing amongst those farmers trying to use the pesticide in a more balanced manner. One particular example of the social cost of induced resistance to pesticides is when this occurs in disease vectors, as has happened with mosquitoes and organochlorines

M. Hebblethwaite

as a result of these pesticides being used against crop pests. The reduced risk of such an occurrence can be an important category of social benefit of pheromones.

The human health and environmental pollution issues can affect farmers' own operations and people working within them, hence constituting private costs. However, when these take the form of pesticide residues they are potential hazards to the general population and to the environment. As such, they constitute important categories of social costs which the adoption of pheromones would avoid.

13.3.2 Implications for adoption of pheromones

The increasing concern within society over the social costs of the use of some types of pesticides, as well as a greater understanding of the risks of deteriorating pest situations emerging through unwise pesticide use, have helped to support the case for pheromone development and promotion. However, the common dichotomy between social benefits of pheromones and private decision-making responsibility represents a potential problem area.

One strategy to assist adoption of BMCs would be to demonstrate clear private benefits from their use. The most important of such benefits would be reduced pest control costs in comparison with pesticides and increased crop revenues through better pest control. The latter might be attainable in a situation of persistent misuse of pesticides, with pest control having broken down. There is one view that the IPM cycle has to run its course, whereby integrated control involving approaches such as pheromones is only adopted in crisis situations. Then the private benefits of a recovery of yield levels might be a sufficient incentive for private decision-makers to adopt pheromones.

However, it is to be hoped that the pesticide-induced crisis stage of the IPM cycle can be foreseen and avoided. In this case, if society is responsible for pest control actions, whereby the social benefits of pheromones are valued, their adoption may proceed in the absence of clearly perceived gains through reduced control costs. On the other hand, as more typically farmers have this responsibility, widespread adoption seems only likely to take place if such gains are offered.

13.4 THE USA, EGYPT AND PAKISTAN AS CONTRASTING SITUATIONS FOR PHEROMONE USE

The foregoing review of the characteristics of farming systems and of the benefits of pheromones, and how these influence pheromone adoption, can be brought into sharper focus by considering the USA, Egypt and Pakistan: three countries in which pheromones are being used for pest control.

13.4.1 The USA: first adoption

Pheromones for field crop pests were first used commercially in 1976 in the USA, for controlling pink bollworm (*Pectinophora gossypiella*) on cotton in Arizona and California. Starting in that year with 6500 application hectares of a hollow-fibre formulation, cumulative use had reached just over half a million application hectares by 1982, and a laminated flake formulation had also come into commercial use, as well as an 'attract and kill' formulation of the pink bollworm pheromone (incorporating pesticide at very low application rates per hectare) and a similar system for boll weevil control.

The reasons why use of pheromones for crop pest control first occurred in the USA are obvious. Companies able to produce and formulate pheromones are located there. The USA is the biggest national market for pest control products, and the country has a large cotton production industry, under increasing pressure from pink bollworm, which reached California in 1965. In California's Imperial Valley, annual costs attributable to this pest have been estimated to vary between 12% and 80% of total cotton crop value during the 1970s (Burrows *et al.*, 1982).

Farming systems characteristics also favoured pink bollworm pheromone use in the south-west USA. Cotton is grown in large and contiguous areas, under high input/output regimes. Use of pheromones for pest control requires more ability to monitor and interpret pest populations compared with calendar spraying of insecticides, and established public and private sector agencies assist this.

Variation exists among farmers in resources, attitudes and practices in cotton pest control, such that simultaneous adoption throughout a farming community would not occur spontaneously. However, the large farm size (indicated in Table 13.1) makes area-wide use feasible at the individual farm

TABLE 13.1 USA: characteristics of early commercial use of pink bollworm pheromone in Blythe, California, 1977 (Brooks *et al.*, 1979)

Area treated[1]	
Average of 4 growers	86 ha
Range of 4 growers	15–206 ha
Average % reduction in pest control cost, compared with conventional practice[2]	
Pheromone only	55%
Pheromone plus insecticide	23%
Pheromone plus insecticide and *Bacillus thuringiensis*	16%
Average yield of the 4 growers compared with conventional practice[3]	133%

[1]Treated areas constituted a more or less contiguous block, and were reasonably well isolated from immigration of mated female pink bollworm moths.
[2]Average cost of conventional practice was £113 per hectare.
[3]4 growers average yield 1212 kg of lint per hectare; conventional practice 914 kg per hectare as average of entire Palo Verde Valley.

or local community level. This is reinforced in the desert south-west of the USA, where cotton-growing areas have a degree of isolation. After farmer initiatives in California's Imperial Valley, this isolation facilitated the mandatory adoption of pink bollworm pheromone use within a manageable area (35 000 ha in 1980). Another exception to individual farmer responsibility for pest control decisions was the use of private consultants, which could have assisted area-wide adoption of pheromones within a locality.

Nevertheless, subject to the above qualifications, farmers have basic control over pest management practices. The companies promoting pheromones in the USA must therefore base their marketing efforts on the private benefits to be expected from their adoption, notwithstanding the strong environmental lobby whose influence comes mainly in registration. The promotion of pink bollworm pheromone therefore emphasizes more cost-effective control, through preservation of beneficial insects, reduction of secondary infestations, avoidance of increasing pesticide dependence hence lower pesticide use, lower control costs and improved yield and quality. Direct benefits to the farmer in the form of higher profits are claimed.

At the outset of commercial promotion of pink bollworm pheromone in 1977, price policy had to permit more cost-effective control, if the above claims were to be fulfilled. Experience is shown in Tables 13.1 and 13.2. The very fact that cost and yield data were highlighted reflects the company's need to focus on private benefits. The experience presented from this first promotion, while somewhat mixed (Table 13.2), did offer support to this focus when reference was made to one area where isolation from external pink bollworm pressure was good (Table 13.1). In subsequent years, commercial promotion of pink bollworm pheromone in the USA has continued with improved cost-effectiveness as the central feature of marketing efforts.

13.4.2 Egypt: state decision-making

Applied research and field trials on the use of pheromones for control of pink bollworm and the Egyptian cotton leafworm (*Spodoptera littoralis*) began

TABLE 13.2 USA: cost trends in early commercial use of pink bollworm pheromone in Arizona and California, 1977 (Brooks *et al.*, 1979)

	Compared with conventional insecticide use, areas with cost:	
	Decrease	Increase
Number	6	4
Combined pheromone-treated area (ha)	2152	1490
Average cost decrease/increase (%)	19.5	41.4

in Egypt in 1978, leading to the involvement of three commercial companies in large-scale trials of pink bollworm formulations. In 1984 the first sales were made, sufficient to treat about 4000 ha with four applications, divided between two governorates. These sales were also the first of a pheromone for pest control in field crops in any developing country. This is explained partly by the great importance attached to pink bollworm in Egypt's cotton crop of around 400 000 ha, with 85% of pesticide costs on cotton being for control of bollworms, principally pink bollworm (Hebblethwaite, 1984).

The characteristics of the farming systems also favoured use of pink bollworm pheromone, despite Egyptian agriculture being dominated by small landowners: 95% of landowners have holdings of less than 2.1 ha (5 feddans) which constitute 53% of the area owned, and three-quarters of the area comprises holdings of less than 8.4 ha (20 feddans) (Table 13.3). However, agriculture is concentrated in the Nile valley and delta, and the small areas of cotton planted by each farmer are consolidated into contiguous blocks (with a crop rotation). This is done to enable government control concerning the issuing and checking of cotton production quotas allocated to supervised cooperatives.

Importantly, this contiguous block system also enables government-organized pest control, with early-season use of 1.5–2.0 million children on manual collection of leafworm egg masses, and mid–late-season application of pesticides, mainly by aircraft.

The high input/output levels of Egyptian cotton production (Table 13.4) are similar to those of California (Table 13.1). The high yield potential favours substantial expenditure on pest control, which is estimated to represent about one-fifth of cotton production and harvesting costs. Cotton accounts for about 70% of pesticide use in Egyptian agriculture, and Table 13.5 shows the rapidly expanding use during the 1950s and 1960s, at a time when cotton was slowly declining.

TABLE 13.3 Egypt: distribution of land ownership[1] 1980 (Central Agency for Public Mobilisation and Statistics, 1983)

Size category		No. of land-owners (×1000)	Area owned (×1000 feddans)	Percentage	
Feddans	ha			Land-owners	Area owned
0–5	0–2.1	3487	2934	95.3	52.9
5–10	2.1–4.2	92	595	2.5	10.7
10–20	4.2–8.4	44	558	1.2	10.1
20–50	8.4–21.0	24	620	0.7	11.2
50–100	21.0–42.0	6	398	0.2	7.2
100 +	42.0+	2	440	0.1	7.9
Totals		3655	5545	100.0	100.0

[1]Not including state land, desert and land in process of distribution.

TABLE 13.4 Egypt: average cotton production costs and break-even yield, 1982[1] (Hebblethwaite, 1984; Shepley, 1984)

Production and harvesting costs (£ per hectare)[2]	652
Of which pest control costs (£ per hectare)[3]	139
Break-even yield level (kg ha^{-1})	2428
Average yield, 1980–1982 (kg ha^{-1} of seed cotton)	2617

[1]Converting at £E1.42 = £1 sterling.
[2]Includes government pest control expenditure; excludes government expenditure on aerial applications of fertilizer and land rent.
[3]Materials and application costs for pesticides, and labour costs for manual collection of cotton leafworm egg masses.

TABLE 13.5 Egypt: total quantity of pesticides used in agriculture, 1953–1982 (Richards, 1982; Central Agency for Public Mobilisation and Statistics, 1983)

Period	Annual average (tonnes)
1953–1954	1 885
1955–1958	9 156
1959–1962	14 246
1963–1966	20 639
1967–1970[1]	27 010
1971–1974	25 841
1975–1978	20 183

[1]Average of 3 years, as data for 1968 unavailable.

Despite obviously low levels of technical skills amongst farmers compared with the USA and despite increasingly fragmented land ownership (Table 13.6), widespread adoption of pesticides at increasing levels of use has been possible because of the transfer of decision-making and control actions from the farmers to the government and its spray contractors. The transfer was applied to limited areas in the 1950s and generally from 1963.

State control of pest management practices was particularly vital to the pilot-scale adoption of pink bollworm pheromone in two ways. First, it made possible the essential area-wide application. Second, concern over the social costs and hazards of increasing pesticide dependence could be directly fed into decisions on pest control policy.

Such concerns helped to launch a cooperative programme involving the Ministry of Agriculture, Ain Shams University, Cairo, the UK's Overseas Development Administration and industry (BASF, ICI Agrochemicals and Sandoz). In the initial years of the programme application costs were borne by the Ministry of Agriculture and the materials were provided as experimental products by the companies. After four years of field research further pheromone product was purchased by the Egyptian government.

TABLE 13.6 Egypt: size distribution of land by ownership and area farmed, 1950, 1961 and 1975 (Richards, 1982)

Size category		% area owned[1]		% area farmed		
Feddans	ha	1950	1961	1950	1961	1975
0–5	0–2.1	35.4	52.1	23.2	37.8	65.9
5–50	2.1–21.0	29.4	32.7	37.7	40.7	33.3
50+	21.0+	34.2	15.2	39.1	21.5	1.8

[1]Data for 1975 not available.

Since 1984, a price tendering system has applied to pink bollworm pheromone in the same way as, but separately from, the traditional pesticide tender system in Egypt. Thus suppliers can hope to take a share of the market in accordance with their tendered price.

Sales of pheromone in Egypt, unlike the USA, are not based on expectations of greater cost-effectiveness through lower control costs and higher crop yield, although these may be achieved. Rather, this new technology has provided an opportunity to introduce a different control technique and chemical mode of action into the control programme, which in Egypt has been based very successfully on the principle of integrating different approaches.

13.4.3 Pakistan: private decision-making

Applied research and field trials with pheromones for the control of pink and also spiny and spotted bollworms (*Earias insulana* and *E. vittella*) began in Pakistan in 1983. At an early stage of this work, a consideration of social and economic factors was especially useful.

In Pakistan 34% of farms are less than 2 ha, and these constitute only 8% of the cultivated area; farms of between 2 and 10 ha account for 57% by number and 56% of the cultivated area; land-holding amongst the remaining 9% of farms is divided roughly equally between under 20 and over 20 ha sizes (Table 13.7). In contrast to Egypt (Table 13.6), there has been a trend towards polarization of farm size, with owners repossessing land from tenants and farming this directly (Hussain, 1983).

Input/output levels are low, especially for an irrigated agricultural system (Table 13.8). Yield potential of the majority of cotton-growing farmers is low owing to a variety of deficiencies in production practices and institutional support. Availability of capital and technical knowledge at farm level is poor, with the exception of some large farmers.

Whereas Egypt has been using pesticides comprehensively on cotton for twenty-five years, mostly under government management, it is not surprising from Pakistan's low input/output levels that pesticide use has

TABLE 13.7 Pakistan: number and area of private farms, classified by size, 1980 (Ministry of Food, Agriculture and Co-operatives, 1983)

Farm size (ha)	Farms[1] No. (million)	%	Farm area ha (million)	%	Cultivated area[2] ha (million)	%
Under 2.0	1.39	34.2	1.35	7.1	1.23	7.7
2.0–5.0	1.60	39.3	5.20	27.3	4.79	30.2
5.0–10.0	0.71	17.4	4.70	24.7	4.12	26.0
10.0–20.0	0.26	6.4	3.39	17.8	2.78	17.5
20.0–60.0	0.10	2.3	2.80	14.7	2.03	12.8
60.0 and above	0.01	0.3	1.62	8.5	0.92	5.8
Total farms[3]	4.07	99.9	19.06	100.1	15.87	100.0

[1]Includes farms under all forms of tenancy (see also Table 13.10) and therefore does not represent distribution of land ownership.
[2]Of the 15.87 million ha cultivated in total 14.78 million ha were irrigated, and 91% of the irrigated area was in Punjab and Sind, which are the cotton-growing provinces; hence the farm size distribution may be approximately equated with distribution of irrigated area by farm size, and also with farm size distribution for cotton production, as cotton is one of the two principal summer crops in the irrigated farming system, irrespective of farm size.
[3]Percentage figures may not add up to 100.0 owing to rounding.

TABLE 13.8 Pakistan: cotton production costs under normal and improved small-farmer management, and break-even yields, 1982/3[1] (estimate based on Gill *et al.*, 1983)

	Management level	
	Normal	Improved
Production and harvesting costs (£ per hectare)[2]	100	204
Of which pest control costs (£ per hectare)	0	52
Break-even yield level (kg seed cotton per hectare)	450	917
Average national yield 1980–1982 (kg seed cotton per hectare)	1027	

[1]Converting at Rs 20.1 = £1.
[2]Includes all field operations, and payments to government of water rates and land revenue; excludes any payments for land rent.

been much less widespread. Table 13.9 shows the Pakistan government's extent of entry into pest control operations in the 1970s. At the peak in 1976/7, 30% of the rice area and 14% of the cotton area was aerially sprayed, but by the early 1980s government aerial spray operations were largely confined to sugarcane. In contrast to Egypt, the low yield potential and absence of a centralized marketing system resulted in government being unable to recover spray costs and so finance the programme.

Pesticide imports (in the absence of domestic production and including a substantial quantity for non-agricultural use) fell from a peak of 16 000 tons in 1976/7 to 5500 tons in 1981/2. This contrasts strongly with Egypt's level

TABLE 13.9 Pakistan: extent of government crop spraying by aircraft, 1972/3–1981/2 (Ministry of Food, Agriculture and Cooperatives, 1983)

	1972/3	1973/4	1974/5	1975/6	1976/7[1]	1977/8	1978/9	1979/80	1980/1	1981/2
Rice										
Cropped area ('000 ha)	1480	1512	1604	1710	1749	1899	2026	2035	1933	1976
Area sprayed ('000 ha)	154	328	378	468	516	322	377	60	60	80
%	10.4	21.7	23.6	27.4	29.5	17.0	18.6	2.9	3.1	4.1
Spray hectares ('000)	284	613	819	916	1006	516	377	60	60	60
Cotton										
Cropped area ('000 ha)	2010	1845	2031	1852	1865	1843	1891	2081	2108	2214
Area sprayed ('000 ha)	84	145	148	160	261	163	155	78	0	0
%	4.2	7.9	7.3	8.6	14.0	8.8	8.2	3.7	0	0
Spray hectares ('000)	234	323	397	478	873	502	246	94	0	0
Sugarcane										
Cropped area ('000 ha)	534	645	673	700	1412	823	753	719	825	947
Area sprayed ('000 ha)	134	124	96	141	168	87	87	87	108	77
%	25.1	19.2	14.3	20.1	11.9	10.6	11.6	12.1	13.1	8.1
Spray hectares ('000)	281	258	260	338	280	202	157	155	183	104
Other crops										
Spray hectares ('000)	38	54	109	93	37	95	108	73	7	7
Total spray hectares ('000)	837	1248	1585	1825	2196	1315	888	382	250	192

[1] Data for maize included with that for sugarcane in 1976/7; included in other crops for other years.

of use (Table 13.5), especially when taking into account Pakistan's much greater summer crop area (4.2 million ha of rice and cotton in Pakistan, and 1.5 million ha of rice, cotton and maize in Egypt). Following the privatization of insecticide imports in 1980 and therefore greater private sector promotion of their use, 20–30% of the cotton area was said to have been ground sprayed in 1982/3. While this was an increase, minimal or zero pest control remained normal, except for large farmers.

Important factors in determining whether Pakistan will adopt pheromones include the distribution of control over pest management actions, and the private and social benefits offered. It seems unlikely that government will resume aerial spraying of cotton as the non-contiguous nature of cotton growing in Pakistan and the presence of obstacles make aerial spraying inefficient, adding to the financing problem already mentioned. Comprehensive government-controlled ground spraying seems equally impracticable.

Responsibility is therefore likely to remain with farmers. In this situation, unlike in Egypt, the decision-makers will place little value on any claims of social benefits arising from pheromone adoption.

It seems likely that, as in the USA, marketing strategies for pheromones will have to hold out the prospect of private gains (lower costs or better control) to individual farmers in comparison with pesticides which, although not generally used, are better known.

Farmer responsibility for decision-making also raises the important issue of land tenure, in two aspects: first, the relative influence of landlords and tenants on pest control decision-making; second, the way in which tenurial arrangements influence the distribution of costs and benefits in the adoption of pest control innovations. The fact that pure tenants or owner-tenants account for almost half the national farm area (Table 13.10) emphasizes the importance of considering tenure.

Evidence from a survey in Khanewal subdivision of Punjab in 1979 (Nabi, 1983) is that landlords influence tenants' decisions regarding use of modern

TABLE 13.10 Pakistan: number, area and average size of private farms, classified by tenure, 1980 (Ministry of Food, Agriculture and Cooperatives, 1983)

Form of tenure	Farms		Farm area		
	No. (million)	%	ha (million)	%	Average size (ha)
Owner-cultivator	2.23	55	9.93	52	4.5
Owner-cum-tenant	0.79	19	5.02	26	6.4
Tenant	1.05	26	4.11	22	3.9
Total farms	4.07	100	19.06	100	4.7

inputs (specifically tractor-ploughing and fertilizer) and that the threat of eviction felt by tenants acts to ensure compliance with landlords' wishes. This survey indicated that landlords and tenants share equally the cost of modern inputs such as fertilizer and seed of modern high-yielding varieties; animal draught power and labour are traditionally provided by the tenants against the landlords' contribution of land. This cost-sharing is especially relevant to wider adoption of improved pest control practices, including pheromones, as pest control operations account for more than half the costs of moving from normal small farmer management to improved management of cotton in Pakistan (Table 13.8). Although money rents based on potential crop yield are reported to be increasing in occurrence (Hussain, 1983), share-cropping is still much more common, tenants and landlords traditionally taking equal shares of the crop output. Provided that pheromone use was not to be especially labour-intensive, the equal sharing of cash input costs in pest control and of additional revenue generated indicates that renting of land should not erode incentives to adopt pheromones.

A conventional imported commercial approach to pheromone adoption in Pakistan is likely to regard large owner-operators as the market, offering the possibility of area-wide use and sufficient technical and managerial skills. However, if the objective of both government and of companies is to bring pheromones to most farmers and most of the cultivated area, the 48% of tenant farmers must be included in the approach, as well as the small owner-cultivators. Given tenant farmers' common cash shortages, landlord contributions to financing inputs—especially if accompanied by assistance in procuring supplies without onerous credit terms—could be very significant in bringing pheromones into use in the tenant farm sector. Furthermore, joint ownership of several farms by one land-owner is common (although not shown by Table 13.7). As much as 50% of the tenant-operated farm area at the 1972 census was rented out by owners of 60 ha or more of land.

Thus a more imaginative strategy might regard large land-owners not only as potential pheromone users but also as intermediaries in encouraging adoption by smaller tenant farmers. If land rented out to several tenants is in contiguous fields, working through landlords would offer the important possibility of an area-wide approach. This strategy may be more in keeping with the realities of a country such as Pakistan, with fewer extension and credit agencies for the small farmers, with possible difficulties in encouraging direct cooperation amongst neighbouring tenant farmers, and with landlord–tenant relations often being dominant. However, any such strategy should not rely on 'trickle-down' of new practices from large farmers, but should be positively directed to small farmers, within the social and economic realities as outlined. Different approaches are likely to be needed for large farmers, tenant farmers and small owner-cultivators.

13.5 CONCLUSIONS

Some conditions favouring adoption are common to pheromones and pesticides, while others are more specific to pheromones. Experience to date indicates that pheromones will entail pest control costs similar to pesticides, in which case high input/output farming systems are necessary for their adoption. If pheromones are more demanding than pesticides in terms of technical skills needed, this must be reflected either in the capabilities of individual pest managers or supporting public and private agencies.

To be effective in control, pheromones require area-wide use, the minimum area being dependent on pest and ecological situation. Area-wide use may be achieved at the level of individual farms in countries such as the USA having large farms, although even there mandatory use by all farmers within a locality has featured prominently.

Area-wide use will exceed the size of individual farms in many countries. In these situations, the normal adoption curve of an innovation within a farming community over time, arising from inter-farm differences in resources, goals and institutional access, will be unacceptable. Hence the normal extension approach to individual farmers will be invalid.

This problem is avoided in those few countries such as Egypt where the state controls pest management decisions and actions. With the state as pest manager, the social benefits of pheromones may be valued, reducing the pressure to offer improved cost-effectiveness with pheromones in comparison with pesticides. In this situation, the state may be content to pay the same price for the two control methods, once convinced of equal effectiveness alongside the social benefits of pheromones.

In most countries, the state is not responsible for pest management decisions and actions, either because the farm structure precludes efficient centralized control programmes, because of limited institutional capability in the public sector, or because of difficulties of recovering costs of centralized control programmes. If the use of pheromones for pest control is to have a chance of success in these countries, simultaneous adoption amongst groups of farmers will be necessary. The degree of economic, social, political and cultural homogeneity and harmony will largely determine the feasibility of farmer cooperation, and these aspects must be studied on a case-by-case basis. Functioning agricultural producer cooperatives may already exist, in which case area-wide use should be more easily attainable. In countries such as Pakistan in which land is commonly rented out, the presence of land-owners might be turned to advantage by regarding these as targets for pheromone promotion, not only in the sense of individuals with large farms, more wealth and more skills, but also as joint financiers and joint decision-makers with their several tenants.

The use of pesticides will be fairly well known in countries where pheromones might be promoted. To encourage their adoption by farmers in the absence of state-controlled pest management, prospects of improved cost-effectiveness must be offered with pheromones. Farmers as individuals or in small groups will attach little or no value to the social benefits of pheromone use. If pesticides are not generally used, the promotion of pheromones at a price which promises more cost-effective pest control than pesticides would offer the prospect of shortening the IPM cycle. On the other hand, if such a price policy is not practised, it is likely that adoption of pheromones would only follow a pesticide-induced crisis, taking into account the additional demands associated with area-wide use.

Companies promoting pheromones may have a choice. They can limit the focus to three categories of countries: those with large, high input/output farms; those having efficient state-managed pest control; and those which have already suffered a pesticide crisis. Alternatively, a wider range of countries to include those not falling into one of these categories can be aimed at, through a price policy which offers more cost-effective control to farmers before a pesticide crisis is reached, and through developing pheromone promotion programmes well adapted to the social and economic structure of agriculture. Such a BMC price policy which is more flexible and more competitive with pesticides would depend on the cost structure for pheromones. To the extent that pheromones are marketed by pesticide companies it would also depend on companies' views on the ease and profitability of promoting pheromones within pesticide-dominated programmes.

Collaboration between commercial companies and public sector research and development institutions in developed and developing countries may be helpful. This could introduce an understanding of social and economic aspects of farming communities into pheromone development programmes, in a manner more thorough and at less cost than is possible within the commercial operations of companies working alone.

13.6 REFERENCES

Brooks, T. W., Doane, C. C. and Staten, R. T. (1979). Experience with the first commercial pheromone communication disruptive for suppression of an agricultural insect pest. In *Chemical Ecology: Odour Communication in Animals*, F. J. Ritter (Ed.), Elsevier/North Holland, Amsterdam, pp. 375–388.

Burrows, T. M., Sevacherian, V., Browning, H. and Baritelle, J. (1982). History and cost of the pink bollworm (Lepidoptera: Gelechiidae) in the Imperial Valley. *Bull. Entomol. Soc. Am.*, **28**, 286–290.

Central Agency for Public Mobilisation and Statistics, Arab Republic of Egypt (1983). *Statistical Yearbook 1952–1982*, CAMPUS, Cairo.

Feder, G., Just, R. E. and Zilberman, D. (1982). *Adoption of Agricultural Innovation in Developing Countries*, Staff Working Paper 542, World Bank, Washington, DC.

Gill, I., Ahmed, Z. and Ashraf, M. (1983). Some impressions of production factors limiting cotton yields in Multan district. *Pakistan Cottons*, **27**, 33–43.

Hebblethwaite, M. J. (1984). *An Economic Assessment of Pheromones in Cotton Pest Control in Egypt*, report R1224(R), Tropical Development and Research Institute, London.

Hussain, A. (1983). *The Land Reforms in Pakistan*, Group 83 Series, Lahore.

Ministry of Food, Agriculture and Co-operatives, Government of Pakistan (1983). *Agricultural Statistics of Pakistan 1982*, Printing Corporation of Pakistan, Lahore.

Nabi, I. (1983). *Contracts, Resource Use and Productivity in Sharecropping*, Discussion Paper 34, University of Warwick (DERC), Warwick.

Richards, A. (1982). *Egypt's Agricultural Development 1800–1980*, Westview Press, Boulder.

Shepley, S. C. (1984). *Egyptian Cotton Production Economics and Farmer Response to Government Price Intervention and Meat Import Policies*, USAID project 263-0031, Ministry of Agriculture, Cairo.

Insect Pheromones in Plant Protection
Edited by A. R. Jutsum and R. F. S. Gordon
©1989 John Wiley & Sons Ltd

14

The Further Understanding of Pheromones: Biological and Chemical Research for the Future

C. J. Sanders
Great Lakes Forest Research Centre,
Sault Ste Marie, Ontario, Canada

14.1 INTRODUCTION

The dependence of insects on chemical stimuli for successfully carrying out essential activities in their life cycles has not escaped the attention of the economic entomologist. Manipulation of chemical stimuli presents a tempting opportunity to modify insect behaviour and so to regulate survival and reproduction. Advances in analytical and synthetic chemistry over the past twenty years, coupled with sophisticated behavioural research, has led to the identification of many chemicals which can be used to modify insect behaviour, and pheromone products for control of insects are now commercially available. The use of pheromones in the regulation of insect populations has been fully reviewed on several occasions in the past few years (see Birch, 1974; Shorey and McKelvey, 1977; Ritter, 1979; Roelofs, 1979; Mitchell, 1981; Nordlund *et al.*, 1981; Silverstein, 1981; Kydonieus and Beroza, 1982; Leonhardt and Beroza, 1982; and Campion, 1985, for an excellent review which includes good coverage of areas other than North America).

Progress in the identification and use of pheromones and other behaviour-modifying chemicals has been uneven. Most advances have been in research on the sex pheromones of Lepidoptera and Coleoptera, the aggregating pheromones of Coleoptera, and also epideictic pheromones which influence oviposition and other behaviours. The following discussion will concentrate principally on the use of lepidopteran sex pheromones, focusing on what has been learned of the principles involved in their use and, from this, what can be recommended for future research. It should be noted, however, that rapid progress is being made in our understanding of the role of chemicals in parasite ecology (Nordlund et al., 1981) and in the interactions between insect and host plants (Hedin, 1983). Work in these areas is expanding rapidly, and it is difficult to chart the future. Certainly the pay-off from research on behaviour-modifying chemicals in terms of the potential for population regulation is high among all insect groups, and among all aspects of insect behaviour.

Insect pheromones have been utilized in the control of insect pests in two principal ways: trapping and mating disruption. In trapping the insects are lured by the pheromone to a trap where they are captured, either to remove sufficient insects from the population to reduce subsequent population densities (mass trapping, Chapter 4) or to provide information on the presence and numbers of insects for interpretation and use by pest managers (detection or monitoring, Chapter 3). In disruption (Chapter 5), the pheromone is released into the atmosphere to disrupt normal mating behaviour, so preventing mating, with a resultant reduction in population density.

Fundamental to making the optimum use of pheromones for both trapping and disruption is the correct identification of the complete pheromone blend.

14.2 PHEROMONE IDENTIFICATION

Most, if not all, lepidopteran sex pheromones are blends of two or more chemicals (see Inscoe, 1982, for a listing of identified pheromones by species), and there is increasing evidence that incomplete blends do not elicit all behavioural steps, or the same intensity of response as do the natural blends (Linn and Roelofs, 1985). As a corollary, it can be concluded that if a blend does not evoke the same response as a calling female then the blend is incomplete (Rothschild, 1982; Vetter and Baker, 1983, 1984; Sanders, 1984), although it is possible that other features of the natural pheromone plume, such as a fluctuating signal produced by pulsed release of pheromone by calling females, may also affect response (Cardé et al., 1984).

Certainly re-analysis of the pheromone blends of most species should be carried out, and the identification of additional minor components is likely in most of them. The capability for identification of trace quantities of material has improved considerably over the past few years. Not only are analytical techniques more sensitive, but techniques for the collection of volatiles emitted by insects have also improved (Baker et al., 1980; Weatherston et al., 1981). In addition research on biosynthetic pathways has helped to narrow the field and to provide clues as to the likely identity (Bjostad and Roelofs, 1984; Morse and Meighen, 1984).

Among the Noctuidae, where detailed studies have been carried out on the presence of trace chemicals in *Heliothis* spp. (Klun *et al.*, 1979), it has been determined that the pheromone of *H. zea* Boddie is made up of four components (Klun *et al.*, 1980a) and that of *H. virescens* F. of seven components (Klun *et al.*, 1980b). But more recent studies on the responses of male *H. zea* and *H. virescens* in a wind tunnel by Vetter and Baker (1983, 1984) have shown that omission of some of the minor components from the blends had no effect on male behaviour, although in neither species did the identified blend elicit the same response as a calling female. In another noctuid, *Trichoplusia ni* Hbn. where six components have been identified (Bjostad *et al.*, 1984), behavioural studies have shown that some of the minor components produce similar effects, and can be substituted for each other, suggesting a form of redundancy (Linn *et al.*, 1984). Redundancy and substitution may also occur where geographical variation in blends are found within the same species, as in *Spodoptera littoralis* Boisd. and *Agrotis segetum* Schiff (Campion *et al.*, 1980; Arn *et al.*, 1981).

Determining the role of the various components in mating behaviour is not easy, but significant advances have been made. Generalizations are dangerous, but already it appears that among the Tortricidae maximum responses are evoked at all stages of mating behaviour (arousal, locomotion, courtship and copulation) by the complete pheromone blend, with the components present in the same ratios as in the natural pheromone (Baker *et al.*, 1976; Baker and Cardé, 1979; Linn and Roelofs, 1983). In the Noctuidae, however, different stages in the mating sequence are modulated by different components of the blend (Nakamura and Kawasaki, 1977; Kawasaki, 1981; Linn and Gaston, 1981; Bradshaw *et al.*, 1983). Similar long- and short-range roles for different pheromone components as are found in the Noctuidae have also been reported in *Phthorimaea operculella* Zell. (Toth *et al.*, 1984).

Further significant advances in determining the role of the various components of sex pheromone blends can be anticipated in the future, following along the lines of the pioneering work of Cardé, Baker and their co-workers. Ultimately this should lead to the identification of blends which will be competitive with, or out-compete, the virgin female moths.

14.3 TRAPPING

The rapid increase in the identification and manufacture of sex pheromones and their analogues has enabled research into their potential use as lures in traps to proliferate. Numerous articles on sex pheromone trapping are now found each month in the scientific literature.

14.3.1 Types of trap

Traps can be divided into two types: those that capture the insects on a sticky surface, and those that kill the attracted insects, either in a liquid such as a mixture of alcohol and ethylene glycol, or by an insecticide such as dichlorvos impregnated into a plastic strip. The sticky traps used for Lepidoptera, although cheap and easy to use, have only limited capacity since the sticky surface deteriorates over time and also becomes saturated with trapped insects, which reduces further catches (Riedl, 1980; Houseweart et al., 1981; Brown, 1984). Saturation has been less of a problem in trapping bark beetles (Bedard and Wood, 1981; Peacock et al., 1981), partly because the beetles are small and partly because trap size can be increased as necessary. Liquid and insecticidal traps are somewhat more difficult to handle and deploy. Their capacity is, however, limited only by their size (Granett, 1973; Ramaswamy and Cardé, 1982), although there is evidence that the accumulation of previously caught, decaying insects may reduce subsequent catches of both Lepidoptera (Sanders, 1978) and bark beetles (Lie and Bakke, 1981).

Special problems occur in designing traps for beetles. Similar traps to those used for Lepidoptera have been found effective for trapping boll weevils (Mitchell and Hardee, 1974) and Japanese beetles (Klein, 1981). But for bark beetles and ambrosia beetles the silhouette of the tree trunk itself is an important visual component in attraction (Vité and Bakke, 1979; Lindgren, 1983; Birch, 1984). As a result, black 'drainpipe' traps, baited with pheromone, which mimic tree boles were used successfully in Scandinavia to control a recent outbreak of Ips typographus L. (see Chapter 4).

14.3.2 Trapping efficiency

The term 'efficiency' has often been used synonymously with rate of capture, the most efficient trap being considered the one which catches most insects. But rate of capture depends upon several processes involving different aspects of insect behaviour, and lumping them together into a single measurement obscures which features of a trap are satisfactory and those which can be improved (Lewis and Macauley, 1976; Mastro et al., 1977; Charmillot, 1980; Lingren et al., 1980; Raulston et al., 1980; Hartstack and Witz, 1981). Rate of capture depends upon the numbers of insects attracted to the vicinity of the trap, the numbers entering the trap and the numbers retained in the trap—attraction, entry and retention. Future research on trap design should be carried out in such a way as to make it quite clear which of these processes is being affected. Maximizing the efficiency of one attribute may decrease the efficiency of another: traps which permit easy entry often permit easy escape. Unfortunately in the past, few investigators have made these distinctions, but there are examples of the types of studies that are required. Lewis and Macauley (1976) and Angerilli and McLean (1984) used smoke to simulate pheromone plumes from various types of trap and found the highest catches in traps which produced discrete, well-structured plumes, presumably because these enabled more moths to orient successfully to the trap entrance. Mastro *et al.* (1977) and Elkinton and Childs (1983) observed the behaviour of male gypsy moths in the vicinity of pheromone traps, which led to suggestions for modifying trap design to increase the proportion of moths entering the traps once they had been attracted to the traps' vicinity.

Catches of both Lepidoptera and Coleoptera are affected by the location of the traps. Many empirical studies have shown that height above the ground or proximity to host plants affects catch (e.g. McNally and Barnes, 1981; Bakke and Rie, 1982; Riedl *et al.*, 1979; Aliniazee, 1983; Liebhold and Volney, 1984); but few observations have been made of insect activity to maximize probabilities of capture. An exception is the observation of Richerson *et al.* (1976) that male gypsy moths orient to tree boles. As a result, traps placed close to the trunks of trees catch more moths than those further away (Elkinton and Childs, 1983). Much can be learned about trap design by observing the response of insects to traps in wind tunnels, but ultimately any modification in design must be proven in the field. For day-flying insects all that is required is time and patience, but for crepuscular and night-flying insects, observations will depend upon new technologies, such as night viewing devices (Lingren *et al.*, 1980, 1982) or the use of infrared illumination and sensing equipment (Murlis *et al.*, 1982; Schaefer and Bent, 1984).

14.3.3 Pheromone dispensers

Responses of insects to traps are also influenced by quality and quantity of the pheromone released, which in turn depends upon the attributes of the dispenser or lure. These affect the longevity of the release of pheromone, the uniformity of the release rate over time, the stability of the pheromone, and the rate of pheromone release, which, along with the shape of the plume, dictate the active space within which male moths are activated and attracted to the source. While it is probable that catches can be significantly increased by improvements in trap efficiency (attraction, entry and retention) and by optimizing pheromone blends, significant gains are likely to arise also from future research into the performance of dispensers used to release the pheromone. As with trap design, numerous experiments have been carried out comparing catches in traps baited with different types of dispenser: cigarette filters, hollow fibres, plastic laminates, plastic pellets and rubber septa (see Chapter 10). The criteria by which a dispenser is judged are the protection of the active ingredients from degradation and the release rate measured in absolute terms and in comparative rates over time—the more uniform the release rate the better. In general, the evidence indicates that many types of dispenser are adequate for the purpose. An exception occurs, however, when the active ingredient contains one or more aldehydes. Steck *et al.* (1979) showed the need for caution in using rubber septa, and more recently Dunkelblum *et al.* (1984) have pointed out the effects of polymerization of aldehydes on the potency of lures. Certainly stability of pheromones is a constant source of concern and should be checked before any type of dispenser is chosen for operational use, not only in the laboratory but also in the field. As more complex molecules are considered for operational use, such as the trienes and tetraenes, identified as pheromone components in the Geometridae (Roelofs *et al.*, 1982; Wong *et al.*, 1984) and Arctiidae (Hill and Roelofs, 1981; Hill *et al.*, 1982), additional problems can be anticipated.

Another consideration is that the dispenser should release the components of the pheromone blend in the optimum ratios. This requires that, first, the optimum pheromone blend has been correctly identified (see Section 14.2). Second, it requires that the dispenser is actually releasing the pheromone components in the correct proportions. Different compounds may be released at different rates, meaning that the chemical blend emitted by the dispenser may be different from the blend with which the dispenser was loaded. Consequently, it is important that the blend emitted by the dispenser is analysed, and that the appropriate quantities of material are incorporated into the lure to achieve the required release ratio (McDonough and Butler, 1983).

14.4 DISRUPTION

A listing of all the research papers among Lepidoptera alone which relate to the disruption of mating behaviour for the purpose of regulating insect numbers would run to several hundred. Yet the number of successes, as indicated by the number of instances where pheromones and other behaviour-modifying chemicals are being used in the operational control of insects, is very limited. In North America pheromones have been registered by the US Environmental Protection Agency for use as disruptants against few pests: pink bollworm, tobacco budworm, oriental fruit moth, artichoke plume moth, gypsy moth and pine shoot borer. In other countries pheromones are being used operationally against the cotton leafworm (Israel and Egypt), ambrosia beetle (Canada), spruce bark beetle (Scandinavia), as well as the pink bollworm in many cotton-producing countries. These are the successes—why are there not more like them?

Certainly part of the reason that the success rate has not been higher is due to the fact that early attempts were naive and strictly empirical. Arbitrary quantities of chemicals were released at unknown rates, usually with unknown effects on behaviour. Treatments were usually dictated more by the attributes of the formulation than by the biology of the target pest. Results were measured in terms of reductions in trap catch, which were often dramatic, but it was soon discovered that large reductions in trap catch did not necessarily mean similar reductions in mating frequency. As a result of such disappointments there has been a period of retrenchment in which researchers are now looking more critically at the experimental techniques to try and determine the causes of failure, and thereby to increase the chances of success.

14.4.1 Mechanisms of disruption

Perhaps the most fundamental factor affecting success is an understanding of the mechanisms by which disruption works, so that treatments can be designed to yield maximum effects. Attraction involves a series of processes beginning with the reception of pheromone molecules at the sensillae, ending with transmission of an output to the muscles which results in the stimulated insect moving towards the pheromone source (see Bell and Cardé, 1984). Disruption of any of these processes will reduce mating success, but in most instances little if anything is known of which process is being modified. Clearly, if it is known which process is involved then treatments could be targeted more effectively.

Adaptation and habituation of the nervous system may both play a role in disruption, but the relative roles of the two are difficult to unravel (Bartell, 1982). Adaptation, or the reduction in output from the insect's olfactory receptors with constant exposure to pheromone, is unlikely to be of great importance since disadaptation occurs quickly once the stimulus is removed. Habituation, which occurs in the central nervous system, and is represented by a decline in response due to prolonged exposure, persists for longer periods and could therefore contribute significantly towards disruption (Bartell and Lawrence, 1977a, b). Exposure to constant stimuli does not produce such long-lasting habituation as does exposure to pulses of pheromone (Bartell and Lawrence, 1977a). It is therefore important to know, not just what concentration of pheromone is present in the atmosphere, but also how the concentration fluctuates and, if such fluctuations are important, how they can be enhanced to increase levels of mating disruption.

One mode of action by which mating may be disrupted is the luring of males away from calling females along artificial or false odour plumes. Here it is important that the sources of the plumes are at least as attractive as a calling female. Also it is essential that the attracted insects are removed from the pool of reproductive individuals, either by being so stimulated by the source that they remain in its vicinity, or by being killed when they arrive at the source in the so-called male annihilation technique (Flint *et al.*, 1976; Burkholder and Shapas, 1978). Again, as with research for maximizing trap catch, the emphasis is then on the identification and release rates of the natural pheromone blend, and determining the role of the various components to determine how to maximize attraction and to maximize the length of time that the attracted insects remain at the pheromone source.

An alternative to disruption by false-trail following is the masking of the pheromone plumes produced by calling females by raising the ambient background concentration of synthetic pheromone to a point where the males cannot differentiate the natural plumes against this background. Since the pheromone emitted by the females becomes dispersed by turbulence, concentration decreases with distance. Raising the background concentration therefore, in effect, truncates the active space of the pheromone plume: in theory, the higher the background concentration, the smaller the active space and the smaller the number of males attracted to the calling female. But, complete masking of the plumes to within a few centimetres of the calling females may require impracticably high concentrations of pheromone. Males passing within a few centimetres of a calling female may still be able to locate and successfully mate with her, particularly if other cues such as vision play a role in close-range behaviour. Critical topics to be addressed here are: the size of the active space; insect response thresholds; and the orientation mechanisms of insects (see reviews by Kennedy, 1983; Bell and Cardé, 1984). Basic to these is a knowledge of the structure and shapes of odour plumes as they are borne away from the source by air movement, and the actual concentrations of synthetic pheromones in the air space (not merely the amounts released over time) and their persistence and fate over time (Murlis and Jones, 1981; David *et al.*, 1982, and review by Elkinton and Cardé, 1984). The limitation of time-averaged data for describing pheromone plumes is becoming increasingly evident. As Cardé (1984) has stated, this remains 'a fertile area for investigation'. The use of wind tunnels and flight simulators over the past few years has resulted in significant advances in our understanding of the in-flight behaviour of moths (e.g. Miller and Roelofs, 1978; Cardé and Hagaman, 1979; Kennedy *et al.*, 1981; Kuenen and Baker, 1982a, b; Preiss and Kramer, 1983). To complete our understanding of male moth orientation to odour plumes the need now is for studies which integrate information on the fine-scale structure of the plumes with the behaviour of the insects (see Chapter 2).

In addition to using the synthetic pheromone itself to mask the natural pheromone plume, there are a number of other techniques which may be effective. Masking of the plumes may be achieved by releasing one or more components or mimics of the blend which cause a distortion in the pheromone blend perceived by the insect. An added advantage of this technique is that disruption may be achieved by using chemicals that are cheaper and more stable than the natural pheromone, such as formates in place of aldehydes (Mitchell *et al.*, 1975; Beevor *et al.*, 1981). Masking may also be targeted at the close-range behavioural components of courtship, including male-released aphrodisiacs (Birch, 1974), as well as female-emitted chemicals. Little attention has been given to the possibilities of disrupting behaviour at this level, but detailed studies of courtship behaviour certainly indicate that the possibilities exist and warrant further investigation.

14.4.2 Formulations

Although much research has been carried out to determine release rates of various formulations under laboratory conditions, it is essential that these are also checked under field conditions (Campion *et al.*, 1981). Release rates are influenced by environmental factors such as relative humidity and temperature: the higher the temperature and the lower the relative humidity the higher the release rate (Bierl-Leonhardt *et al.*, 1979). The peak rates of release of most female Lepidoptera occur in the evening or night when temperatures are at their minimum. To achieve similar or higher rates of release to out-compete the calling females the formulations waste large quantities of pheromone during the warmer parts of the day when release rates are at a maximum. Synthetic pheromones are generally expensive, therefore any modifications to the formulations which would bring about a reduction in emission rates during the heat of the day will result in large savings.

The fate of the chemicals once they have left the emitter has been little studied. Pheromones are adsorbed by soil and water, and re-released into the atmosphere (Caro, 1982; Shaver, 1983). Of potential significance is the adsorption of pheromone by the vegetation. Wall and Perry (1983) have shown that male pea moths are attracted to vegetation in the vicinity of a pheromone-baited trap long after the trap has been removed, which indicates that the vegetation has adsorbed and is re-releasing significant amounts of pheromone. Not only will the average concentration of pheromone in the air space be affected, but such adsorption and re-release may affect the different components of a pheromone blend differently and so distort the blend in the surrounding air. Pheromone concentrations may be considerably higher in the immediate vicinity of the foliage, and insects may perceive the pheromone directly, by contacting the contaminated foliage. The effects of such exposure on insects resting on the foliage is not known, and little attention has been paid to them in the assessment of disruption trials. Measurements of actual pheromone concentrations in the atmosphere (Plimmer *et al.*, 1978; Caro *et al.*, 1980; Wiesner *et al.*, 1980; Meighen *et al.*,

1982) are clearly a significant improvement over calculations based solely on release rates from the formulation. Measurement of such parameters should be included in all future disruption trials, to provide meaningful interpretation of dose/response relationships, as well as the necessary background information for observations on insect behaviour.

14.4.3 Alternative methods of disruption

The focus of this discussion thus far has been primarily on the disruption of attraction, i.e. the orientation process by which an insect arrives in the vicinity of an odour source. As mentioned previously most research has focused on attraction, partly for historical reasons, but partly because it is a clear-cut, prolonged, easily discernible behaviour, which is amenable to experimental analysis and, by inference, to disruption by man. But further crucial steps occur in behavioural sequences after attraction. Many of these steps are also mediated by chemicals and are therefore targets for behavioural modification. As discussed earlier, the information content of the different chemical components of lepidopteran sex pheromones as they affect attraction is only just beginning to be appreciated. Even less is known of the factors governing mate recognition and courtship. In some species the male releases 'aphrodisiacs' which elicit a response in the female, or at least acceptance of the male by the female (reviewed by Birch, 1974). Most such chemicals are highly volatile, and must be present at high concentrations to elicit a response. Therefore the potential for their use in modifying behaviour is not great. Nevertheless, they should not be neglected. Further research is necessary to determine their role in mating behaviour before the full potential for their use in modifying behaviour can be assessed.

Once the male insect has arrived in the vicinity of the female, he still has to locate her. In the oriental fruit moth, *Grapholitha molesta* (Busck.), the male 'calls' the female to him by puffing air over his hair pencils towards her (Baker and Cardé, 1979), but in all other species for which information is available, contact depends on the males actively searching for the female. Vision may play a role in this, but ultimately the male must contact the female with his antennae in order for further courtship to proceed (Grant and Brady, 1975; Colwell *et al.*, 1978; Baker and Cardé, 1979; Sanders, 1979; Castrovillo and Cardé, 1980; Ellis and Brimacombe, 1980). Presumably at this stage he recognizes the female as a potential partner, but what the required stimulus is that he perceives is not known. Certainly body and/or wing scales appear to be an important factor (Ono, 1977, 1981; Sanders, 1979; Inoko, 1982), but whether the stimulus is chemical or not is not known. If it is, then again there is a potential role for behaviour-modifying chemicals.

In several species of Lepidoptera it has also been shown that female moths perceive their own pheromone (Mitchell *et al.*, 1972; Birch, 1977; Palaniswamy and Seabrook, 1978), although the biological significance of this is not fully understood. High titres of pheromone may cause the females to become more mobile, thus reducing the probability of their mating. It has also been suggested, but not proven, that dispersal of female moths may be triggered by high concentrations of pheromone (Palaniswamy and Seabrook, 1978; Trematerra and Battaini, 1987), a convenient way for the species to monitor population density. Consideration should be given to the possibility that the pheromones or components of them can also be targeted at the females as well as the males.

14.4.4 Experimental design

The success of pheromone disruption trials have also been affected by experimental design. In earlier experiments, assessment was frequently based on how effectively the treatment depressed catches in pheromone-baited traps compared to traps in untreated check plots. Unfortunately, impressive reductions in trap catch were not always associated with significant reductions in mating based on observations of moth behaviour at mating stations or on the mating status of tethered females or captured feral females. The problem was often magnified by the use of traps baited with laboratory-reared female moths of lower potency than the feral females or with synthetic lures of unknown potency compared to virgin females. The shortcomings of evaluating disruption experiments by reduction in trap catch alone are now well recognized.

In future, monitoring the effects of disruption trials should include:

(1) measuring the incidence of mating in the treated areas, either indirectly on 'mating tables' or directly by capturing feral females, to determine the proportion of females still mating;

(2) observing the behaviour of both male and female insects to determine which aspects of reproduction are affected by the treatment; and

(3) measuring the impact on the following generation and on crop damage to determine the economic effect of the treatment.

Another confounding factor is the possibility of egg-bearing female moths moving into the treated area from surrounding untreated areas. This may take the form of local movement into treated fields or orchards by female moths (Rothschild, 1981; Doane and Brooks, 1981; Madsen and Peters, 1976; Van der Kraan and Van Deventer, 1982) or by female boll weevils (Johnson *et al.*, 1976), or the longer-range displacement of whole populations of egg-laden moths, as occurs among some noctuids (Rainey, 1979; Riley *et al.*, 1983) and forest tortricids (Greenbank *et al.*, 1980; Schönherr, 1980). In other instances wind-borne larvae can repopulate treated areas, as with the gypsy moth (Mason and McManus, 1980) and tussock moths (Mitchell, 1979).

Although dispersal is an ever-present concern in the operational use of pheromones, it can be avoided as a confounding factor in experimental trials in one of several ways:

(1) by creating buffer zones so that dispersing insects do not penetrate into the experimental area;
(2) by selecting experimental plots well away from other known populations;
(3) by caging the treated plots to keep out invading insects;
(4) by quantifying dispersal and allowing for its effects.

The first three methods have all been tried, but none is without problems. Protecting the experimental plots by leaving a buffer zone wider than the dispersal range of the insects is appropriate and valid if the range is known. Unfortunately there is no generally accepted technique for trapping female moths engaged in local redistribution during oviposition in order to find out how far they move. Extrapolations made from trapping male insects (e.g. Baltensweiler and Fischlin, 1979) are of dubious value. A reliable technique for trapping female moths would be of great value. For this reason the report that mated *Adoxophyes* females are attracted by acrylic acid (Tamaki *et al.*, 1984) is of considerable interest and should be further explored.

The selection of experimental plots well beyond the dispersal range of other known populations is clearly a viable option. To be successful this again assumes that dispersal distances are known. Where long-range dispersal has been studied in detail (e.g. Schaefer, 1976; Greenbank *et al.*, 1980; Riley *et al.*, 1983) it has been shown that insects may travel tens or hundreds of kilometres in a single flight. If the insects spread out in all directions then their numbers would be diluted, and the problem would be less serious. However, it appears that unless wind speeds are negligible, individuals orient in the same direction, usually downwind. Large numbers of insects therefore travel together, and numbers can be further concentrated by wind convergence. Unfortunately the study of long-range dispersal is not easy: it requires sophisticated equipment, such as radar, and an extensive interdisciplinary research team including biologists, physicists and meteorologists. Thus far, the necessary effort has been applied to only a few insect pests, but clearly more research is needed, not only to assist the development of pheromones, but for the implementation of any control programme. As stated by Kennedy and Way (1979) in summing up a conference on insect dispersal: 'The problem of movement is a bottleneck for pest management around the world.'

In field trials, caging the experimental populations is the most reliable way of avoiding contamination by invaders (e.g. Marks *et al.*, 1981; Beevor *et al.*, 1981; Carpenter *et al.*, 1982; Lingren *et al.*, 1982). For stored-product insects the problem is less serious, since these insects are in a sense already caged (Burkholder, 1973; Vick *et al.*, 1978; Hagstrum and Davis, 1982). The problem for others who are caging part of a larger universe is to make the cages large enough to allow full expression of the natural behaviour of the insects. Ironically, the dispersal behaviour of the insects within the cage is one of the confounding factors since, where dispersal is a natural part of behaviour, caging prevents the emigration of the caged population, forcing the would-be emigrés to oviposit inside the cages.

The ultimate solution to the problem of contamination by invading females lies in the fourth option: quantifying dispersal and allowing for its effects. As is evident from the previous discussions, however, current knowledge is not sufficient for this to be a reliable option for any insect, and clearly much further research is needed. Meanwhile the influence that dispersal can have both on experiments and on operational treatments should not be underestimated.

14.4.5 Criteria for success of mating disruption

Essential for developing the use of pheromones for mating disruption is an understanding of the population dynamics of the pest species, since survivorship curves are the baseline on which all interventions are superimposed. For instance, as discussed above, dispersal is of particular concern; invasion by egg-bearing females from untreated areas into the treated plot can completely negate the effects of the treatment. Quantifying the effects of dispersal is therefore of high priority.

An integrated view of pest management also helps to identify situations which favour the use of a particular management tool. In the case of pheromones aimed at mating disruption these are:
(1) where the target species is a pest of concern and can be treated at low densities;
(2) where the pest species has a cryptic lifestyle which protects it from treatment by conventional insecticides;
(3) where the treated area is unlikely to be reinvaded by egg-bearing females.

The level of success in mating disruption should be assessed in the context of the overall pest management programme (e.g. Klassen *et al.*, 1982). For instance, although the 73% reduction in mating of gypsy moth achieved in trials in Massachusetts was considered not enough by itself to reduce populations (Schwalbe *et al.*, 1983), it may be a very significant contribution when viewed in conjunction with other control techniques. If, in a population of sexually reproducing insects with a fecundity of 200, 10% survive to reproductive age, then disruption alone will have to reduce mating success by at least 90% to bring about a decline in the density of the insect population from one generation to the next. If, however, another agent applied prior to the pheromone treatment reduced the adult population by one-half, then a subsequent 80% reduction in mating would bring about a decline in population numbers. On the other hand, the effect of disruption can be overestimated if the longevity of the insects is not allowed for in the calculations. For instance, as Kiritani and Kanoh (1984) have pointed out, a 90% reduction in mating each night still results in 65% of the females mating if they each live for 10 days. However, even a delay in mating may contribute to a reduction in total fecundity since it shortens the female oviposition period and also exposes the insects to a longer period of predation before oviposition begins.

Too often, in all the various disciplines of pest control, different control techniques are looked at in isolation. As a result, a promising technique may be undervalued. The potential of any mortality factor in insect control can be determined only after obtaining an accurate picture of the survival curves for the species. The effects of additional factors can then be assessed, as can the equally important attributes required of an experimental treatment for it to contribute to population regulation. Thus a benchmark is provided, against which success can be measured.

14.4.6 Advantages and disadvantages of pheromones as control agents

Many authors have pointed out the potential advantages of pheromones over more conventional pesticides for regulating insect pests. These include: specificity to the target organism; small amounts of material required; negligible toxicity to other organisms; and, in some cases, increased efficacy. Let us consider the last point first. Because of their cryptic lifestyle, many insect pests are not susceptible to insecticide or are susceptible for only short periods. Good examples are stem borers such as the families Sessiidae and Cossidae (Lepidoptera), shoot, cone and seed borers among the Diptera and in the families Pyralidae and Olethreutidae (Lepidoptera) as well as many Coleoptera, such as weevils and bark beetles. In other instances pheromones may be used to increase the efficacy of conventional control methods by aggregating pests where insecticides can be used or where the infested host plant can be destroyed; again, the bark beetles are a good example. In these cases pheromones may be more effective than chemicals. Success rates in the use of pheromones for control, and hence their credibility, might be increased if research were focused on such species. Lepidoptera which feed on exposed plant surfaces are, in general, more readily controlled by conventional pesticides, making the pheromone option less attractive and less urgent. However, where insecticide resistance has developed, or where environmental regulations prevent the use of pesticides, pheromones may have significant opportunities.

Specificity of a control agent to the target organism is generally considered to be a favourable attribute, but it can also be disadvantageous. The aggregating pheromone of bark beetles may influence the behaviour of sibling species of beetle and can also attract predators and parasites (Birch, 1984). Although this is clearly undesirable in a programme involving the mass trapping of bark beetles, the problem has been overcome by designing traps which allow the entry of the beetles but not the predators.

There is evidence that the behaviour of parasites of Lepidoptera are affected by the sex pheromones of their hosts (Lewis *et al.*, 1982; Nordlund *et al.*, 1983). The study is a rapidly expanding field of research (see reviews in Nordlund *et al.*, 1981; Vinson, 1984), and as more becomes known of the chemical interactions among insects it is possible that more instances will be discovered where sex pheromones of one species affect the behaviour of another, either to the advantage of the recipient (kairomones) or to its disadvantage (allomones). Researchers should certainly be on the lookout for such relationships during development of control programmes involving pheromones.

The extreme specificity of pheromones may also be a disadvantage where a crop is afflicted by more than one pest. Little advantage is gained if control of one pest by pheromones leaves other pests to be controlled by conventional insecticides (Rothschild, 1982). One proposal for circumventing this is to use a blend of chemicals which disrupts the mating of more than one species (Mitchell, 1975). This is particularly appealing where several pest species of one crop share the same pheromone components, as occurs among Tortricidae in orchards (Roelofs *et al.*, 1976) and Noctuidae in field crops (Klun *et al.*, 1979; Steck *et al.*, 1980). This strategy has not been tested extensively in the case of the Noctuidae because of problems with formulations of the pheromones (McLaughlin *et al.*, 1981). In the case of the Tortricidae it is due to discouraging results in early tests and the realization that more basic information on disruption of the individual species is required before multi-species disruption can be adequately assessed (Roelofs and Novak, 1981). Certainly, the use of compromise blends will be less effective if complete natural pheromone blends are necessary for maximizing disruption. The strategy may prove more successful where the pheromones of the species involved have different chemical components (e.g. Roehrich *et al.*, 1979; Henneberry *et al.*, 1981).

An oft-heard disadvantage in the use of pheromones is their cost. The active ingredients themselves are expensive relative to conventional insecticides, although unquestionably the economics of scale will reduce costs once the materials are being produced for large-scale use. The use of relatively sophisticated formulations and application equipment in some cases for dissemination of pheromones also raises costs. Pheromones are used in extremely low quantities of active ingredient per hectare. While this is an advantage in terms of cost and environmental pollution, it becomes a disadvantage in the economics of production and profit-making. It is therefore no surprise that the greatest success in the operational use of pheromones in pest management has been in the control of the pink bollworm, a multivoltine pest affecting extensive areas in many countries (Doane and Brooks, 1981; Critchley et al., 1983) (see Chapter 5). For some pests, however, pheromones may well become the control technique of choice because of their negligible effect on non-target organisms. It is difficult to put a cost figure on such considerations, but it is probable that the scope for using pheromones, even when not cost-competitive, will increase in view of the increasing concerns over environmental pollution caused by some conventional insecticides.

To make pheromones more competitive, attention in the future should be given to reducing costs. Formulations which do not require expensive specialized equipment should be used when possible. Behavioural research is also needed to determine if disruption can be achieved using cheaper, more stable pheromone analogues (parapheromones).

A major concern over the extensive use of sex pheromones is that of selection for resistant strains (Pimentel et al., 1984). Evidently selection pressures during the natural evolution of pheromone communication systems have been strong (Cardé and Baker, 1984), resulting in numerous instances of sibling species in which sex pheromones are the major barriers to hybrid matings (e.g. Roelofs and Brown, 1982). Considerable variation may exist among individuals within a species in the quantity of pheromones released, in blend ratios (which may be polymorphic) and in male response (Baltensweiler et al., 1978; Cardé et al., 1978; Flint et al., 1979). But where selection pressure has been removed, as in laboratory rearings and in the introduction of a species into another continent, the resulting genetic drift has been but slight (Klun et al., 1975; Cardé and Webster, 1979; Flint et al., 1979; Wallner et al., 1984), although Tompkins and Hall (1984) found that male Drosophila were able to distinguish females from different laboratory populations by pheromone. In the only extensive field investigation thus far, Haynes et al. (1984) found no evidence of resistance in pink bollworm in cotton fields treated with gossyplure for 3–5 years. As Haynes et al. point out, in an excellent review of the problem, treatments with pheromone are unlikely to be frequent enough or intensive enough to exert strong selection pressures, even when applied repeatedly to chronic agricultural pests. If shifts in the composition of pheromone blends and responses of the male did occur, adjustments in the treatments could compensate. The danger of selection causing a change in pheromone blend or concentration appears therefore to be slight, but it is still a possibility, and it should be continually monitored.

More serious is the possibility that resistance will arise by the selection of individuals which rely on cues other than pheromones for locating mates. If high concentrations of sex pheromone in the atmosphere cause males to be more active then, even if long-range orientation is disrupted, there remains the possibility that males will come within visual range of a female.

In several species of Lepidoptera it has been shown that males can locate a female from several centimetres away by vision (Shorey and Gaston, 1970; Hidaka, 1972; Carpenter and Sparks, 1982). In other Lepidoptera, vision plays a role only at very short range (Traynier, 1968; Richerson, 1977; Colwell, *et al.* 1978; Baker and Cardé, 1979) but the relative importance of vision compared to other cures has not been determined, and the possibility exists that selection could lead to the development of strains in which vision plays a dominant role. However, such individuals would be at a distinct disadvantage when population density declined. The establishment of such strains is therefore unlikely.

Another concern is that parthenogenetic strains may develop. Parthenogenesis has been reported in *Lymantria*, *Orgyia* and *Bombyx* (in Wigglesworth, 1950). Nielsen and Cantwell (1973) quote several references which claim to have demonstrated parthenogenesis in *Galleria*, but they themselves found no evidence of parthenogenesis in some 280 000 *Galleria* eggs laid by virgin females. Thus although parthenogenesis remains a possibility, it is not of sufficient concern at present to warrant extensive research in relation to the use of pheromones.

14.5 ALTERNATIVE PHEROMONES AS POTENTIAL CONTROL AGENTS

The concentration of this book on sex and aggregation pheromones has been deliberate. However, research in recent years has demonstrated a real prospect of alternative pheromones being developed as control agents. Two examples are considered below: oviposition-deterring pheromones and alarm pheromones.

14.5.1 Oviposition deterring-pheromones

There is a clear, theoretical opportunity for a chemical which, when applied to a crop, reduces egg-laying on the crop by the pest species, or redirects egg-laying to less susceptible parts of the crop. Several examples are known where the insects themselves deposit a pheromone with their eggs, whose action is to reduce the likelihood of other females of that species laying eggs nearby.

The case of the fruit flies (*Rhagoletis* spp.) has been researched in detail in Switzerland and the USA. Hurter (1987) reported the identification of an oviposition-deterring pheromone in *R. cerasi* L., the European cherry fruit fly. The compound, *N*(15-(beta glucopyranosyl) oxy-8-hydroxy-palmitoyl) taurine, was discovered in faecal samples of *R. cerasi* females, and is deposited on the fruit surface at the time of egg-laying. Field trials with faecal extracts (Katsoyannos and Boller, 1976) demonstrated the action of the deterrent in reducing oviposition on treated fruit. Averill and Prokopy have researched the oviposition-deterring pheromone of *R. pomonella* Walsh, the apple maggot fly. Egg-laying *R. pomonella* showed a marked avoidance of fruit infested by larvae and of fruit marked with a water extract of oviposition-deterring pheromone (Averill and Prokopy, 1987).

The cabbage butterfly, *Pieris brassicae* L., deposits an oviposition-deterring pheromone (ODP) with its egg masses. Field tests (Klijnstra and Schoonhoven, 1987) showed that, in a no-choice cage test, there was no difference in the total numbers of eggs laid between cages containing only ODP-treated and only untreated brussel sprouts. However, in the treated cages most eggs were laid on plants at the edges of the cages, suggesting an avoidance response.

Messina *et al.* (1987) investigated the oviposition-deterring pheromone of *Callosobruchus maculatus* F., the spotted cowpea bruchid. ODPs were extracted by washing glass beads carrying eggs in ether. Subsequent egg-laying was significantly less on seeds that had been dipped in the extract than on those dipped in ether alone. Removal of both antennae of the female *C. maculatus* reduced the insects' ability to distinguish between ODP-treated and untreated seeds.

14.5.2 Alarm pheromones

When aphids are attacked they release an alarm pheromone, (*E*)-beta-farnesene (Bowers *et al*, 1972) to which nearby aphids respond by stopping feeding, moving away and sometimes dropping from the plant. Attempts to use the alarm pheromone alone for aphid control met with failure. However, Griffiths *et al.* (1986) reported some successful experiments with alarm pheromone/insecticide mixtures in which the pheromone increased the aphid mortality achieved by permethrin and bendiocarb. The prospect exists of using alarm pheromones to reduce not only direct aphid feeding damage but also the spread of aphid borne plant viruses.

Work on mite alarm pheromones (e.g. Kuwahara *et al.*, 1987) should yield similar, exciting opportunities for novel control methods or the enhancement of conventional acaricide activity.

14.6 SUMMARY AND CONCLUSIONS

The successful application of pheromones, whether as attractants or disruptants, depends upon a thorough understanding of (1) their chemistry, (2) behaviour of the target insect, and (3) integration of the two into a pest management programme. In the past, many field trials have gone ahead even though information is incomplete, resulting in failures which cannot be satisfactorily explained.

Although attractants are known for many insects, complete pheromone blends have been identified for only a fraction of these, yet the evidence indicates that the natural blends are the most effective for modifying behaviour. This may not be a serious drawback for survey trapping where maximum catches are not essential, but for detection, mass trapping and disruption the true potential of pheromones in pest management cannot be realized unless the optimum blends are identified, and appropriate formulations for their release developed. Current technology appears capable of achieving this. The most serious bottleneck in the use of pheromones in pest management is likely to be our lack of understanding of the biological effects of pheromone compounds and how they can be utilized most effectively. Although laboratory experiments in wind tunnels have provided valuable insight into insect behaviour and should continue to do so, ultimately behaviour must be studied in the field.

Finally, as with any management tool, the operational use of pheromones must be considered in the context of an integrated pest management system. This is as true for traps designed to provide early warning of a problem as for formulations designed to disrupt mating. Viewed in isolation, too much may be expected and too much promised, and if expectations are not met, support for further essential research may be withdrawn.

14.7 REFERENCES

Aliniazee, M. T. (1983). Monitoring the filbertworm, *Melissopus latiferreanus* (Lepidoptera: Olethreutidae), with sex attractant traps: Effect of trap design and placement on moth catches. *Environ. Entomol.* **12**, 141–146.

Angerilli, N. and McLean, J. A. (1984). Wind tunnel and field observations of western spruce budworm, *Choristoneura occidentalis* responses to pheromone-baited traps. *J. Ent. Soc. BC*, **81**, 10–16.

Arn, H., Baltensweiler, W., Bues, R., Buser, H. R., Esbjerg, P., Guerin, P., Mani, E., Rauscher, S., Szocs, G. and Toth, M. (1981). Refining lepidopteran sex attractants. In *Les Médiateurs Chimiques Agissant sur les Comportement des Insectes*. Symp. Internat. Versailles, 16–20 Nov., INRA, Paris, pp. 261–265.

Averill, A. L. and Prokopy, R. J. (1987). Residual activity of oviposition-deterring pheromone in *Rhagoletis pomonella* (Diptera: Tephritidae) and female response to infested fruit. *J. Chem. Ecol.* **13**, 167–177.

Baker, T. C., Cardé, R. T. and Roelofs, W. L. (1976). Behavioral response of male *Argyrotaenia velutinana* (Lepidoptera: Tortricidae) to components of its sex pheromone. *J. Chem. Ecol.*, **2**, 333–352.

Baker, T. C. and Cardé, R. T. (1979). Analysis of pheromone-mediated behaviors in male *Grapholitha molesta*, the oriental fruit moth (Lepidoptera: Tortricidae). *Environ. Entomol.*, **8**, 956–968.

Baker, T. C., Cardé, R. T. and Miller, J. R. (1980). Oriental fruit moth *Grapholitha molesta*, pheromone component emission rates measured after collection by glass surface adsorption. *J. Chem. Ecol.*, **6**, 749–758.

Bakke, A. and Rie, L. (1982). The pheromone of the spruce bark beetle, *Ips typographus* and its potential use in the suppression of beetle populations. In *Insect Suppression with Controlled Release Pheromone Systems* (Vol. II), A. F. Kydonieus and M. Beroza (Eds), CRC Press, Boca Raton, Florida, pp. 3–15.

Baltensweiler, W., Priesner, E., Arn, H. and Delucchi, V. (1978). Unterschiedliche sexuallockstoffe bei Lärchen- und Arvenform des Graven Lärchenwicklers (*Zeiraphera diniana* Gn. Lepidoptera: Tortricidae). *Mitt. Schweiz. Entomol. Ges.*, **51**, 133–142.

Baltensweiler, W. and Fischlin, A. (1979). The role of migration for the population dynamics of the larch bud moth, *Zeiraphera diniana* Gn. (Lepidoptera: Tortricidae). *Mitt. Schweiz. Entomol. Ges.*, **52**, 259–271.

Bartell, R. J. (1982). Mechanisms of communication disruption by pheromone in the control of Lepidoptera: a review. *Physiol. Entomol.*, **7**, 353–364.

Bartell, R. J. and Lawrence, L. A. (1977a). Reduction in responsiveness of male apple moths, *Epiphyas postvittana*, to sex pheromone following pulsed pheromonal exposure. *Physiol. Entomol.*, **2**, 1–6.

Bartell, R. J. and Lawrence, L. A. (1977b). Reduction in responsiveness of male light-brown apple moths, *Epiphyas postvittana*, to sex pheromone following pulsed pre-exposure to pheromone components. *Physiol. Entomol.*, **2**, 89–95.

Bedard, W. D. and Wood, D. L. (1981). Suppression of *Dendroctonus brevicomis* by using a mass-trapping tactic. In *Management of Insect Pests with Semiochemicals: Concepts and Practice*, E. R. Mitchell (Ed.), Plenum Press, New York, pp. 103–114.

Beevor, P. S., Dyck, V. A. and Arida, G. S. (1981). Formate pheromone mimics as mating disruptants of the striped rice borer moth, *Chilo suppressalis* (Walker). In *Management of Insect Pests with Semiochemicals: Concepts and Practice*, E. R. Mitchell (Ed.), Plenum Press, New York, pp. 305–311.

Bell, W. J. and Cardé, R. T. (Eds) (1984). *Chemical Ecology of Insects*, Chapman & Hall, London, 524 pp.

Bierl-Leonhardt, B. A., DeVilbiss, E. D. and Plimmer, J. R. (1979). Rate of release of disparlure from laminated plastic dispensers. *J. Econ. Ent.*, **72**, 319–321.

Birch, M. C. (Ed.) (1974). *Pheromones*, North-Holland Research Monographs, Frontiers of Biology (Vol. 32), 495 pp.

Birch, M. C. (1977). Responses of both sexes of *Trichoplusia ni* (Lepidoptera: Noctuidae) to virgin females and to synthetic pheromone. *Ecol. Entomol.*, **2**, 99–104.

Birch, M. C. (1984). Aggregation in bark beetles. In *Chemical Ecology of Insects*, W. J. Bell and R. T. Cardé (Eds), Chapman & Hall, London, pp. 331–353.

Bjostad, L. B., Linn, C. E., Du, J-W. and Roelofs, W. L. (1984). Identification of new sex pheromone components in *Trichoplusia ni* predicted from biosynthetic precursors. *J. Chem. Ecol.*, **10**, 1309–1323.

Bjostad, L. B. and Roelofs, W. L. (1984). Biosynthesis of sex pheromone components and glycerolipid precursors from carbon-14 labelled sodium acetate in red-banded leaf roller moth, *Argyrotaenia velutinana*. *J. Chem. Ecol.*, **10**, 681–692.

Bowers, W. S., Nault, L. R., Webb, R. E. and Dutky, S. R. (1972). Aphid alarm pheromone: isolation, identification, synthesis. *Science* **177**, 1121–1122.

Bradshaw, J. W. S., Baker, R. and Lisk, J. C. (1983). Separate orientation and release components in a sex pheromone. *Nature*, **304**, 265–267.

Brown, M. W. (1984). Saturation of pheromone sticky traps by *Platynota idaeusalis* (Walker) (Lepidoptera: Tortricidae). *J. Econ. Ent.*, **77**, 915–918.

Burkholder, W. E. (1973). Black carpet beetle. Reduction of mating by megatomoic acid, the sex pheromone. *J. Econ. Ent.*, **66**, 1327–1328.

Burkholder, W. E. and Shapas, T. J. (1978). Use of entomopathogen with pheromones and attractants in pest management systems for stored-product insects. In *Microbial Control of Insect Pests: Future Strategies for Pest Management Systems*, G. E. Allen, C. M. Ignoffo and R. P. Jacques (Eds), University of Florida, Gainesville, 236 pp.

Campion, D. G. (1985). Survey of pheromone uses in pest control. In *Techniques in Pheromone Research*, H. E. Hummel and T. A. Miller (Eds), Springer-Verlag, New York, pp. 405–449.

Campion, D. G., Hunter-Jones, P., McVeigh, L. J., Hall, D. R., Lester, R. and Nesbitt, B. F. (1980). Modification of the attractiveness of the primary pheromone component of the Egyptian cotton leafworm, *Spodoptera littoralis* Lepidoptera: Noctuidae, by secondary pheromone components and related chemicals. *Bull. Ent. Res.*, **70**, 417–434.

Campion, D. G., McVeigh, L. J., Hunter-Jones, P., Hall, D. R., Lester, R., Nesbitt, B. F., Marrs, G. J. and Alder, M. R. (1981). Evaluation of microencapsulated formulations of pheromone components of the Egyptian cotton leafworm in Crete. In *Management of Insect Pests with Semiochemicals: Concepts and Practice*, E. R. Mitchell (Ed.), Plenum Press, New York, pp. 253–265.

Cardé, R. T. (1984). Chemo-orientation in flying insects. In *Chemical Ecology of Insects*, W. J. Bell and R. T. Cardé (Eds), Chapman & Hall, London, p. 119.

Cardé, R. T. and Baker, T. C. (1984). Sexual communication with pheromones. In *Chemical Ecology of Insects*, W. J. Bell and R. T. Cardé (Eds), Chapman & Hall, London, pp. 355–383.

Cardé, R. T., Dindonis, L. L., Agar, B. and Foss, J. (1984). Apparency of pulsed and continuous pheromone to male gypsy moth, *Lymantria dispar*. *J. Chem. Ecol.*, **10**, 335–348.

Cardé, R. T. and Hagaman, T. E. (1979). Behavioral responses of the gypsy moth in a wind tunnel to air-borne enantiomers of disparlure. *Environ. Entomol.*, **8**, 475–484.

Cardé, R. T., Roelofs, W. L., Harrison, R. G., Vawter, A. T., Brussard, P. F., Mutuura, A. and Munroe, E. (1978). European corn borer: Pheromone polymorphism or sibling species. *Science*, **199**, 555–556.

Cardé, R. T. and Webster, R. P. (1979). Variation in attraction of individual male gypsy moths to (+) and (−) disparlure. *J. Chem. Ecol.*, **5**, 935–939.

Caro, J. H. (1982). The sensing, dispersion and measurements of pheromone vapors in air. In *Insect Suppression with Controlled Release Pheromone Systems* Vol. I, A. F. Kydonieus and M. Beroza (Eds), CRC Press, Boca Raton, Florida, pp. 145–158.

Caro, J. H., Glotfelty, D. E. and Freeman, H. P. (1980). (Z)-9-Tetradecen-1-ol formate: Distribution and dissipation in the air within a corn crop after emission from a controlled-release formulation. *J. Chem. Ecol.*, **6**, 229–239.

Carpenter, J. E. and Sparks, A. N. (1982). Effects of vision on mating behavior of the male corn earworm, *Heliothis zea*. *J. Econ. Ent.*, **75**, 248–250.

Carpenter, J. E., Sparks, A. N. and Gueldner, R. C. (1982). Effects of moth population density and pheromone concentration on mating disruption of the corn earworm, *Heliothis zea*, in large screened cages. *J. Econ. Ent.*, **75**, 333–336.

Castrovillo, P. J. and Cardé, R. T. (1980). Male codling moth (*Laspeyresia pomonella*) orientation to visual cues in the presence of pheromone and sequences of courtship behaviors. *Ann. Ent. Soc. Am.*, **73**, 100–105.

Charmillot, P. J. (1980). Efficiency of codling moth, *Laspeyresia pomonella* trap baited with synthetic sex attractant. *Ann. Zool. Ecol. Anim.*, **11**, 587–598.

Colwell, A. E., Shorey, H. H., Gaston, L. K. and Van Vorhis Kay, S. E. (1978). Short range precopulatory behavior of males of *Pectinophora gossypiella* (Lepidoptera: Gelechiidae). *Beh. Biol.*, **22**, 323–335.

Critchley, B. R., Campion, D. G., McVeigh, L., Hunter-Jones, P., Hall, D. R., Nesbitt, B. F., Marrs, G. J., Jutsum, A. R., Hosny, M. M. and Nasr, E. A. (1983). Control of pink bollworm, *Pectinophora gossypiella* (Lepidoptera: Gelechiidae) in Egypt by mating disruption using an aerially applied microencapsulated pheromone formulation. *Bull. Ent. Res.* **73**, 289–300.

David, C. T., Kennedy, J. S., Ludlow, A. R., Perry, J. N. and Wall, C. (1982). A reappraisal of insect flight towards a distant point source of wind-borne odor. *J. Chem. Ecol.*, **8**, 1207–1215.

Doane, C. C. and Brooks, T. W. (1981). Research and development of pheromones for insect control with emphasis on the pink bollworm. In *Management of Insect Pests with Semiochemicals: Concepts and Practice*, E. R. Mitchell (Ed.), Plenum Press, New York, pp. 285–303.

Dunkelblum, E., Kehat, M., Klug, J. T. and Shani, A. (1984). Trimerization of *Earias insulana* sex pheromone, (*E,E*)-10,12-hexadecadienal, a phenomenon affecting trap efficiency. *J. Chem. Ecol.*, **10**, 421–428.

Elkinton, J. S. and Cardé, R. T. (1984). Odor dispersion. In *Chemical Ecology of Insects*, W. J. Bell and R. T. Cardé (Eds), Chapman & Hall, London, pp. 73–91.

Elkinton, J. S. and Childs, R. D. (1983). Efficiency of 2 gypsy moth (Lepidoptera: Lymantria) pheromone baited traps. *Environ. Entomol.*, **12**, 1519–1525.

Ellis, P. E. and Brimacombe, L. C. (1980). The mating behaviour of the Egyptian cotton leafworm moth, *Spodoptera littoralis* (Boisd.). *Anim. Behav.*, **28**, 1239–1248.

Flint, H. M., Smith, R. L., Bariola, L. A., Horn, B. R., Forey, D. E. and Kuhn, S. J. (1976). Pink bollworm: Trap tests with gossyplure. *J. Econ. Ent.*, **69**, 535–538.

Flint, H. M., Balasubramanian, M., Campero, J., Strickland, G. R., Ahmad, Z., Barral, J., Barbosa, S. and Khail, A. F. (1979). Pink bollworm. Response of mature males to ratios of Z,Z- and Z,E-isomers of gossyplure in several cotton growing areas of the world. *J. Econ. Ent.*, **72**, 758–762.

Granett, J. (1973). A disparlure-baited box trap for capturing large numbers of gypsy moths. *J. Econ. Ent.*, **66**, 359–363.

Grant, G. G. and Brady, U. E. (1975). Courtship behavior of phycitid moths. I. Comparison of *Plodia interpunctella* and *Cadra cautella* and role of male scent glands. *Can. J. Zool.*, **53**, 813–826.

Greenbank, D. O., Shaefer, G. W. and Rainey, R. C. (1980). Spruce budworm (Lepidoptera: Tortricidae) moth flight and dispersal: New understanding from canopy observations, radar, and aircraft. *Mem. Ent. Soc. Can.*, **110**, 49.

Griffiths, D. C., Dawson, G. W., Pickett, J. A. and Woodcock, C. (1986). Use of the aphid alarm pheromone and derivatives. *Proc. Brit. Crop. Prot. Conf.*

Hagstrum, D. W. and Davis, L. R., Jr (1982). Mate seeking behavior and reduced mating by *Ephestia cautella* in a sex pheromone permeated atmosphere. *J. Chem. Ecol.*, **8**, 507–516.

Hartstack, A. W. and Witz, J. A. (1981). Estimating field populations of tobacco budworm, *Heliothis virescens* moths, from pheromone trap catches. *Environ. Entomol.*, **10**, 908–914.

Haynes, K. F., Gaston, L. K., Mistrot Pope, M. and Baker, T. C. (1984). Potential evolution of resistance to pheromones: Interindividual and interpopulational variation in chemical communication system of pink bollworm moth. *J. Chem. Ecol.*, **10**, 1551–1565.

Hedin, P. A. (Ed.) (1983). Plant resistance to insects. *Amer. Chem. Soc.*, *Symp. Series*, 208.

Henneberry, T. J., Bariola, L. A., Flint, H. M., Lingren, P. D., Gillespie, J. and Kydonieus, A. F. (1981). Pink bollworm and tobacco budworm mating disruption studies on cotton. In *Management of Insect Pests with Semiochemicals*, E. R. Mitchell (Ed.), Plenum Press, New York, pp. 267–283.

Hidaka, T. (1972). Biology of *Hyphantria cunea* Drury (Lepidoptera: Arctiidae) in Japan. XIV. Mating behavior. *Appl. Ent. Zool.*, **7**, 116–132.

Hill, A. S. and Roelofs, W. L. (1981). Sex pheromone of the saltmarsh caterpillar, *Estigmene acrea*. *J. Chem. Ecol.*, **7**, 655–668.

Hill, A. S., Kovaley, B. G., Nikolaeva, L. N. and Roelofs, W. L. (1982). Sex pheromone of the fall webworm moth, *Hyphantria cunea*. *J. Chem. Ecol.*, **8**, 383–396.

Houseweart, M. W., Jennings, D. T. and Sanders, C. J. (1981). Variables associated with pheromone traps for monitoring spruce budworm populations (Lepidoptera: Tortricidae). *Can. Ent.*, **113**, 527–537.

Hurter, J., Boller, E. F., Staedtler, E., Blattmann, B., Buser, H. R. and Bosshard, N. U. (1987). Oviposition-deterring pheromone in *Rhagoletis cerasi* L.: purification and determination of the chemical constitution. *Experientia*, **43**, 157–164.

Inscoe, M. N. (1982). Insect attractants, attractant pheromones, and related compounds. In *Insect Suppression with Controlled Release Pheromone Systems* (Vol. II), A. F. Kydonieus and M. Beroza (Eds), CRC Press, Boca Raton, Florida, pp. 201–312.

Inoko, H. (1982). Role of tactile signals in courtship behavior of *Bombyx mori* Lepidoptera: Bombycidae. *Jpn. J. Appl. Entomol. Zool.*, **26**, 10–14.

Johnson, W. L., McKibben, G. H., Rodriguez, V. J. and Davich, T. B. (1976). Bollweevil: Increased longevity of grandeme using different formulations and dispensers. *J. Econ. Ent.*, **69**, 263–265.

Katsoyannos, B. I. and Boller, E. F. (1976). First field application of oviposition-deterring marking pheromone of European cherry fruit fly. *Environ. Entomol.*, **5**, 151–152.

Kawasaki, K. (1981). A functional difference of the individual component of *Spodoptera litura* (F.) (Lepidoptera: Noctuidae) sex pheromone in the attraction of flying male moths. *Appl. Ent. Zool.*, **16**, 63–70.

Kennedy, J. S. (1983). Zigzagging and casting as a programmed response to wind borne odour: A review. *Physiol. Entomol.*, **8**, 109–120.

Kennedy, J. S. and Way, M. J. (1979). Summing up the conference. In *Movement of Highly Mobile Insects: Concepts and Methodology in Research*, R. L. Rabb and G. G. Kennedy (Eds), North Carolina State University, Raleigh, pp. 446–456.

Kennedy, J. S., Ludlow, A. R. and Sanders, C. J. (1981). Guidance of flying male moths by wind-borne sex pheromone. *Physiol. Entomol.* **6**, 395–412.

Kiritani, K. and Kanoh, M. (1984). Influence of delay in mating on the reproduction of the oriental tea tortrix *Homona magnanima* (Lepidoptera: Tortricidae) with reference to pheromone based control. *Prot. Ecol.*, **6**, 137–144.

Klassen, W., Ridgway, R. L. and Inscoe, M. (1982). Chemical attractants in integrated pest management programs. In *Insect Suppression with Controlled Release Systems* (Vol. I), A. F. Kydonieus and M. Beroza (Eds), CRC Press, Boca Raton, Florida, pp. 13–130.

Klein, M. G. (1981). Mass trapping for suppression of Japanese beetles. In *Management of Insect Pests with Semiochemicals: Concepts and Practice*, E. R. Mitchell (Ed.), Plenum Press, New York, pp. 183–190.

Klijnstra, J. W. and Schoonhoven, L. M. (1987). Effectiveness and persistence of the oviposition deterring pheromone of *Pieris brassicae* in the field. *Ent. Exp. Appl.*, **45**, 227–235.

Klun, J. A. *et al.* (1975). Insect sex pheromones: Intraspecific pheromonal variability of *Ostrinia nubilalis* in North America and Europe. *Environ. Entomol.*, **4**, 891–894.

Klun, J. A., Plimmer, J. R., Bierl-Leonhardt, B. A., Sparks, A. N. and Chapman, O. L. (1979). Trace chemicals: The essence of sexual communication systems in *Heliothis* species. *Science*, **204**, 1328–1330.

Klun, J. A., Plimmer, J. R., Bierl-Leonhardt, B. A., Sparks, A. N., Primiani, M., Chapman, O. L., Lee, G. H. and Lepone, G. (1980a). Sex pheromone chemistry of the female corn earworm moth, *Heliothis zea*. *J. Chem. Ecol.*, **6**, 165–175.

Klun, J. A., Bierl-Leonhardt, B. A., Plimmer, J. R., Sparks, A. N., Primiana, M., Chapman, O. L., Lepone, G. and Lee, G. H. (1980b). Sex pheromone chemistry of the female tobacco budworm moth, *Heliothis virescens*. *J. Chem. Ecol.*, **6**, 177–183.

Kuenen, L. P. S. and Baker, T. C. (1982a). Optomotor regulation of ground velocity in moths during flight and sex pheromone at different heights. *Physiol. Entomol.*, **7**, 193–202.

Kuenen, L. P. S. and Baker, T. C. (1982b). The effects of pheromone concentration on the flight behaviour of the oriental fruit moth, *Grapholitha molesta*. *Physiol. Entomol.*, 423–434.

Kuwahara, Y., Akimoto, K., Leal, W. S., Nakao, H. and Suzuki, T. (1987). Isopiperitenone: a new alarm pheromone of the acarid mite, *Tyrophagus similis* (Acarina, Acaridae). *Agric. Biol. Chem.*, **51**, 3441–3442.

Kydonieus, A. F. and Beroza, M. (Eds) (1982). *Insect Suppression with Controlled Release Pheromone Systems* (2 vols), CRC Press, Boca Raton, Florida.

Leonhardt, B. A. and Beroza, M. (Eds) (1982). *Insect Pheromone Technology: Chemistry and Application*, ACS Symposium Series 190, American Chemical Soc., Washington, DC, 229 pp.

Lewis, T. and Macauley, E. D. M. (1976). Design and evaluation of sex-attractant traps for pea moth, *Cydia nigricana* (Steph.), and the effects of plume shape on catches. *Ecol. Ent.*, **1**, 175–187.

Lewis, W. J., Nordlund, D. A., Gueldner, R. C., Teal, P. E. A. and Tumlinson, J. H. (1982). Kairomones and their use for management of entomophagous insects. XIII. Kairomonal activity for *Trichogramma* species of abdominal tips, excretion and a synthetic sex pheromone blend of *Heliothis zea* (Boddie) moths. *J. Chem. Ecol.*, **8**, 1323–1331.

Lie, R. and Bakke, A. (1981). Practical results from the mass-trapping of *Ips typographus* in Scandinavia. In *Management of Insect Pests with Semiochemicals: Concepts and Practice*, E. R. Mitchell (Ed.), Plenum Press, New York, pp. 175–181.

Liebhold, A. M. and Volney, W. J. A. (1984). Effect of foliage proximity on attraction of *Choristoneura occidentalis* and *C. retiniana* (Lepidoptera: Tortricidae) to pheromone sources. *J. Chem. Ecol.*, **10**, 217–227.

Lindgren, B. S. (1983). A multiple funnel trap for scolytid beetles (Coleoptera). *Can. Ent.*, **115**, 299–302.

Lingren, P. D., Burton, J., Shelton, W. and Raulston, J. R. (1980). Night vision goggles: For design, evaluation and comparative efficiency determination of a pheromone trap for capturing live adult male pink bollworms. *J. Econ. Ent.*, **73**, 622–630.

Lingren, P. D., Raulston, J. R., Sparks, A. N. and Wolf, W. W. (1982). Insect monitoring technology for evaluation of suppression via pheromone systems. In *Insect Suppression with Controlled Release Pheromone Systems* (Vol. I), A. F. Kydonieus and M. Beroza (Eds), CRC Press, Boca Raton, Florida, pp. 171–193.

Linn, C. E., Jr and Gaston, L. K. (1981). Behavioral function of the components and the blend of the sex pheromone of the cabbage looper, *Trichoplusia ni*. *Environ. Entomol.*, **10**, 751–755.

Linn, C. E. Jr and Roelofs, W. L. (1983). Effect of varying proportions of the alcohol component on sex pheromone blend discrimination in male oriental fruit moths. *Physiol. Entomol.*, **8**, 291–306.

Linn, C. E. Jr and Roelofs, W. L. (1985). Response specificity of male pink bollworm moths to different trends and dosages of sex pheromone. *J. Chem. Ecol.*, **11**, 1583–1590.

Madsen, H. F. and Peters, F. E. (1976). Pest management: Monitoring populations of *Archips argyrospilus* and *Archips rosanus* (Lepidoptera: Tortricidae) with sex pheromone traps. *Can. Ent.*, **108**, 1281–1284.

Marks, R. J., Hall, D. R., Lester, R., Nesbitt, B. F. and Lambert, M. R. K. (1981). Further studies on mating disruption of the red bollworm, *Diparopsis castanea* Hampson (Lepidoptera: Noctuidae) with a microencapsulated mating inhibitor. *Bull. Ent. Res.*, **71**, 403–418.

Mason, C. J. and McManus, M. L. (1980). The role of dispersal in the natural spread of the gypsy moth. In *Dispersal of Forest Insects: Evaluation, Theory and Management Implications*, Proc. IUFRO Conf., Sandpoint, Idaho.

Mastro, V. C., Richerson, J. V. and Cameron, E. A. (1977). An evaluation of gypsy-moth pheromone baited traps using behavioral observations as a measure of trap efficiency. *Environ. Entomol.*, **6**, 128–132.

McDonough, L. M. and Butler, L. I. (1983). Insect sex pheromones: Determination of half lives from formulations by collection of emitting vapor. *J. Chem. Ecol.*, **9**, 1491–1502.

McLaughlin, J. R., Mitchell, E. R. and Cross, J. H. (1981). Field and laboratory evaluation of mating disruptants of *Heliothis zea* and *Spodoptera frugiperda* in Florida. In *Management of Insect Pests with Semiochemicals: Concepts and Practice*, E. R. Mitchell (Ed.), Plenum Press, New York, pp. 243–251.

McNally, P. S. and Barnes, M. M. (1981). Effects of codling moth pheromone trap placement, orientation and density on trap catches. *Environ. Entomol.*, **10**, 22–26.

Meighen, E. A., Slessor, K. N. and Grant, G. G. (1982). Development of a bioluminescence assay for aldehyde pheromones of insects. I. Sensitivity and specificity. *J. Chem. Ecol.*, **8**, 911–921.

Messina, F. J., Barmore, J. L. and Renwick, J. A. A. (1987). Oviposition deterrent from eggs of *Callosobruchus maculatus*: spacing mechanism or artifact? *J. Chem. Ecol.*, **13**, 219–226.

Miller, J. R. and Roelofs, W. L. (1978). Gypsy moth responses to pheromone enantiomers as evaluated in a sustained-flight tunnel. *Environ. Entomol.*, **7**, 42–44.

Mitchell, E. R. (1975). Disruption of pheromone communication among co-existent pest insects with multichemical formulations. *Bio-Science*, **25**, 493–499.

Mitchell, E. R. (Ed.) (1981). *Management of Insect Pests with Semiochemicals: Concepts and Practice*, Plenum Press, New York, 514 pp.

Mitchell, E. R. and Hardee, D. D. (1974). In-field traps: A new concept in survey and suppression of low populations of boll weevils. *J. Econ. Ent.*, **67**, 506–508.

Mitchell, E. R., Jacobson, M. and Baumhover, A. H. (1975). *Heliothis* spp: Disruption of pheromonal communication with (*Z*)-9-tetradecen-1-ol formate. *Environ. Entomol.*, **4**, 577–579.

Mitchell, E. R., Webb, J. C. and Hines, R. W. (1972). Capture of male and female cabbage loopers in field traps baited with synthetic sex pheromone. *Environ. Entomol.*, **1**, 525–526.

Mitchell, R. G. (1979). Dispersal of early instars of the Douglas-fir tussock moth. *Ann. Ent. Soc. Am.*, **72**, 291–297.

Morse, D. and Meighen, E. (1984). Detection of pheromone biosynthetic and degradative enzymes *in vitro*. *J. Biol. Chem.*, **259**, 475–480.

Murlis, J., Bettany, B. W., Kelley, J. and Martin, L. (1982). The analysis of flight paths of male Egyptian cotton leafworm moths, *Spodoptera littoralis*, to a sex pheromone source in the field. *Physiol. Entomol.*, **7**, 435–441.

Murlis, J. and Jones, C. D. (1981). Fine scale structure of odour plumes in relation to insect orientation to distant pheromone and other attractant sources. *Physiol. Entomol.*, **6**, 71–86.

Nakamura, K. and Kawasaki, K. (1977). The active space of the *Spodoptera litura* (F.) sex pheromone and the pheromone component determining this space. *Appl. Ent. Zool.*, **12**, 162–177.

Nielsen, R. A. and Cantwell, G. E. (1973). The question of parthenogenesis in the greater wax moth. *J. Econ. Ent.*, **66**, 37–38.

Nordlund, D. A., Jones, R. L. and Lewis, W. J. (Eds) (1981). *Semiochemicals: Their Role in Pest Control*, Wiley, New York, 306 pp.

Nordlund, D. A., Lewis, W. J. and Gueldner, R. D. (1983). Kairomones and their use for management of entomophagous insect. XIV. Response of *Telenomus remus* to abdominal tips of *Spodoptera frugiperda*, (*Z*)-9-tetradecen-1-ol acetate and (*Z*)-9-dodecen-1-ol acetate. *J. Chem. Ecol.*, **9**, 695–701.

Ono, T. (1977). The scales as a releaser of the copulation attempt in Lepidoptera. *Naturwiss.*, **64**, 386–387.

Ono, T. (1981). Factors releasing the copulation attempt in three species of Phycitidae (Lepidoptera: Phycitidae). *Appl. Ent. Zool.*, **16**, 24–28.

Palaniswamy, P. and Seabrook, W. D. (1978). Behavioral responses of the female eastern spruce budworm, *Choristoneura fumiferana* (Lepidoptera: Tortricidae) to the sex pheromone of her own species. *J. Chem. Ecol.*, **4**, 649–655.

Peacock, J. W., Cuthbert, R. A. and Lanier, G. N. (1981). Deployment of traps in a barrier strategy to reduce populations of the European elm bark beetle, and the incidence of Dutch elm disease. In *Management of Insect Pests with Semiochemicals: Concepts and Practice*, E. R. Mitchell (Ed.), Plenum Press, New York, pp. 155–174.

Pimentel, D., Glenister, C., Fast, S. and Gallahan, D. (1984). Environmental risks of biological pest controls. *Oikos*, **42**, 283–290.

Plimmer, J. R., Caro, J. H. and Freeman, H. P. (1978). Distribution and dissipation of aerially-applied disparlure under a woodland canopy. *J. Econ. Ent.*, **71**, 155–157.

Preiss, R. and Kramer, E. (1983). Stabilization of altitude and speed in tethered flying gypsy moth males: Influence of (+) and (−)-disparlure. *Physiol. Entomol.*, **8**, 55–68.

Rainey, R. C. (1979). Dispersal and redistribution of some Orthoptera and Lepidoptera by flight. *Mitt. Schweiz. Entomol. Ges.*, **52**, 125–132.

Ramaswamy, S. B. and Cardé, R. T. (1982). Nonsaturating traps and long-life attractant lures for monitoring spruce budworm males. *J. Econ. Ent.*, **75**, 126–129.

Raulston, J. R., Sparks, A. N. and Lingren, P. D. (1980). Design and comparative efficiency of a wind-oriented live trap for capturing *Heliothis* spp. *J. Econ. Ent.*, **73**, 586–589.

Richerson, J. V. (1977). Pheromone-mediated behavior of the gypsy moth. *J. Chem. Ecol.*, **3**, 291–308.

Richerson, J. V., Cameron, E. A. and Brown, E. A. (1976). Sexual activity of the gypsy moth. *Am. Midl. Nat.*, **95**, 299–312.

Riedl, H. (1980). The importance of pheromone trap density and trap maintenance for the development of standardized monitoring procedures for the codling moth (Lepidoptera: Tortricidae). *Can. Ent.*, **112**, 655–663.

Riedl, H., Hoying, S. A., Barnett, W. W. and DeTar, J. E. (1979). Relationship of within-tree placement of the pheromone trap to codling moth catches. *Environ. Entomol.*, **8**, 765–769.

Riley, J. R., Reynolds, D. R. and Farmery, M. J. (1983). Observations of the flight behaviour of the armyworm moth, *Spodoptera exempta*, at an emergence site using radar and infra-red optical techniques. *Ecol. Ent.*, **8**, 395–418.

Ritter, F. J. (Ed.) (1979). *Chemical Ecology: Odour Communication in Animals*, Elsevier/North-Holland Biomedical Press, Amsterdam, pp. 427.

Roehrich, R., Carles, J. P., Trésor, C. and de Vathaire, M. A. (1979). Essais de confusion sexuelle contre les Tordeuses de la grappe l'Eudemis, *Lobesia botrana* Den. et Schiff. et la Cochylis, *Eupoecilia ambiguella* Tr. *Ann. Zool. Ecol. Anim.*, **11**, 659–675.

Roelofs, W. L. (Ed.) (1979). Establishing efficacy of sex attractants and disruptants for insect control. *Entomol. Soc. Am.*, 97.

Roelofs, W. L. and Brown, R. L. (1982). Pheromones and the evolutionary relationships of the Tortricidae. *Ann. Rev. Ecol. Syst.*, **13**, 395–442.

Roelofs, W. L., Cardé, R. T., Taschenberg, E. F. and Weires, R. W., Jr (1976). Pheromone research for the control of lepidopterous pests in New York. In *Pest Management with Insect Sex Attractants*, M. Beroza (Ed.), Amer. Chem. Soc. Symp. Series, **23**, pp. 75–87.

Roelofs, W. L., Hill, A. S., Linn, C. E., Meinwald, J., Jain, S. C., Herbert, H. J. and Smith, R. F. (1982). Sex pheromone of the winter moth, *Operophtera brumata*, a geometrid with unusually low temperature precopulatory responses. *Science*, **217**, 657–659.

Roelofs, W. L. and Novak, M. A. (1981). Small-plot disorientation tests for screening potential mating disruptants. In *Management of Insect Pests with Semiochemicals: Concepts and Practice*, E. R. Mitchell (Ed.), Plenum Press, New York, pp. 229–242.

Rothschild, G. H. L. (1981). Mating disruption of Lepidopterous pests: Current status and future prospects. In *Management of Insect Pests with Semiochemicals: Concepts and Practice*, E. R. Mitchell (Ed.), Plenum Press, New York, pp. 207–228.

Rothschild, G. H. L. (1982). Suppression of mating in codling moths with synthetic sex pheromone and other compounds. In *Insect Suppression with Controlled Release Pheromone Systems* (Vol. II), A. F. Kydonieus and M. Beroza (Eds), CRC Press, Boca Raton, Florida, pp. 117–134.

Sanders, C. J. (1978). Evaluation of sex attractant traps for monitoring spruce budworm populations (Lepidoptera: Tortricidae). *Can. Ent.*, **110**, 43–50.

Sanders, C. J. (1979). Mate location and mating in eastern spruce budworm. *Can. Dep. Env., Bi-mon. Res. Notes*, **35**, 2–3.

Sanders, C. J. (1984). Sex pheromone of the spruce budworm (Lepidoptera: Tortricidae): Evidence for a missing compound. *Can. Ent.*, **116**, 93–100.

Schaefer, G. W. (1976). Radar observations of insect flight. In *Insect Flight*, R. C. Rainey (Ed.), Symp. Royal Ent. Soc. 7, pp. 157–197.

Schaefer, G. W. and Bent, G. A. (1984). An infra-red remote sensing system for the active detection and automatic determination of insect flight trajectories (IRADIT), *Bull. Ent. Res.*, **74**, 261–278.

Schönherr, J. (1980). Ausbreitung des Europäischen Tannentriebwicklers, *Choristoneura murinana* Hb. (Lepidoptera: Tortricidae). In *Dispersal of Forest Insects: Evaluation Theory and Management Implications*, A. A. Berryman and L. Safranyik (Eds), Proc. IUFRO Conference.

Schwalbe, C. P., Paszek, E. C., Bierl-Leonhardt, B. A. and Plimmer, J. R. (1983). Disruption of gypsy moth, *Lymantria dispar* (Lepidoptera: Lymantriidae), mating with disparlure. *J. Econ. Ent.*, **76**, 841–844.

Shaver, T. N. (1983). Environmental fate of Z-11-hexadecenal and Z-9-tetradecenal, components of a sex pheromone of the tobacco budworm, *Heliothis virescens* (Lepidoptera: Noctuidae). *Environ. Entomol.*, **12**, 1802–1804.

Shorey, H. H. and Gaston, L. K. (1970). Sex pheromones of noctuid moths XX. Short-range visual orientation by pheromone-stimulated males of *Trichoplusia ni*. *Ann. Ent. Soc. Amer.*, **63**, 829–832.

Shorey, H. H. and McKelvey, J. J., Jr (Eds.) (1977). *Chemical Control of Insect Behavior: Theory and Application*, Wiley, London, 414 pp.

Silverstein, R. M. (1981). Pheromones: Background and potential for use in insect pest control. *Science*, **213**, 1326–1332.

Steck, W. F., Bailey, B. K., Chisholm, M. D. and Underhill, W. E. (1979). 1,2-Dianilinoethane, a constituent of some red rubber septa which reacts with aldehyde components of insect attractants and pheromones. *Environ. Entomol.*, **8**, 732–733.

Steck, W., Underhill, E. W. and Chisholm, M. D. (1980). Trace components in lepidopterous sex attractants. *Environ. Entomol.*, **9**, 583–585.

Tamaki, Y., Sugie, H. and Hirano, C. (1984). Acrylic acid: An attractant for the female smaller tea tortrix moth (Lepidoptera: Tortricidae). *Jap. J. Appl. Ent. Zool.*, **28**, 161–166.

Tompkins, L. and Hall, J. C. (1984). Sex pheromones enable *Drosophila* males to discriminate between conspecific females from different laboratory stocks. *Anim. Behav.*, **32**, 349–352.

Toth, M., Bellas, T. E. and Rothschild, G. H. L. (1984). Role of pheromone components in evoking behavioral responses from male potato tuberworm moth *Phthorimaea operculella* (Lepidoptera: Gelechiidae). *J. Chem. Ecol.*, **10**, 271–280.

Traynier, R. M. M. (1968). Sex attraction in the Mediterranean flour moth, *Anagasta* (=*Ephestia*) *kuhniella*. *Can. Ent.*, **100**, 5–10.

Trematerra, P. and Battaini, F. (1987). Control of *Ephestia kuehniella* Zeller by mass trapping. *J. Appl. Ent.*, **104**, 336–340.

Van Der Kraan, C. and Van Deventer, P. (1982). Range of action and interaction of pheromone traps for the summer fruit tortrix moth, *Adoxophyes orana* (F.v.R.). *J. Chem. Ecol.*, **8**, 1251–1262.

Vetter, R. S. and Baker, T. C. (1983). Behavioral responses of male *Heliothis virescens* in a sustained-flight tunnel to combinations of seven compounds identified from female sex pheromone glands. *J. Chem. Ecol.*, **9**, 747–759.

Vetter, R. S. and Baker, T. C. (1984). Behavioral responses of male *Heliothis zea* moths in sustained-flight tunnel to combination of 4 compounds identified from female sex pheromone gland. *J. Chem. Ecol.*, **10**, 193–202.

Vick, K. W., Coffelt, J. A. and Sullivan, M. A. (1978). Disruption of pheromone communication in the angoumois grain moth with synthetic female sex pheromone. *Environ. Entomol.*, **7**, 528–531.

Vinson, S. B. (1984). Parasitoid–host relationship. In *Chemical Ecology*, W. J. Bell and R. T. Cardé (Eds), Chapman & Hall, London, pp. 205–233.

Vité, J. P. and Bakke, A. (1979). Synergism between chemical and physical stimuli in host colonization by an ambrosia beetle. *Naturwiss.*, **66**, 528–529.

Wall, C. and Perry, J. N. (1983). Further observations on the responses of male pea moth, *Cydia nigricana*, to vegetation previously exposed to sex-attractant. *Ent. Exp. Appl.*, **33**, 112–116.

Wallner, W. E., Cardé, R. T., Xu-Chonghua, Weseloh, R. M., Sun-Xilin, Yan-Jingjun and Schaefer, P. W. (1984). Gypsy moth (*Lymantria dispar* L.) attraction to disparlure enantiomers and the olefin precursor in the People's Republic of China. *J. Chem. Ecol.*, **10**, 753–757.

Weatherston, J., Golob, M. A., Brooks, T. W., Huang, Y. Y. and Benn, M. H. (1981). Methodology for determining the release rates of pheromones from hollow fibres. In *Management of Insect Pests with Semiochemicals: Concepts and Practice*, E. R. Mitchell (Ed.), Plenum Press, New York, pp. 425–443.

Wiesner, C. J., Silk, P. J., Tan, S.-H. and Fullarton, S. (1980). Monitoring of atmospheric concentrations of the sex pheromone of the spruce budworm. *Choristoneura fumiferana* (Lepidoptera: Tortricidae). *Can. Ent.*, **112**, 333–334.

Wigglesworth, V. B. (1950). *The Principles of Insect Physiology*, Methuen, London, pp. 487.

Wong, J. W., Palaniswamy, P., Underhill, E. W., Steck, W. F. and Chisholm, M. D. (1984). Novel sex pheromone components from the fall cankerworm moth, *Alsophila pometaria*. *J. Chem. Ecol.*, **10**, 463–473.

Chemical Modification of Insect Behaviour in Plant Protection
Edited by A. R. Jutsum and R. F. S. Gordon
©1989 John Wiley & Sons Ltd

Epilogue

A. R. Jutsum

ICI Agrochemicals, Fernhurst, Haslemere, Surrey GU27 3JE, UK

R. F. S. Gordon

ICI Agrochemicals, Jealott's Hill Research Station,
Bracknell, Berks, RG12 6EY, UK

A decade of pheromone research, development and commercialization has passed since publication of Shorey and McKelvey's classic (1977) survey. It has been a decade in which the considerable optimism of the early seventies, fuelled by rapid progress in the identification of insect pest pheromones, gave way increasingly to a realistic view that pheromones and other behaviour-modifying chemicals have a significant, but relatively small, role to play in commercial pest control. There would be no widespread replacement of insecticides by pheromones. Rather, they would have a role in complementing insecticide use in carefully researched, integrated pest management programmes. Furthermore, this role was to be achieved only by the considerable application of skills from many disciplines and a visionary determination to succeed.

It has been our objective in compiling this book to reflect the present realism, the challenges, successes and failures, and above all the breadth and depth of endeavour that has characterized pheromone research over the past decade.

The volume of pheromone research appears to have stabilized at a healthy level with new work building on success in each of the main areas of applicability. The challenge to senior researchers will be to maintain a steady supply of new research graduates in the face of the competitive lure of genetic and biotechnological research fields. The more far-sighted can be relied upon to bring new techniques to bear on old problems and to take pheromone research and exploitation down new paths, assisted by genetics and biotechnological research. .

In retrospect, the advance in analytical chemistry which permitted rapid and reliable identification of hundreds of pheromones was the key enabling technology in the basic research. But we have seen that it is not enough to know the chemical blend emitted by an insect to understand, mimic and disrupt the behaviour elicited by that chemical. The detailed accounts of formulation development may surprise those who are not formulation chemists by the difficulties that were faced and largely overcome. Experience with trap design and (still too rare) detailed behavioural field experiments amply demonstrate not only the scope to be totally confused by laboratory or field results, but also the opportunities to improve the performance of pheromone dispensers and traps.

Several factors conspire against the more widespread commercial development of pheromones in plant protection. These factors tend to feed on one another. Thus, for example, the relatively small market size expected for a pheromone product, because of its specificity, engenders relatively high pheromone active ingredient (a.i.) prices. High a.i. prices result in expensive proving trials, in the case of mating disruption, because of the necessarily large treatment areas. The scale of even initial pheromone trials and the natural desire to minimize costs at such a time, when the likelihood of not proceeding is highest, are incompatible with evaluating a representative range of pheromone blends and formulation types. The same pressure to minimize initial costs hinders the acquisition of the type of insect behavioural data which might show a way out of the vicious circle by indicating a more cost-effective trials procedure.

In our view, these inhibitory factors are being increasingly counter-balanced. The threat and the reality of insecticide resistance, the ever increasing cost and duration of new insecticide registration, the demands of the increasingly widespread, integrated pest management schemes for new techniques to complement conventional insecticides, and the demands for ever safer pest control products, are all factors which enhance the need for pheromonal control agents. Importantly, these factors specifically encourage pheromone research by creating markets within the most cost-sensitive sectors, for control of the major pests (*Heliothis*, boll weevil, pink bollworm, *Plutella*, aphids, mites, etc).

Competition for some of these markets may be expected from microbial pesticides, in particular, which are likely to share with pheromones the advantage of relatively short and inexpensive registration procedures. But pheromone attributes such as the ability to control pests in which the feeding stage is hidden, will ensure the potential for technical differentiation from competitive methods of control.

Beyond simple pest control, pheromones will continue to offer the 'service' of monitoring. The demand for detailed and costly field studies in monitoring by pheromone traps (here, to optimize and calibrate trap catches) is probably greater than with the other pheromone applications. But the opportunity exists, given adequate funding, to target insecticide applications more judiciously on the basis of pheromone trap data.

To counter the oft-heard complaint that no major agrochemical company has shown a commitment to pheromones, we cite the fact that four major agrochemical companies (BASF, CIBA-GEIGY, ICI Agrochemicals, and Sandoz) have openly committed significant resources over the last decade to cooperative and individual developments in the use of pheromones for pest control. It is our expectation that these and other companies will support future developments alongside colleagues in universities and government agencies.

We conclude with some questions which arose for us from the contributions to the book, and which may see some answers in the coming years.

The extreme specificity of sex pheromones is both a strength and a weakness in pest control. To what extent can the weakness be eliminated by the search, perhaps using molecular modelling and electrophysiological techniques, for multi-species specific pheromone analogues?

In how many cases can female sex pheromones be detected by females of the same species? How much do we know of their effects on others of the same sex? Is the result of broadcast pheromone application in 'mating disruption' solely a function of a modification in the males' behaviour? If the females also respond to the pheromone in such a way as to enhance the apparent mating disruption, can this effect be optimized through chemical, formulation or application changes?

In which pest situations can a phase of beneficial insect activity be clearly identified as crucial, for instance, in the restriction of a secondary pest? In which pesticide situations are certain products or specific application dates seen to be threatened or implicated in pesticide resistance? In these situations pheromones have had an important role to play and further opportunities will be found.

Will registration authorities adopt pheromone monitoring techniques as a route to ever more judicious use of insecticides?

We expect that a spirit of enquiry and visionary determination will live on in pheromone research and that these and the readers' own questions will find some stimulating and useful answers.

REFERENCE

Shorey, H. H. and McKelvey, J. J. Jr. (1977) *Chemical Control of Insect Behaviour. Theory and Application*. Wiley, London.

Species Index

Subject Index